RISK ASSESSMENT FOR CHEMICALS IN DRINKING WATER

Edited by

ROBERT A. HOWD, Ph.D.
Chief, Water Toxicology Section
Office of Environmental Health Hazard Assessment
California Environmental Protection Agency

ANNA M. FAN, Ph.D.
Chief, Pesticide and Environmental Toxicology Branch
Office of Environmental Health Hazard Assessment
California Environmental Protection Agency

BICENTENNIAL 1807 WILEY 2007 BICENTENNIAL

WILEY-INTERSCIENCE

A JOHN WILEY & SONS, INC., PUBLICATION

Published by John Wiley & Sons, Inc., Hoboken, New Jersey.
Published simultaneously in Canada.

Wiley Bicentennial Logo: Richard J. Pacifico

Library of Congress Cataloging-in-Publication Data:

Howd, Robert A.
Risk assessment for chemicals in drinking water / Robert A. Howd, Anna M. Fan.
p. cm.
Includes bibliographical references.
ISBN 978-0-471-72344-8 (cloth)
1. Drinking water—Contamination—Health aspects. 2. Health risk assessment. I. Fan, Anna M., 1949- II. Title.
RA591.H69 2007
613.2′87—dc22

2007007076

Printed in the United States of America

10 9 8 7 6 5 4 3 2 1

CONTENTS

CONTRIBUTORS

George V. Alexeeff, Scientific Affairs Division, Office of Environmental Health Hazard Assessment, California Environmental Protection Agency, Oakland, California

Steven A. Book, Division of Drinking Water and Environmental Management, California Department of Public Health, Sacramento, California

Joseph P. Brown, Air Toxicology and Epidemiology Branch, Office of Environmental Health Hazard Assessment, California Environmental Protection Agency, Oakland, California

Richard Carrier, Water, Air, and Climate Change Bureau, Health Canada, Ottawa, Ontario, Canada

Vincent James Cogliano, *IARC Monographs* Programme, International Agency for Research on Cancer, Lyon, France

Joyce Morrissey Donohue, Office of Science and Technology, Office of Water, U.S. Environmental Protection Agency, Washington, DC (M.S. Nutrition Research; Ph.D. Biochemistry)

Anna M. Fan, Pesticide and Environmental Toxicology Branch, Office of Environmental Health Hazard Assessment, California Environmental Protection Agency, Oakland, California

Richard C. Hertzberg, Department of Environmental and Occupational Health, Emory University, Atlanta, Georgia

Robert A. Howd, Water Toxicology Section, Office of Environmental Health Hazard Assessment, California Environmental Protection Agency, Oakland, California

Kannan Krishnan, Department of Occupational and Environmental Health, Université dc Montréal, Montréal, Québec, Canada

John C. Lipscomb, National Center for Environmental Assessment, U.S. Environmental Protection Agency, Cincinnati, Ohio

Bruce A. Macler, U.S. Environmental Protection Agency, San Francisco, California

Melanie A. Marty, Air Toxicology and Epidemiology Branch, Office of Environmental Health Hazard Assessment, California Environmental Protection Agency, Oakland, California

Wynne Maynor Miller, Office of Ground Water and Drinking Water, U.S. Environmental Protection Agency, Washington, DC (M.S. Environmental Science and Policy)

Glenn E. Rice, National Center for Environmental Assessment, Office of Research and Development, U.S. Environmental Protection Agency, Cincinnati, Ohio

Richard Sedman, Water Toxicology Section, Office of Environmental Health Hazard Assessment, California Environmental Protection Agency, Oakland, California

Jane E. Simmons, National Health and Environmental Effects Research Laboratory, Office of Research and Development, U.S. Environmental Protection Agency, Research Triangle Park, North Carolina

David P. Spath, Division of Drinking Water and Environmental Management, California Department of Public Health, Sacramento, California

Linda K. Teuschler, National Center for Environmental Assessment, Office of Research and Development, U.S. Environmental Protection Agency, Cincinnati, Ohio

David Ting, Pesticide and Environmental Toxicology Branch, Office of Environmental Health Hazard Assessment, California Environmental Protection Agency, Oakland, California

J. Michael Wright, National Center for Environmental Assessment, Office of Research and Development, U.S. Environmental Protection Agency, Cincinnati, Ohio

FOREWORD

People have a right to expect that the water they drink, the food they eat, the air they breathe, and the environments where they live and work promote the highest possible level of health. They rely on their health agencies to identify hazards in these environments and to prevent avoidable exposures that are inconsistent with this objective.

Public health systems work best when they prevent hazardous exposures without waiting for epidemiologic studies to measure the adverse effects. This is possible through consideration of experimental studies and models that can identify health risks before they can be observed in humans. This means, however, that risk assessment models often cannot be validated by direct observation, as can models in other fields such as demographics, economics, or weather.

Accordingly, the methods of risk assessment are as important as the results of any one risk assessment. Continuous examination is necessary to ensure that risk assessment methods reflect current scientific understanding and benefit from new experimental systems and models. At the same time, public health agencies are facing new demands, for example, to evaluate the cumulative effects of multiple hazards on susceptible populations and life stages. Risk assessors are meeting this challenge by developing methods that go beyond single-chemical, general-population scenarios to address more complex, but also more realistic, situations.

This volume, which examines current risk assessment methods for chemicals in drinking water, should facilitate understanding and improvement of these methods. It includes perspectives from scientists who are grappling with contemporary risk issues at the California EPA, Health Canada, and the U.S. EPA's program, regional, and research organizations.

The existence of vigorous, independent risk assessment programs in many countries and also in state agencies is essential to the public health infrastructure. These programs can be viewed as laboratories where innovations in risk assessment methods are developed, implemented, and tested. The best of these ideas receive wider discussion en route to refinement and adoption by other state, national, and international agencies. Such innovation and examination ensures that risk assessment methods continue to reflect emerging scientific understanding and to meet the needs of health agencies worldwide.

The California risk assessors who have edited this book have a unique and valuable perspective in that California has committed to an independent risk assessment of all regulated chemicals in drinking water. In an effort to share their knowledge gained through years of experience in drinking water risk assessment, they have assembled a stellar list of co-authors to address critical regulatory and risk assessment issues. Although not every important subject can be covered in depth in a single volume, this book represents an important compilation of observations and documentation of risk assessment methods, plus a useful guide to the rest of that voluminous literature.

VINCENT JAMES COGLIANO

Head, IARC Monographs Programme
International Agency for Research on Cancer
Lyon, France

PREFACE

Risk assessment for chemicals in drinking water has much in common with risk assessment for other purposes, together with some elements that are unique. This book is intended to cover both aspects, to provide an integrated source of information on the current principles and practices. It is based on many years of experience in the practice of risk assessment, by the editors and the authors. The perspective taken is that of public health protection, as practiced by federal and state governments, mainly within the United States. The most important source of risk assessment guidance available is the United States Environmental Protection Agency (U.S. EPA). However, information relevant to risk assessment of chemicals in drinking water is scattered across dozens if not hundreds of publications, some not readily available, spanning over the last twenty years. For this book we have attempted to assemble and summarize this information to provide a more comprehensible and up-to-date resource.

In taking on the task, we have also attempted to capture current thinking on major risk assessment issues, uncertainties, and ongoing controversies. We acknowledge that our perspectives do not encompass the entire spectrum of toxicology and risk assessment opinion and practices, and we stand by the use of health-protective assumptions in risk assessment. That is a basic requirement for a public health agency. Our intent in pointing out the uncertainties and controversies is to address the health protectiveness of current practice as well as to indicate areas where current practice might be improved by obtaining information to more adequately address or reduce these uncertainties.

However, when the uncertainties in risk assessment of chemicals in drinking water are acknowledged, risk assessors may face certain criticisms. The general public dislikes being told about uncertainty in protecting their health; the purveyors of drinking water who want to assure the public that their water is safe to drink are not receptive to hearing about how much we do not know; the chemical producers or users often tend to think that the uncertainties about chemical hazards are being vastly overstated. In some cases, there are also community organizations that believe the chemical risks are being understated. The lack of complete or adequate information and the need for methodology development to parallel the generation of new data leave room for future resolution of currently existing scientific issues and conflicts. The responsibility of risk assessors for public health purposes is to examine carefully all the available information, describe and interpret it fairly, and conduct risk assessments that are sufficiently health protective.

Most of the risk assessment experience of the editors has been with the state of California. Our department, the Office of Environmental Health Hazard Assessment (OEHHA) in the California Environmental Protection Agency (Cal/EPA), is the principal risk assessment group for California and has been developing guidance on acceptable levels of chemicals in drinking water for about twenty years. The current system was created with the formation of Cal/EPA in 1991, which incorporated the then-existing risk assessment responsibilities from the Department of Health Services. The program was further strengthened with the passage of the California Safe Drinking Water Act in 1996 (HSC 116350-116415). In California, OEHHA provides the risk assessment for chemicals in drinking water, while the responsibility for regulation of chemicals resides in the Division of Drinking Water and Environmental Management of the Department of Public Health. This is consistent with the guidance in the classic reference, *Risk Assessment in the Federal Government: Managing the Process* (National Research Council, 1983), which recommended separation of risk assessment and risk management. The federal government and many states have a similar system whereby the risk assessment and regulatory functions are kept at arm's length. The U.S. EPA practice is explained in detail in Chapters 1 and 2.

Although microbiological hazards are a major factor in providing safe and potable drinking water, this discussion focuses on the chemicals that may be found as drinking water contaminants. This is largely because microbiological contaminants are addressed in different ways, with different risk assessment methods, and often, by separate governmental agencies or departments. The exclusion is not meant to imply that microbial contaminants are any less important. In fact, development of safe drinking water supplies was initiated and sustained by the need to protect against microbial contamination. That this has led to secondary problems with chemical contaminants formed in the disinfection process is a fact of life for chemical risk assessors, and should not be taken as a source of conflict between those whose task is to manage microbial contamination and those whose focus is on the chemical contaminants. The editors hope that this book may be of interest and use to both groups.

The discussions of risk assessment practices in this book describe the present state of the field and are also intended to reveal directions in which it might be improved. The current practices are under continuous reevaluation and critique. However, advances in risk assessment practices do not occur through the efforts of any central committee, nor by a single systematically organized process, but rather through an avenue of open discussion and input for developing a reasonable level of consensus. Any thoughtful scientist can undertake the initial steps, by pointing out an issue and proposing how it might logically be addressed. This book is intended to support this larger interest group, because the larger the audience of concerned citizens, the more rapidly the issues can be identified and addressed. A majority of senior professionals currently involved in the practice of risk assessment have developed the specialty during their careers. Because the work involves multiple disciplines, they have a special appreciation for how a diversity of backgrounds has enriched the present practice of risk assessment, and wish to see this process continue.

The basic issue is that risk management is best carried out by regulatory agencies, while risk assessment should be driven by science. However, science considerations are often intertwined with the social and economic aspects, and thus may be caught up in the political process. Perhaps this is more likely to occur at the national level, where the results of a decision will have a greater impact, as compared to the state or regional levels. This may lead to situations in which a state is in a better position to address important issues than those who are nominally the national leaders. This is an underlying theme in some of the chapters, but not necessarily made explicit in them.

The California Office of Environmental Health Hazard Assessment has the largest state organization for risk assessment and therefore has been in a unique position to provide an independent viewpoint for risk assessment, with the resources necessary to provide the scientific support for it. The drinking water program at OEHHA has the legislative mandate to provide independent reviews of all regulated chemicals in drinking water. The state law specifies that California standards (maximum contaminant levels) can be equal to or lower than the federal standards. In addition, California can develop regulations for chemicals not regulated at the federal level. In several cases OEHHA risk assessments for emerging chemicals have been finalized earlier than those of U.S. EPA, and California regulations were subsequently developed earlier than national standards. This has not necessarily put us at odds with U.S. EPA scientists, with whom we are likely to be in agreement, but rather, we have occasionally been the standard bearers for new concepts. In some cases California has been first to implement risk assessment practices first described and endorsed by U.S. EPA.

This perspective of the entrenched outsider—the loyal opposition, if you will—was a major factor leading the editors to develop this book. While often finding ourselves not totally hand in hand with the progress at the national level, we press forward, sometimes with the support and encouragement of U.S. EPA staff, sometimes not. This book might be considered to be a showcase for these efforts as a whole. That is, we present here, with the assistance of several U.S. EPA authors and other leaders in risk assessment practice, an overview of the field both as we see it and as we would like it to become—through the combined efforts of those who wish it to be carried forward. Our overall goal is to promote and encourage the science of risk assessment, particularly for exposure to drinking water contaminants.

The book first covers the major concepts and considerations of risk assessment, including how the present practice has evolved and is evolving. We wish to highlight major ongoing efforts, such as the influence of a better understanding of toxicological mechanism on risk assessment, the improved cross-species extrapolation that can be achieved by considering the basic physiological processes of the test species compared to humans, and the sources of variations in toxicological responsiveness. For the latter consideration, major efforts are being put into documentation of changes associated with the different human life stages, from the fetus to the elderly and frail population. Eventually, these present efforts will revolutionize risk assessments, and we can only hope to capture a snapshot of these efforts in passing.

The chapters on risk assessment practices are followed by descriptions of risk assessments of specific chemicals, which are used to illustrate a theme or problem. These chapters illustrate some of the interesting problems of risk assessment, and it should not be inferred that risk of all, or even a majority of the regulated chemicals, is controversial or poses some quandary to the risk assessors (or risk managers). In fact, almost the opposite is true. Most chemical risk assessments are rather straightforward. Needless to say, those are not discussed in detail here. But with the issues and discussions presented, we hope that something else shines through in this lengthy tome—that risk assessment can be intellectually stimulating, and even fun. Most of us like our jobs and enjoy the challenges provided by this risk assessment profession. We hope this is noticeable.

The two final chapters of the book more explicitly describe risk assessment needs and propose directions for the future. You will learn about some frustrations, but also about goals and dreams. Despite our immersion in the day-to-day problems of deadlines, data interpretations, and bureaucracy, it is important to step back once in a while and look around at where we are—or should be—going. This book has provided us the opportunity to do that, for which we are grateful.

Robert A. Howd
Anna M. Fan

1

INTRODUCTION TO DRINKING WATER RISK ASSESSMENT

ROBERT A. HOWD
California Environmental Protection Agency, Oakland, California

The need for a clean and safe drinking water supply for centers of population has been recognized for over 2000 years. The early Romans recognized that human activities and effluent were a major source of water pollution, and that providing water from relatively unpopulated areas was a solution to the problem. In 312 B.C. the Romans under Appius Claudius began development of an aqueduct system to deliver water taken from the Tiber River upstream of the city, thus improving the quality and quantity of their water supply (Okun, 2003). It has been said that the availability of a good water supply through their extensive aqueduct system enabled the rise of Rome as a center of civilization—and it has also been speculated that the use of lead for water pipes helped lead to its downfall, through slow poisoning of the population. This has been disputed, with evidence that terra-cotta was a preferred piping material, resulting in better-tasting drinking water. Thus, the maintenance of drinking water quality has been a major quest throughout the development of modern civilization.

However, it was not until the efforts of John Snow in 1854, analyzing a cholera epidemic in London, that specific diseases were shown to be associated with drinking waters that looked and tasted clean. For those who may not have heard the story, John Snow, a London doctor, noticed that many of the people who died of cholera in that summer's epidemic had a common factor; they all obtained their drinking water through the Broad Street well. He had the pump handle removed and the epidemic faded away. For this analysis and his subsequent publications,

Risk Assessment for Chemicals in Drinking Water, Edited by Robert A. Howd and Anna M. Fan

John Snow is credited as being the father of epidemiology. An excellent summary of these events is available at the Web site of the University of California–Los Angeles, at http://www.ph.ucla.edu/epi/snow.html.

If the slow progress of development of safe drinking water supplies from early Roman times until the mid-nineteenth century seems strange to us today, we should recall that the "germ theory" of disease wasn't elucidated by Louis Pasteur until two decades later, in the late 1870s. Recognition that bacteria were major causes of diseases, that these bacteria could be distributed in drinking water, and that removing the bacteria would protect the population from important diseases such as cholera and typhoid eventually followed.

In the United States, water quality was at first maintained in exactly the same way as in ancient Rome, primarily by transporting clean water through pipes and canals from sparsely populated regions. The need and purpose were exactly the same: to protect the drinking water supply from sewage contamination. However, transporting water over large distances is expensive, and obtaining water from nearby rivers and streams was seen by many municipalities as a preferred option. Filtration through sand was instituted in the late nineteenth century to clarify the water and decrease the bacterial contamination. This step decreased the incidence of cholera, but it soon became obvious that this was not adequate. The incidence of waterborne illnesses such as cholera and typhoid was observed to correlate with the source of the drinking water supply in major American cities, even after filtration was instituted (Okun, 2003; Pontius, 2003). Removal of bacteria by chemical disinfection began to be evaluated.

Chlorination of drinking water for bacteriological control was begun in the United States in 1908 (in Boonton, New Jersey), although it had been studied extensively before that time in both Europe and the United States (Baker, 1948). The treatment was quickly demonstrated to make a tremendous difference in disease transmission. The discoveries leading to the technique are considered to be one of the greatest public health breakthroughs of all time, preventing millions of illnesses and deaths.

DEVELOPMENT OF DRINKING WATER REGULATIONS

The first regulations for drinking water purity were primarily for bacteriological control, beginning with the U.S. Public Health Standards of 1914. These first standards applied only to water used in interstate commerce. However, eventually all 50 states adopted comparable standards for their public water supply systems (U.S. EPA, 1999). Drinking water standards for chemicals were introduced in the U.S. Public Health Standards amendments in 1925, which included standards for lead, copper, and zinc. A few more metals were added in the amendments of 1942. By 1962, the 28 constituents or properties listed in Table 1 were regulated by the U.S. Public Health Service (U.S. DHEW, 1969).

Information on the potential health effects of contaminants in drinking water, particularly those derived from the developing chemical industries, accumulated

TABLE 1. Contaminants Regulated Under the 1962 Public Health Service Standards

Alkyl benzene sulfonate	Lead
Arsenic	Manganese
Barium	Nitrate
Beta and photon emitters	Phenols
Cadmium	Radium-226
Carbon chloroform extract	Selenium
Chloride	Silver
Chromium	Strontium-90
Color	Sulfate
Copper	Threshold odor number
Cyanide	Total coliform
Fluoride	Total dissolved solids
Gross alpha emitters	Turbidity
Iron	Zinc

Source: Adapted from U.S. EPA (1999).

through the 1960s and early 1970s. Hueper (1960) reported that cities in Holland that obtained drinking water from rivers had higher cancer rates than did cities that used groundwater. Nobel laureate Joshua Lederberg pointed out in a *Washington Post* column that disinfection with chlorine was likely to form mutagenic compounds (Lederberg, 1969). A U.S. Environmental Protection Agency (EPA) study (U.S. EPA, 1972) identified 36 organic chemicals in finished drinking water from a New Orleans water treatment plant, accompanied by many more unidentified compounds. Page et al. (1974, 1976) then reported that cancer rates were higher in Louisiana cities that obtained their drinking water from the Mississippi River. These concerns led to passage of the federal Safe Drinking Water Act (SDWA) in 1974 "to assure that water supply systems serving the public met minimum national standards for protection of public health."

The SDWA authorized the EPA to set national health-based standards for drinking water to protect against both naturally occurring and human-made contaminants in drinking water (U.S. EPA, 1999, 2004). The SDWA, especially after further amendments in 1986 and 1996, requires many actions to protect drinking water and its sources (rivers, lakes, reservoirs, springs, and wells). The SDWA applies to every public water system in the United States, but does not regulate private wells that serve fewer than 25 people. The act sets up a system under which the EPA, states, and water systems work together to make sure that the standards are met.

Originally, the SDWA focused primarily on water treatment to ensure safe drinking water. The 1996 amendments expanded the law by recognizing source water protection, operator training, funding for water system improvements, and providing information to the public as important components of the drinking water delivery system. The National Primary Drinking Water Regulations implemented by the EPA under the SDWA provide for national science- and public health–based standards for drinking water, considering available technology and costs. The regulations set enforceable maximum contaminant levels (MCLs) for contaminants in drinking water or required ways to treat water to remove contaminants. In addition to setting the standards, the EPA provides detailed guidance and public information about drinking water issues, compiles drinking water data, and oversees state drinking water programs.

Drinking water supply systems are regulated directly by state drinking water programs. The states applied to the EPA for the authority to implement the SDWA within their jurisdictions *(primacy)*, which required the adoption of standards at least as stringent as the EPA's, as well as a supporting inspection and regulatory system. All states and territories except Wyoming and the District of Columbia have received primacy. The responsible agency, called the *primacy agent*, makes sure that water systems test for contaminants, reviews plans for water system projects, conducts inspections and sanitary surveys, and provides training and technical assistance. The primacy agent is also responsible for taking action against water systems that are not meeting the standards.

To aid in the development of national standards for tap water, the EPA prioritizes contaminants for potential regulation based on risk and how often they occur in water supplies. The EPA conducts a risk assessment for each chemical and sets a maximum contaminant-level goal (MCLG) based on health risk (including risks to sensitive subpopulations, e.g., infants, children, pregnant women, the elderly, and the immunocompromized). The agency also performs a cost–benefit analysis for each standard and obtains input from interested parties to help develop feasible standards. The EPA sets the MCL for the contaminant in drinking water (or a required treatment technique) as close to the health goal as they judge to be feasible.

States then adopt the new standards and are given two or more years to bring their regulated water systems into compliance. The provision in the law that state standards may be more stringent if deemed appropriate is intended primarily to allow a higher purity standard if it is economically feasible in a given region. For example, the federal standard for arsenic in drinking water was set at 10 ppb (a very high cancer risk level) based on high groundwater levels of arsenic in a few states. Cleanup to a more protective standard was judged to be cost-prohibitive in these areas. However, states with less serious arsenic problems are free to set lower, more health-protective standards. New Jersey, for example, has chosen to set the arsenic standard at 5 ppb, based on their local cost–benefit calculation (New Jersey DEP, 2004). In addition, states may decide to develop their own MCLs without waiting for the federal mandate, because the federal process is quite slow.

As of 2006, federal primary standards (MCLs, action levels, or maximum residual disinfectant levels) have been established for 80 chemicals in drinking water (see Table 2). Microbiological contaminants (e.g., cryptosporidium, total coliforms, heterotrophic plate counts) are regulated by treatment standards, as are a few other contaminants or conditions (e.g., acrylamide, epichlorhydrin, turbidity). The federal standards for lead and copper are somewhat unique. These

TABLE 2. Drinking Water Contaminants Regulated by the EPA, with Their Critical Effects and Regulatory Levels

Contaminant	Critical Effects	MCL (ppb)
Acrylamide	Nervous system or blood problems; increased risk of cancer	TT[a]
Alachlor	Eye, liver, kidney or blood problems, anemia; increased risk of cancer	2
Antimony	Increased blood cholesterol, decreased blood sugar	6
Arsenic	Skin damage or circulatory system problems; increased risk of lung, bladder, or skin cancer	10
Asbestos	Increased intestinal polyps	7 MFL[b] $> 10\,\mu M$
Atrazine	Cardiovascular system or reproductive problems	3
Barium	Increased blood pressure	2,000
Benzene	Anemia; decrease in blood platelets; increased risk of cancer	5
Benzo[*a*]pyrene	Reproductive difficulties; increased risk of cancer	0.2
Beryllium	Intestinal lesions	4
Bromate	Increased risk of cancer	10
Cadmium	Kidney damage	5
Carbofuran	Blood, nervous system, or reproductive problems	40
Carbon tetrachloride	Liver problems; increased risk of cancer	5
Chloramines (as Cl_2)	Eye/nose irritation; stomach discomfort, anemia	4,000[c]
Chlordane	Liver or nervous system problems; increased risk of cancer	2
Chlorine (as Cl_2)	Eye/nose irritation; stomach discomfort	4,000[c]
Chlorine dioxide (as ClO_2)	Anemia; nervous system effects in infants and young children	800[c]
Chlorite	Anemia; nervous system effects in infants and young children	1,000
Chlorobenzene	Liver or kidney problems	100
Chromium (total)	Allergic dermatitis	100
Copper	Short-term exposure: gastrointestinal distress; long-term exposure: liver or kidney damage	1,300[d]
Cyanide (free)	Nerve damage or thyroid problems	200
Dalapon	Minor kidney changes	200
Dibromochloropropane	Reproductive difficulties; increased risk of cancer	0.2

(*continued overleaf*)

TABLE 2. (*continued*)

Contaminant	Critical Effects	MCL (ppb)
Dichlorobenzene, 1,2-	Liver, kidney, or circulatory system problems	600
Dichlorobenzene, 1,4-	Anemia; liver, kidney, or spleen damage; changes in blood	75
Dichloroethane, 1,2-	Increased risk of cancer	5
Dichloroethylene, 1,1-	Liver problems	7
Dichloroethylene, *cis*-1,2-	Liver problems	70
Dichloroethylene, *trans*-1,2-	Liver problems	100
Dichloromethane	Liver problems; increased risk of cancer	5
Dichlorophenoxyacetic acid, 2,4-D	Kidney, liver, or adrenal gland problems	70
Dichloropropane, 1,2-	Increased risk of cancer	5
Di(2-ethylhexyl)adipate	Weight loss, liver problems, or possible reproductive difficulties	400
Di(2-ethylhexyl)phthalate	Reproductive difficulties, liver problems; increased risk of cancer	6
Dinoseb	Reproductive difficulties	7
Diquat	Cataracts	20
Endothall	Stomach and intestinal problems	100
Endrin	Liver problems	2
Epichlorhydrin	Increased cancer risk, and over a long period of time, stomach problems	TT
Ethylbenzene	Liver or kidney problems	700
Ethylene dibromide	Problems with liver, stomach, reproductive system, or kidneys; increased risk of cancer	0.05
Fluoride	Bone fluorosis; mottled teeth in children	4,000
Glyphosate	Kidney problems; reproductive difficulties	700
Gross alpha activity	Increased risk of cancer	15 pCi/L[e]
Gross beta activity	Increased risk of cancer	50 pCi/L[e]
Haloacetic acids, total	Increased risk of cancer	60
Heptachlor	Liver damage; increased risk of cancer	0.4
Heptachlor epoxide	Liver damage; increased risk of cancer	0.2
Hexachlorobenzene	Liver or kidney problems; reproductive difficulties; increased risk of cancer	1
Hexachlorocyclopentadiene	Kidney or stomach problems	50

TABLE 2. ***(continued)***

Contaminant	Critical Effects	MCL (ppb)
Lead	Infants and children: delays in physical or mental development; children could show slight deficits in attention span and learning abilities; adults: kidney problems; high blood pressure	15[d]
Lindane	Liver or kidney problems	0.2
Mercury (inorganic)	Kidney damage	2
Methoxychlor	Reproductive difficulties	40
Nitrate	Infants < 6 months old could become seriously ill, and, if untreated, may die; symptoms include shortness of breath and blue baby syndrome	10,000 (as N)
Nitrite	See above	1,000 (as N)
Nitrate + nitrite	See above	10,000 (as N)
Oxamyl	Slight nervous system effects	200
Pentachlorophenol	Liver or kidney problems; increased risk of cancer	1
Picloram	Liver problems	500
Polychlorinated biphenyls	Skin changes; thymus gland problems; immune deficiencies; reproductive or nervous system difficulties; increased risk of cancer	0.5
Radium-226 and -228	Increased risk of cancer	5 pCi/L
Selenium	Hair or fingernail loss, finger or toe numbness, circulatory problems	50
Simazine	Problems with blood	4
Strontium-90	Increased risk of cancer	8 pCi/L (now covered by gross beta)
Styrene	Liver, kidney, or circulatory system problems	100
2,3,7,8-TCDD (dioxin)	Reproductive difficulties; increased risk of cancer	0.00003
Tetrachloroethylene	Liver problems, increased risk of cancer.	5
Thallium	Hair loss; blood changes; kidney, intestine, or liver problems	2
Toluene	Nervous system, kidney, or liver problems	1,000
Toxaphene	Kidney, liver, or thyroid problems; increased risk of cancer	3
TP, 2,4,5- (Silvex)	Liver problems	50
Trichlorobenzene, 1,2,4-	Changes in adrenal glands	70

(continued overleaf)

TABLE 2. ***(continued)***

Contaminant	Critical Effects	MCL (ppb)
Trichloroethane, 1,1,1-	Liver, nervous system, or circulatory problems	200
Trichloroethane, 1,1,2-	Liver, kidney, or immune system problems	5
Trichloroethylene	Liver problems; increased risk of cancer	5
Trihalomethanes (total)	Liver, kidney, or CNS problems, cancer	80
Tritium	Increased risk of cancer	20,000 pCi/L (now covered by gross beta)
Uranium	Increased risk of cancer; kidney toxicity	30
Vinyl chloride	Increased risk of cancer	2
Xylenes	Nervous system damage	10,000

Source: U.S. EPA (2006).
[a]TT, treatment technology standard.
[b]MFL, million fibers per liter.
[c]Maximum residual disinfectant level.
[d]Action level.
[e]Picocuries per liter.

standards, known as *action levels*, are measured at the tap rather than at the source (the drinking water plant). This difference is based on the fact that a home or business plumbing system can be a major source of lead and copper, leaching from pipes, solder, and fixtures.

Secondary standards increase the total list of constituents of concern. These standards, for such chemicals as aluminum, iron, and manganese, are commonly based on taste, odor, or appearance of the water. Some chemicals have both primary and secondary standards (e.g., fluoride), and the secondary standards may be set higher or lower than the MCLs and the federal secondary standards. The secondary standards are also often based on local conditions (*U.S. Code of Federal Regulations*, 2002). Delivery of municipal water exceeding the secondary standards is allowed, but discouraged.

THE RISK ASSESSMENT PROCESS

Determination of safe levels for contaminants in drinking water requires a comprehensive system to evaluate the risk of adverse effects from exposure to chemicals and other contaminants. Considering the hundreds of contaminants that can be found using present analytical techniques, the system requires considerable resources and expertise. Simple prohibition of chemical contaminants from drinking water is not feasible, because water is a very good solvent, and analytical techniques are exquisitely sensitive. Parts per trillion (1×10^{-12}, or 1 drop in 1000 backyard swimming pools) can now be quantitated for many chemicals in drinking water. The basic premise of the risk assessment process is that the probability and severity of possible toxic effects of chemicals is related to dose. Therefore, the question is always: How clean must the water be to provide a negligible risk? The science of toxicology has evolved to address this issue.

We can think of toxicology as having started with observations of poisoning of people by plants and animal venoms. As humans developed mines and other industries, occupational poisons such as lead and mercury were documented by their obvious ill effects. About 400 B.C., Hippocrates wrote extensively of poisons and toxicology principles, and 800 years later, the Romans used poisons extensively for political gain. Paracelsus, a doctor and alchemist (1493–1541), is credited with providing the basis for the modern science of toxicology. Although he contributed much to a systematic understanding of toxicologic principles, his observation, summarized as, "the dose makes the poison," is best known.

This approach resulted in a gradual, painstaking accumulation over the next 400 years of observational knowledge about toxic chemicals and their effects on animals and humans. However, it was not until the 1940s that the need for rapid development of antimalarial drugs during World War II led to the development of formal toxicity test procedures using animal models (Gallo, 1996). At first, this involved straightforward testing of single doses of chemicals in animals to determine acute toxic effects and doses. Protocols were then extended for repeated dosing to evaluate the cumulative effects of chemicals. This was strictly for development of drugs, not to characterize toxic effects of chemicals per se. However, the value of the animal tests was self-evident, and they were quickly applied to the development of insecticides. Relative toxicity of chemicals was compared using simple criteria such as the LD_{50}, the dose that will kill 50% of the test animals within a specified period of time.

In addition to potency, studies in animals are designed to assess toxic mechanisms or concerns, such as acute or longer-term effects, effects on growth and development, effects on reproductive processes, and cancer. The most relevant data are considered to derive from studies in mammals, especially primates, but all data are potentially useful, including mutagenicity tests in microbes (Ames assays). Data from human accidental or occupational exposures to chemicals may also be relevant, as well as the results of epidemiological investigations. The potential exposure to the chemicals through the drinking water supply and other exposure routes must be considered along with the inherent hazards of the chemicals.

Drinking water risk assessment integrates all the available information into estimates of safe levels of chemicals in water. For most efficiency, it is important that the drinking water risk assessment system be integrated with toxicology evaluation programs set up for other purposes. Therefore, studies used to develop guidance on food tolerances for pesticide residues in foods are used to help determine acceptable concentrations of the pesticides in drinking water. In addition, inhalation studies intended to develop occupational standards for volatile solvents are incorporated in the risk assessment whenever available. Studies on pharmaceutical products may be relevant, especially for perspectives on mechanisms. Occupational and industrial exposures by multiple routes are also considered. In attempting to determine maximal safe doses, the assessments must consider lifetime exposures at low levels for the entire population as well as potentially susceptible subpopulations such as pregnant women and their fetuses, and the elderly.

A complex risk assessment process was developed in a relatively short time in response to the regulatory mandates. Risk assessment as a specialty barely predates the Safe Drinking Water Act of 1974. The Society for Risk Analysis was chartered in 1980 to serve the growing risk assessment community (Thompson et al., 2005). The regulatory risk assessment process was first discussed at some length in a groundbreaking report of the National Academy of Sciences (NAS) in 1983 (NAS, 1983). This report, entitled, *Risk Assessment in the Federal Government: Managing the Process* is remembered primarily for its recommendation for separation of risk assessment and risk management, so that risk assessment can be maintained as a scientific process, while risk management brings in the practical and political considerations necessary to resolve a problem. This helped lead to the field of risk assessment as a separate, scientifically driven activity.

Another influential point in the 1983 NAS report was the definition of risk assessment as a four-step process, comprised of hazard identification, dose–response assessment, exposure assessment, and risk characterization (with an accompanying rationale for this separation into parts). These definitions were also important in providing direction to the growing field of risk assessment.

As the requirements for determination of acceptable maximum exposure levels have grown, so has the list of conventions and assumptions used in risk assessment. One important convention is the separation of risk assessment into cancer and noncancer methods. Cancer risk is estimated quantitatively and the risk of noncancer effects is protected against by using uncertainty or safety factors. This separation derives from the assumption that cancer risks can be estimated using models that are linear through zero dose, whereas noncancer effects are subject to a threshold. Whether or not these assumptions are true, or should be assumed to be true for the protection of public health, is subject to much debate and is discussed at greater length in later chapters.

Current practice for cancer risk assessment involves application of a mathematical model to extrapolate measurable cancer risks at a large dose to a negligible risk level. The negligible risk level is generally considered to be in the range of 1 case in 10,000 to 1 in a million (usually expressed as 10^{-4} to 10^{-6} risk) over a lifetime of exposure. The standard animal study design for cancer evaluation, using 50 animals per dose group, has a sensitivity level of about 10% for statistical significance. That is, a 10% greater incidence of a particular type of tumor in a treated group than in the control group is required to obtain statistical significance at the $p < 0.05$ level (1 chance in 20 that the observation represents random variations rather than a toxic effect caused by the chemical treatment).

It is assumed that one-tenth the dose level cited above will result in a 1% tumor incidence, one-hundredth of the dose will result in a 0.1% tumor incidence, and so on. Obviously, whether this assumption is true cannot be determined from the tumor data. A very large study, using hundreds of animals per group, would improve the statistical significance level by only a few-fold. Such a study would cost millions of dollars and still be very far from providing information on toxicity in the dose range of greatest interest (i.e., the relatively low-level environmental exposures). The prediction of tumorigenicity might be extended to a lower dose level by using a biochemical marker that correlates with tumorigenicity, such as mutations or cross-linked DNA. However, this approach is limited by the same type of question; that is, does the amount of DNA change really correlate with the number of tumors at a very low dose level?

Epidemiological studies are potentially much more sensitive, because they can involve many thousands of people. However, these studies are also more prone to uncontrollable bias, multiple interacting risk factors, high background tumor rates, and difficulty in estimating chronic doses. For all these reasons, cancer risk assessment typically requires extrapolation of doses by several orders of magnitude to estimate human population risks on the order of 1 cancer case in 10,000 to 1 million exposed people.

For noncarcinogens, the 10% incidence threshold for statistical significance still applies if studies use 50 animals. However, among common study types, the number of animals recommended varies widely and may be as low as four per group for some studies in dogs. In addition, the effects may be observed as a significant change in a measurable parameter rather than as a yes/no variable such as the presence or absence of a tumor. A greater than 10% change in a measured value (e.g., body or organ weight, hormone level, or biochemical marker of a toxic effect) is often the approximate level for a statistically significant change from control values. A 10% change has also been utilized as an assumed threshold of biological or toxicological significance, although this is not really correct; the degree of perturbation of a system required to produce toxic effects varies according to a host of factors, including the duration of the change induced.

The concept of a threshold for toxic effects recognizes that some changes are too small to be of concern because the organism adapts to the stress. In addition, small amounts of damage can readily be repaired. Doses below this threshold would therefore be tolerable. However, the threshold dose is expected to vary among different people (as well as among different species and strains of animals) because of inherent differences in sensitivity and/or preexisting conditions such as sex, age, pregnancy, or disease. Variations in rate of *absorption, distribution, metabolism, and excretion* (ADME) of a chemical are expected. The amount of variation can be demonstrated in some physiological systems and estimated in others. Known and unknown variations can be expressed through an uncertainty factor. Traditionally, the variation among humans has been expressed as a factor of 10. That is, if one determines the average dose that is without adverse effects in a typical small group of test subjects and divides that number by 10, this dose should be without effect in any person within the entire population. Evaluations of this with real data, such as data on pharmaceutical effect levels (Hattis et al., 1999) or drug half-lives (Ginsberg et al., 2002), show that a factor of 10 is adequate to encompass most of the variability among adults but is not necessarily adequate to protect infants and children.

The same concept is used to account for other sources of variation or uncertainty, such as extrapolation from data in animals to potential effects in humans. The nominal rationale for this is that over a wide range of chemicals tested in humans and in animals, humans are as much as 10 times more sensitive to *some* of them (and less sensitive to others). Thus, if humans have not been tested, dividing the no-effect animal dose by 10 should be adequate to protect against toxic effects in humans.

Additional factors of 10 are used to extrapolate from observed acute effects to potential chronic effect levels, and from observed toxic effect levels to no observed effect levels. For each additional factor, the toxic effect level observed is divided by another factor of 10. However, the uncertainties are not necessarily multiplicative, so this strategy is likely to be overprotective. This is not necessarily inappropriate, because it protects people against unknown effects from poorly tested chemicals and provides a powerful inducement for companies that make or use the chemicals to conduct relevant toxicity tests. The possibility of additive effects or other interactions among the numerous chemicals to which people are exposed is an additional reason for caution. The more that is known about a chemical and its interactions, the smaller the uncertainty factor should be.

Risk assessment has always taken a cautionary approach, to ensure that risks have not been underestimated. However, since the very beginning of our formal risk assessment process, risk assessors have been attempting to refine the process to characterize the parameters more accurately: exposure as well as toxic effects. The presumption has been that if we have more data on chemicals and their effects on the human body, the uncertainty in risk assessment will decrease. This may lead to increases in our estimates of acceptable exposures. In practice, little or no trend toward allowing increased exposures has occurred. Other considerations, such as more and better data on susceptible populations, has tended to offset the contribution of improved estimates of exposure and toxic effects. In addition, tremendous improvements in analytical methodology have continued to reveal more chemicals in drinking water, with a resulting demand that they be evaluated and regulated.

The increase in the number of chemicals monitored has resulted in attempts to consider the effects of mixtures. The effects of chemicals may be additive, greater than additive (synergistic), or antagonistic. Interactions among chemicals may occur at every step within an organism (ADME or effects on the sensitive receptor, tissue, or organ). Such interactions are as yet relatively poorly understood and modeled; this is the subject of intense research, as discussed in a later chapter.

Attempts to better understand and model interactions of chemicals with the body are perhaps the most intense area of research. The development of physiologically based pharmacokinetic models for chemicals has greatly improved our ability to predict the behavior of chemicals in vivo, and thus to replace uncertainty factors with actual data. Credible models of toxicodynamics—the chain of events from interaction of a chemical with a receptor to tissue damage—are still in a relatively primitive stage of development. The basic risk assessment paradigm, as well as the areas of active research, are described in more detail in subsequent chapters.

PUBLIC PERCEPTIONS AND THE PRECAUTIONARY PRINCIPLE

The risk assessment system must protect public health in the face of uncertainty and recognize the public demand for a clean and safe drinking water supply. Ideally, the assessment of risks should be separated from the process for management of risks, which depends on technical and economic feasibility. Maintaining an independent risk assessment process helps assure the public that true estimates of risk will not be hidden while justifying an economic consideration—although risk–benefit trade-offs are always necessary. The public must be kept informed of the efforts made to ensure the safety of their water as well as of any problems that might occasionally arise.

As we have seen particularly in California in recent years, if people question the safety or quality of their municipal water supply, they will decline to drink it. In the United States, several billion dollars per year are now being spent on bottled water (Squires, 2006) (although the increase in bottled water consumption is by no means limited to the United States; see Doria, 2006). The fact that water quality standards for bottled water are essentially the same as for tap water seems not to be well known (Bullers, 2002; Raj, 2005; Stossel, 2005). However, safety is only one factor; people also choose to drink bottled water for taste, convenience, and even fashion (Doria et al., 2005).

To some extent, choosing bottled water is an example of the public's use of the precautionary principle. They have heard of contaminants in the tap water, and some may even have read the annual consumer confidence report from their municipal water supply company. However, the most important aspect of the use of bottled water is the exercise of choice—the public chooses what they perceive to be a higher-quality product, even if they have no evidence to substantiate this.

This brings up the ultimate rationale and justification for good risk assessments of chemicals in drinking water: our responsibility as risk assessors to ensure the public that their water supply is safe. Unfortunately, much is still unknown about the interactions of most chemicals with the human body, and many new chemicals are being introduced into commerce, and the environment, every year. With tight budgets and an expanding workload, it will be a struggle to keep up to date. The sustained diligence of risk assessment professionals is required to ensure protection of public health while moving the science forward.

Disclaimer

The opinions expressed in this chapter are those of the author and not necessarily those of the Office of Environmental Health Hazard Assessment or the California Environmental Protection Agency.

REFERENCES

Baker MN. 1948. *The Quest for Pure Water*. American Water Works Association, Denver, CO.

Bullers AC. 2002. Bottled water: better than the tap? *FDA Consum* 36(4): 14–18. Accessed at: http://www.fda.gov/fdac/features/2002/402_h2o.html.

Doria MF. 2006. Bottled water versus tap water: understanding consumer's preferences. *J Water Health* 4(2): 271–276.

Doria MF, Pidgeon N, Hunter P. 2005. Perception of tap water risks and quality: a structural equation model approach. *Water Sci Technol* 52(8): 143–149.

Gallo MA. 1996. History and scope of toxicology. In: *Casarett and Doull's Toxicology: The Basic Science of Poisons*, 5th ed. Klaassen CD, Amdur MO, Doull J, eds. McGraw-Hill, New York, pp. 3–11.

Ginsberg G, Hattis D, Sonawane B, Russ A, Banati P, Kozlak M, Smolenski S, Goble R. 2002. Evaluation of child/adult pharmacokinetic differences from a database derived from the therapeutic drug literature. *Toxicol Sci* 66(2): 185–200.

Hattis D, Banati P, Goble R. 1999. Distributions of individual susceptibility among humans for toxic effects: How much protection does the traditional tenfold factor provide for what fraction of which kinds of chemicals and effects? *Ann N Y Acad Sci* 895: 286–316.

Hueper WC. 1960. Cancer hazards from natural and artificial water pollutants. *Proceedings of the Conference on the Physiological Aspects of Water Quality*. U.S. Public Health Service, Washington, DC.

Lederberg J. 1969. We're so accustomed to using chlorine that we tend to overlook its toxicity. Washington Post, May 3, p. A15.

NAS (National Academy of Sciences). 1983. *Risk Assessment in the Federal Government: Managing the Process*. National Research Council, National Academies Press, Washington, DC.

New Jersey DEP (Department of Environmental Protection). 2004. Safe Drinking Water Act Rules: Arsenic. Adopted November, 4. NJDEP, Trenton, NJ Accessed at: http://www.nj.gov/dep/rules/adoptions/arsenic_rule7-10.pdf.

Okun DA. 2003. Drinking water and public health protection. In: *Drinking Water Regulation and Health*. Pontius FW, ed. Wiley, Hoboken, NJ, pp. 3–24.

Page T, Talbot E, Harris RH. 1974. *The Implications of Cancer-Causing Substances in Mississippi River Water*. Environmental Defense Fund, Washington, DC.

Page T, Harris RH, Epstein SS. 1976. Drinking water and cancer mortality in Louisiana. *Science* 193: 55.

Pontius FW, ed. 2003. *Drinking Water Regulation and Health*. Wiley, Hoboken, NJ.

Raj SD. 2005. Bottled water: How safe is it? *Water Environ Res* 77(7): 3013–3018.

Squires S. 2006. Testing the waters. *Washington Post* July 4, p. HE01. Accessed at: http://www.washingtonpost.com.

Stossel J. 2005. Is bottled water better than tap? Commentary. ABC News, May 6. Accessed at: http://abcnews.go.com.

Thompson KM, Deisler PF, Schwing RC. 2005. Interdisciplinary vision: the first 25 years of the Society for Risk Analysis (SRA), 1980–2005. *Risk Anal* 25(6): 1333–1386.

U.S. Code of Federal Regulations. 2002. CFR part 143, *National Secondary Maximum Contaminant Levels*. 44 FR 42198. Accessed at: www.access.gpo.gov/nara/cfr/waisidx_02/40cfr143_02.html.

U.S. DHEW (Department of Health, Education, and Welfare). 1969. *Public Health Service Drinking Water Standards, 1962*. Public Health Service Publication 956. Reprinted September 1969. U.S. DHEW, Washington, DC. (As cited in U.S. EPA, 1999.)

U.S. EPA (Environmental Protection Agency). 1972. *Industrial Pollution of the Lower Mississippi River in Louisiana*. U.S. EPA, Region VI, Dallas, TX.

______. 1999. *25 Years of the Safe Drinking Water Act: History and Trends*. U.S. EPA, Washington, DC. Accessed at: http://permanent.access.gpo.gov/websites/epagov/www.epa.gov/safewater/sdwa/trends.htm.

______. 2001. Controlling disinfection by-products and microbial contaminants in drinking water. Chapter 2 in: *A review of Federal Drinking Water Regulations in the US* (by James Owens). EPA/600/R-01/110. U.S. EPA, Washington, DC. Accessed at: http://www.epa.gov/NRMRL/pubs/600r01110/600r01110.htm.

______. 2004. *Safe Drinking Water Act 30th Anniversary: Understanding the Safe Drinking Water Act*. EPA/816/F-04/030. U.S. EPA, Washington, DC. Accessed at: http://www.epa.gov/safewater/sdwa/30th/factsheets/understand.html.

______. 2006. *List of Drinking Water Contaminants and MCLs*. Office of Water, U.S. EPA, Washington, DC. Accessed at: http://www.epa.gov/safewater/mcl.html.

2

SUMMARY OF THE DEVELOPMENT OF FEDERAL DRINKING WATER REGULATIONS AND HEALTH-BASED GUIDELINES FOR CHEMICAL CONTAMINANTS

JOYCE MORRISSEY DONOHUE AND WYNNE MAYNOR MILLER
U.S. Environmental Protection Agency, Washington, DC

Concern for the purity of drinking water in the United States, specifically its relationship to infectious disease, has been a matter of public interest for over a century. The discovery that chlorination of drinking water dramatically reduced the incidence of waterborne disease gave rise to the treatment of potable water with chemical agents (AWWA, 2005). In addition, public awareness of source water contamination from industrial wastes discharged into rivers and other source waters increased as the impact on the environment became apparent.

The U.S. Public Health Service (PHS) established the first formal bacteriological public health standards or guidelines for drinking water in 1914. These standards applied only to contagious disease-causing contaminants and to water systems that provided drinking water to interstate carriers (e.g., ships, trains, and buses). In essence, these 1914 standards were set primarily "to protect the health of the traveling public." The PHS subsequently established the first guidelines for chemical contaminants in 1925 and revised, expanded, and/or updated the guidelines in 1946, 1947, and 1962. Although the 1962 final standards for 28 substances were not federal mandates for public drinking water systems, over

Risk Assessment for Chemicals in Drinking Water, Edited by Robert A. Howd and Anna M. Fan

time and with some minor changes, all 50 states adopted these standards as either regulations or guidelines (Knotts, 1999; U.S. EPA, 1999).

By the late 1960s, waste discharges from industry, factories, and mining, as well as runoff from agricultural and urban landscapes, were beginning to affect water sources and drinking water supplies. A 1969 PHS survey of 969 community water supplies found that 41% of water treatment facilities were not meeting the standards set by the PHS. A 1972 survey detected 36 chemicals in treated water from treatment plants that relied on the Louisiana portion of the Mississippi River as a water source. In the early 1970s, because of increasing concerns about the safety of the public's drinking water and a host of other environmental issues, Congress established the U.S. Environmental Protection Agency (EPA) and passed several federal laws to address the release of pollutants into the environment (Knotts, 1999; U.S. EPA, 1999).

The Safe Drinking Water Act (SDWA), which Congress passed in 1974, is the primary law that addresses the safety of public drinking water. As directed by the 1974 SDWA, the EPA issued the first federally mandated regulations for public water systems (PWSs) in 1975, building upon existing Public Health Service standards. These 1975 regulations included interim drinking water standards for six organic chemicals, 10 inorganic chemicals, turbidity, and total coliform bacteria. The EPA set interim standards for radionuclides in 1976 and regulations for total trihalomethanes (TTHMs) in 1979. In addition to authorizing the EPA to set drinking water standards, the 1974 SDWA provided for monitoring the levels of chemical and microbial contaminants in drinking water and for public notification (U.S. EPA, 1999).

Although Congress made slight revisions to the SDWA in 1977, 1979, and 1980, it was the 1986 and the 1996 SDWA reauthorizations that resulted in the most significant changes, refining and expanding the EPA's authority and including risk assessment and risk management practices. With respect to the setting of standards, the 1986 SDWA amendments mandated the EPA to develop maximum contaminant level goals (MCLGs) and maximum contaminant levels (MCLs) for 83 contaminants by 1989. This list of contaminants included the 16 interim standards that EPA issued in 1975. The SDWA required the agency to finalize the interim standards as national primary drinking water regulations (NPDWRs). The EPA promulgated final NPDWRs for 76 of the 83 contaminants by 1992. These regulations applied to organic chemicals, inorganic chemicals, and pathogens. The remaining contaminants specified by the 1986 SDWA included arsenic, radon, radionuclides, and sulfate (U.S. EPA, 1986b, 1999).

Congress also amended the SDWA in 1988 with the Lead Contamination Control Act, which established a program to eliminate lead-containing drinking water coolers in schools (U.S. EPA, 1999), and then again in the 2002 Bioterrorism Act, which mandates certain requirements for community water systems to guard against terrorist attacks or other intentional acts that jeopardize the nation's drinking water supply (U.S. EPA, 2004). The 1996 SDWA amendments contained several key mandates that provided further enhancements to the existing act. The amendments directed the EPA to develop or amend specific drinking water standards for arsenic, radon, disinfection by-products, *Cryptosporidium*, and disinfection requirements for groundwater systems. The EPA proposed a regulation for radon in 1999 and revised and promulgated a final standard for arsenic in 2001. The EPA addressed sulfate with the development of a health advisory, and regulations for *Cryptosporidium* and disinfectants and their by-products (DBPs) have been addressed through a series of regulations between 1998 and 2006. After the 1996 SDWA amendments, the EPA also established final drinking water standards for radionuclides in 2000 (some of which had been interim since 1976). As of 2006, more than 90 microbial, chemical, and radiological contaminants are regulated by the SDWA, guiding the safety of water delivered by public water systems across the United States (U.S. EPA, 1999).

The 1986 SDWA amendments directed the EPA to regulate 25 contaminants every three years. However, this approach made it difficult for the agency to prioritize and target high priority contaminants for drinking water regulation. Hence, the 1996 amendments eliminated this provision and allowed the agency to determine which unregulated contaminants should be regulated with an NPDWR by developing and applying a risk-based contaminant selection approach based on meeting certain SDWA criteria. This process, called the *contaminant candidate list* (CCL) *process* is described in the next section.

SELECTING CANDIDATES FOR REGULATORY CONSIDERATION

As mentioned above, the 1996 SDWA amendments require the EPA to publish a list of contaminants that are known or anticipated to occur in public water systems and may require regulation with a national primary drinking water regulation. In selecting contaminants for the CCL, the SDWA requires the agency to select those of greatest public health concern based on health effects and occurrence. The SDWA directs the agency to publish this list of unregulated contaminants every five years. The agency uses the CCL to guide its research efforts and to set regulatory priorities for the drinking water program. The first CCL was finalized in 1998 and the second in 2005.

The 1996 SDWA amendments also require the EPA to make regulatory determinations for no fewer than five contaminants from the CCL within five years after enactment of the 1996 amendments and every five years thereafter. The criteria established by the SDWA for making a determination to regulate a contaminant (i.e., a positive regulatory determination) include the following:

- The contaminant may have an adverse effect on the health of persons.
- The contaminant is known to occur or there is a substantial likelihood that the contaminant will occur in public water systems with a frequency and at levels of public health concern.
- In the sole judgment of the administrator, regulation of the contaminant presents a meaningful opportunity for health risk reduction for persons served by public water systems.

In order to make a decision to regulate a contaminant, the findings for all three criteria must be affirmative.

To address the statutory criteria for making a regulatory determination, the EPA typically evaluates the health impact of the contaminant, quantifies the dose–response relationship through a formal health risk assessment process, and analyzes occurrence data from public water systems. The agency also evaluates the availability of analytical methods for monitoring and for effective treatment technologies. In addition, the agency considers the potential impact to populations exposed, especially the impacts on sensitive populations (e.g., infants, children, elderly, immunocompromized), the national distribution of the contaminant in public water systems, and other sources of exposure information.

As required by the SDWA, a decision to regulate a contaminant on the CCL commits the EPA to publication of a *maximum contaminant level goal* (MCLG) and promulgation of a *national primary drinking water regulation* (NPDWR) for that contaminant. An MCLG is defined in SDWA Section 1412(b)(4)(A) as "the level at which no known or anticipated adverse effects on the health of persons occur and which allows an adequate margin of safety." Depending on the contaminant, the NPDWR is generally established as either a *maximum contaminant level* (MCL) or a *treatment technique* (TT) regulation. The EPA defines the MCL as the maximum permissible level of a contaminant in drinking water that is delivered to any user of a public water system, and a TT as an enforceable procedure or level of technological performance, that public water systems must follow to ensure control of a contaminant. The agency can also determine that there is no need for a regulation when a contaminant fails to meet one of the statutory criteria.

The first regulatory determination was completed in 2003. The agency issued health advisories for sulfate, sodium, and manganese and provided guidance regarding acanthamoeba for contact lens wearers. Drinking water regulations were not recommended for aldrin, dieldrin, hexachlorobutadiene, metribuzin or naphthalene, based on their low occurrence in public water supplies. The agency is presently in the process of making regulatory determinations from the second CCL. Where the agency determines that regulation of contaminant on the CCL is necessary, SDWA requires the agency to propose the regulation within 24 months of the published regulatory determination and to finalize the regulation within 18 months of the regulation proposed. Although to date, the agency has not identified any candidates for regulation through the CCL and regulatory determination process, the following section provides a general description of the key components that have been evaluated during past regulatory development efforts for chemical contaminants.

KEY COMPONENTS FOR REGULATORY DEVELOPMENT

In developing NPDWRs for chemical contaminants, the agency evaluates several key components that form the underpinnings of a national primary drinking

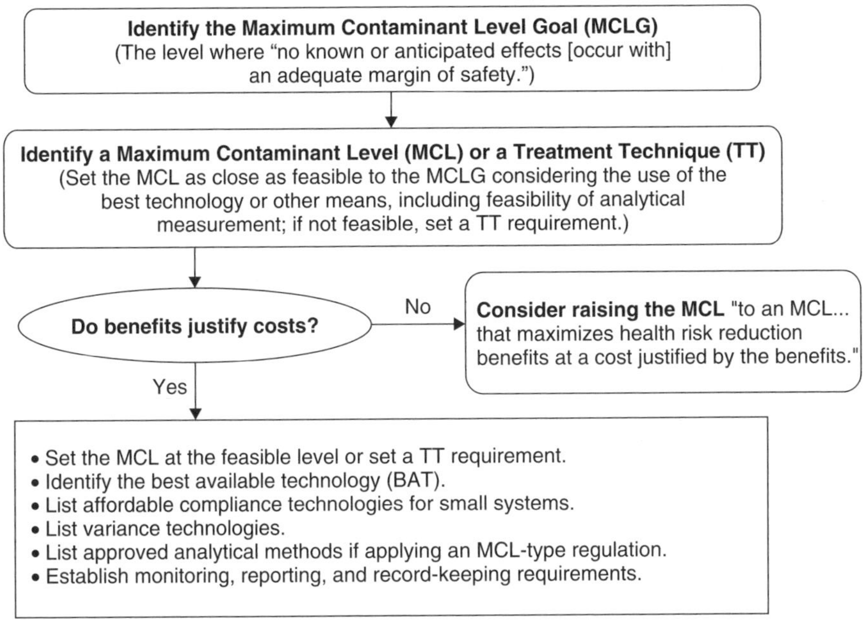

Figure 1. Typical process for establishing a maximum contaminant level goal and a national primary drinking water regulation.

water standard. Figure 1 illustrates how these key components generally interact during the regulatory development process. The key components that are typically evaluated when establishing a drinking water regulation include the following:

- Health effects
- Analytical methods
- Occurrence data
- Treatment technologies
- Cost–benefit information

Although not an exhaustive description, a synopsis of the typical data, information, and/or factors considered for each key regulatory development component is provided in the following sections.

Health Effects

In conducting a risk assessment for the development of an MCLG (the health-based component of a regulation), all available toxicity data are gathered from published literature or other sources, such as those that may have been submitted to regulators as unpublished reports or confidential business information. Data describing cancer effects or data relevant to cancer risk are assessed separately from data describing noncancer effects.

Noncancer Effects For effects other than cancer, the EPA develops an oral *reference dose* (RfD), defined as an estimate (with uncertainty spanning perhaps an order of magnitude) of a daily oral exposure to the human population (including sensitive subgroups) that is likely to be without an appreciable risk of deleterious effects during a lifetime (U.S. EPA, 1993). To develop an RfD, data from human and/or animal studies are evaluated. Oral drinking water studies are preferred, but studies using other routes, such as diet or gavage, may be considered. For each of these studies, the highest dose that causes no adverse effect, the *no-observed-adverse-effect level* (NOAEL), and the lowest dose that produces an adverse effect, the *lowest-observed-adverse-effect level* (LOAEL), are identified for each test species relevant to humans. The effect(s) at the LOAEL are termed *critical effect*(s).

The critical effect(s) are not just the effects observed at the lowest doses tested, but are also preferably effects that increase in severity as doses increase. The existence of a dose-related response supports the conclusion that an effect is due to compound administration. Increasingly, the dose–response relationship for the critical effect(s) is modeled mathematically to identify a response level [*benchmark response* (BMR)] and its associated *benchmark dose* (BMD) at the lower end of the range of observation. The *lower confidence bound on the BMD* (BMDL) is used as the point of departure for the RfD determination. Response levels of 10%, 5%, a one standard deviation change, or a half standard deviation change are those most frequently selected for the point of departure for the quantitative assessment. In cases where the critical effect is apparent only at the highest dose tested or where the dose–response pattern does not fit any of the available models, the NOAEL or, in the absence of an NOAEL, the LOAEL serves as the point of departure for the RfD analysis.

The RfD is estimated by dividing the BMDL, NOAEL or LOAEL by a *composite uncertainty factor*, which accounts for differences in the response to toxicity within the human population as well as differences between humans and animals when animal data are used (Table 1). If the study selected as the basis for the RfD involves an exposure duration other than the lifetime, another factor may be used. Similarly, if an LOAEL is used in estimating the RfD, a factor may be applied to account for the absence of an NOAEL. Professional judgment may suggest the use of an additional uncertainty factor due to an insufficient database for that chemical.

In selecting the uncertainty factor, each area of uncertainty is evaluated and assigned a value of 1, 3, or 10, depending on the strength of the data. A threefold factor is used when data are available to reduce the need to apply a 10-fold unit of uncertainty. For example, a LOAEL that is an early biomarker of toxicity or a nearly complete toxicity data set may required threefold uncertainty factor rather than a factor of 10. An uncertainty factor of 1 is employed when the data are clearly from the most sensitive members of the human population, eliminating the need for intraspecies adjustment. The net uncertainty factor is the product of the individual factors used. Uncertainty factors tend to range from 1 to 3000-fold.

TABLE 1. Uncertainty Factors

Uncertainty Component	Description
UF_H	A factor of 1, 3 (approximately $\frac{1}{2}$ $\log_{10}$ unit), or 10 used to account for variation in sensitivity among members of the human population (intraspecies variation)
UF_A	A factor of 1, 3, or 10 used to account for uncertainty when extrapolating from valid results of long-term studies on experimental animals to humans (interspecies variation)
UF_S	A factor of 1, 3, or 10 used to account for the uncertainty involved in extrapolating from less-than-chronic NOAELs to chronic NOAELs
UF_L	A factor of 1, 3, or 10 used to account for the uncertainty involved in extrapolating from LOAELs to NOAELs
UF_D	A factor of 1, 3 or 10 used to account for the uncertainty associated with extrapolation from the critical study data on some of the key toxic endpoints is lacking, making the database incomplete

Uncertainty factors greater than 3000 may indicate too much uncertainty to have any confidence in the RfD.

Tumorigenic Effects The data for tumorigenic effects are assessed qualitatively (hazard identification) and quantitatively (dose–response assessment). The qualitative evaluation for a carcinogen involves an assessment of the weight of evidence for the chemical's potential to cause cancer in humans and takes into account the mode of action by which chemical exposure leads to cancer (U.S. EPA, 2005). The data considered in the risk assessment include both human epidemiology and animal studies.

The EPA (1986a) *Guidelines for Carcinogenic Risk Assessment* established five alphanumeric cancer categories, as identified in Table 2. Although the revised guidelines (U.S. EPA, 2005) have replaced those of 1986, the 1986 assessments

TABLE 2. Alphanumeric Cancer Classification Categories

Group	Category
A	Human carcinogen
B	Probable human carcinogen: B1 indicates limited evidence from epidemiological studies B2 indicates sufficient evidence from animal studies and inadequate or no data from epidemiological studies
C	Possible human carcinogen
D	Not classifiable as to human carcinogenicity
E	Evidence of noncarcinogenicity for humans

Source: U.S. EPA (1986a).

apply to the majority of chemicals regulated by the agency prior to the 1996 reauthorization of the SDWA. The 2005 cancer guidelines have replaced the alphanumeric system with descriptors and narratives describing a chemical's potential to produce cancer in humans. Under the new guidelines, there are five generalized descriptors for carcinogens:

- Carcinogenic to humans
- Likely to be carcinogenic to humans
- Suggestive evidence of carcinogenic potential
- Inadequate information to assess carcinogenic potential
- Not likely to be carcinogenic to humans

The quantitative assessment of tumorigenic potency is determined by the mode of action. The potency for those chemicals that lead to tumors through a known mutagenic mode of action and those for which a mode of action cannot be determined is determined to be linear and expressed in terms of a slope factor. Those tumors that are the result of nongenotoxic mechanisms (e.g., regenerative hyperplasia) and do not exhibit a linear response to dose are quantified using an RfD-like approach and, where possible, are based on the dose–response relationship for a precursor effect in the mode of action leading to the tumors.

The process for deriving the slope factor for a carcinogen is similar to the benchmark modeling described above. The quantal relationship of the tumors to dose is plotted using the multistage model available in the agency benchmark dose modeling software. The *point of departure* is a dose that falls at the lower end of the range of observation for tumor response. It is generally a dose that is statistically greater than the background (control) response for a specific tumor type or group of related tumors. The lower bound on the point of departure is determined and a straight line is plotted from the lower bound to zero. The *slope factor* is the slope of that line and represents the tumorigenic potency of the chemical. The RfD or cancer slope factor from the health risk assessment provided the foundation for developing the health-based regulatory values described in the next section.

Analytical Methods

In developing an NPDWR, Section 1401(1)(D) of the SDWA directs the EPA to include criteria and procedures to assure a supply of drinking water that complies dependably with the maximum contaminant levels, including accepted methods for quality control and testing procedures to ensure compliance with such levels. Only approved analytical methods may be used for compliance monitoring of drinking water contaminants regulated under an NPDWR.

In promulgating an NPDWR for a drinking water contaminant, Section 1401 of the SDWA directs the agency to specify an MCL or treatment technique. More specifically, Section 1401(1)(C)(I) directs the EPA to set an MCL for NPDWRs if it is "economically and technologically feasible to ascertain the level of a contaminant water in public water systems." Alternatively, if it is not "economically and technologically feasible to so ascertain the level of such contaminant," the SDWA specifies that the EPA may, in lieu of an MCL identify known treatment techniques that reduce the contaminant sufficiently in drinking water [section 1401(1)(C)(ii)]. In deciding whether an analytical method used to ascertain a contaminant in drinking water is economically and technologically feasible, the agency generally considers the following factors:

- Sensitivity of analytical method(s) to address the concentration of concern (i.e., are detection and quantitation sufficient to meet the MCL?)
- Method reliability, precision (or reproducibility), and bias (accuracy or recovery) at the MCL
- Ability to identify the contaminant of concern in the presence of potential interferences (method specificity)
- Methods that are suitable for routine use in compliance monitoring
- Availability of certified laboratories, equipment, and trained personnel sufficient to conduct compliance monitoring
- Cost of the analysis to public drinking water systems

For the first criterion (sensitivity), the EPA typically uses the *method detection limit* (MDL) and *practical quantitation level* (PQL) to estimate the limits of performance of an analytical method to measure chemical contaminants in drinking water. The MDL is defined as "the minimum concentration of a substance that can be measured and reported with 99% confidence that the analyte concentration is greater than zero" (40 CFR Part 136 Appendix B), and the PQL is generally defined by the EPA's Drinking Water Program as "the lowest concentration of an analyte that can be reliably measured within specified limits of precision and accuracy during routine laboratory operating conditions" (U.S. EPA, 1985). Because MDLs can be operator, method, laboratory, and matrix specific, they may not necessarily be reproducible within a laboratory or between laboratories, due to the day-to-day analytical variability that can occur and the difficulty of measuring an analyte at very low concentrations. The EPA considers this analytical variability during regulation development and uses the PQL to estimate the minimum reliable quantitation level that most laboratories can be expected to meet during day-to-day operations. The agency has used the PQL to estimate the feasible level of measurement for most regulated chemical contaminants. The EPA has set the MCL at the feasible level of measurement for approximately 23 carcinogenic and two noncarcinogenic compounds.

Occurrence Data

Data and information on the likely occurrence of a contaminant in public water systems is a key component in the development of an NPDWR. The EPA uses this occurrence data and information to estimate the number of public water systems affected and the potential population exposure at the various regulatory levels under consideration (e.g., a potential MCL). The EPA uses the estimates of the total number of systems and populations affected in its cost–benefit analysis. In evaluating occurrence data for use in regulation development, the EPA typically considers some of the following factors:

- Overall quality and limitations of the data
- Reporting limit and whether the data are useful for evaluating regulatory levels of interest
- Number and types of water systems (e.g., small or large water systems, transient or nontransient)
- Type of water (raw or finished)
- Source water used by the water system (ground or surface)
- Representativeness (national, regional, or local in scope)
- Time period and frequency of sample collection

Over the years, the EPA has used several sources of occurrence data for the various regulatory development efforts. Some of the primary and supplemental sources of occurrence data or information that the agency has used in regulation development have included but were not limited to the following:

- Required unregulated contaminant monitoring (to provide data for CCL regulatory determinations): 1988 to 1992, 1993 to 1997, 2000 to 2003
- National inorganics and radionuclides survey (NIRS): 1984 to 1986
- National organic monitoring survey (NOMS): 1976 to 1977
- Rural water survey (RWS): 1978 to 1980
- Community water system surveys (CWSSs)
- National pesticide survey
- Information collection request (ICR) for disinfection by-products

Treatment Technologies and Best Available Technologies

When establishing an NPDWR, Section 1412(b)(4)(E) of the SDWA requires the EPA to list the treatment technology and other means that are feasible for achieving an MCL and to identify the *best available technology* (BAT). In the process of meeting this requirement, the EPA evaluates data and information to determine if the following criteria are met satisfactorily:

- Documented high removal efficiency
- Full-scale operation testing
- General geographic applicability
- Environmentally safe options for residuals handling
- Compatibility with other water treatment processes

- Broad applicability in achieving water system compliance
- Effects on the distribution system
- Water resource and water reuse options
- Potential environmental quality concerns
- Reasonable cost basis for large and medium-sized systems

Although the EPA is required to list BATs when establishing an MCL, systems are not required to use a listed BAT to comply with the MCL.

In addition to the BAT requirements for medium-sized and large systems, Section 1412(b)(4)(E)(ii) of the SDWA requires the EPA to list affordable technologies for small drinking water systems that achieve compliance with an MCL. The EPA evaluates affordable compliance technologies for three size categories of small public water systems. These three small system size categories include (1) those serving more than 25 but fewer than 500 people, (2) those serving more than 500 but fewer than 3300 people, and (3) those serving more than 3300 but fewer than 10,000 people. To determine whether compliance technologies are affordable for a small system, the EPA compares (for a representative system) the current annual household water bill (or baseline cost) plus the incremental cost of the new regulation to an affordability threshold of 2.5% of the median household income. On March 2, 2006, the EPA proposed changes and is requesting comments on its methodology for determining affordability for small systems (U.S. EPA, 2006a).

The 1996 SDWA amendments allow states to grant variances to small public water systems serving fewer than 10,000 people that cannot afford to comply with an NPDWR. The SDWA specifies that variances cannot be granted for microbial contaminants. States may grant small system variances only for those drinking water standards that the EPA has determined are unaffordable for one or more categories of small systems. In order for states to grant variances on a case-by-case basis, the EPA must find that (1) small systems cannot afford the technology recommended, (2) affordable variance technologies are available, and (3) the variance technologies available are protective of public health (U.S. EPA, 2006a).

Cost–Benefit Information

Prior to 1996, the SDWA did not contain the provision that the costs of a rule had to be supported by the benefits achieved. Costs and benefits were considered, but to a more limited extent than is required by the 1996 SDWA. The 1996 SDWA amendments require the EPA to determine whether or not the quantifiable and nonquantifiable benefits of an MCL justify the quantifiable and nonquantifiable costs based on the health risk reduction and cost analysis (HRRCA) required under SDWA Section 1412(b)(3)(c). The 1996 SDWA amendments also grant the EPA discretionary authority in certain cases to set an MCL that is less stringent than the feasible level if the benefits of an MCL set at the feasible level would not justify the costs [Section 1412(b)(6)].

Assessing the cost–benefit impacts of an NPDWR is complex and involves not only mandates from the SDWA but also other federal acts and executive orders. The EPA typically summarizes the potential impacts of a regulation in an *economic analysis* document. The costs associated with a drinking water regulation include primarily (1) the costs incurred by public water systems to comply with the NPDWR and its monitoring requirements, and (2) the costs incurred by states to implement and enforce the regulation. Some of the key factors that the EPA typically considers in performing a cost analysis include:

- Distribution of contaminant occurrence for various water systems
- Treatment technologies and non-treatment-related decisions that water systems might use to achieve compliance
- Unit costs of the various technologies used for compliance at regulatory levels of interest
- Costs of monitoring, implementation, record keeping, and reporting (by the water system and/or by the state).

The EPA uses this information to estimate national costs for the various regulatory options, system costs (by system size and type), and potential household costs.

The benefits component of the cost–benefit assessment is equally complex. The first step is to consider the adverse health effects that are likely to be reduced as a result of regulation. As specified by the SDWA, when the EPA develops an NPDWR, the agency identifies the quantifiable and nonquantifiable benefits associated with the regulation. Estimates of reduced morbidity (illness) and mortality (death) risks are generally based on (1) estimates of population risks, (2) estimates of change in risks that will result from the regulation, and (3) estimates of the number of adverse health outcomes that will be avoided as a result of the proposed regulation. If possible, the agency attempts to monetize the heath benefits that are identified through the assessment.

For the populations served by an NPDWR, the value of reducing the risk of adverse health effects generally includes two components: avoidance of medical costs and productivity losses associated with illness, and reduction in risk of premature mortality. This conceptual valuation framework goes beyond valuing out-of-pocket medical costs and lost time to include the value that consumers place on avoiding pain and suffering and the risk premium.

DEVELOPMENT OF REGULATORY VALUES

Maximum Contaminant Level Goal

An MCLG is defined as a concentration of a contaminant in drinking water that is anticipated to be without adverse health effects over a lifetime. An MCLG is a nonenforceable value. The methodology used in establishing an MCLG will differ based on the nature of the critical adverse effect and its mode of action. In the case of chemicals with an experimentally supported threshold mode of action, the RfD provides a point of departure for an MCLG calculation. In the case of chemicals that have no threshold for their mode of action, such as mutagenic carcinogens and carcinogens with an unidentified mode of action, the MCLG is zero according to current agency practices.

The following equation is used in deriving a nonzero MCLG:

$$\text{MCLG} = \frac{\text{RfD} \times \text{body weight} \times \text{relative source contribution}}{\text{drinking water intake}} \tag{1}$$

where

RfD = reference rose
body weight = 70 kg (adults)
drinking water intake = 2 L/day
relative source contribution = portion of the total exposure contributed by drinking water

Historically, chemicals have been grouped into three categories (category I, II, or III) for derivation of the MCLG (U.S. EPA, 1991). *Category I chemicals* are those that were categorized as human/known or likely/probable carcinogens, and thus assigned a zero MCLG. Chemicals characterized as possible carcinogens under the 1986 cancer guidelines were placed in *category II*. Rather than treating the possible carcinogens as category I chemicals in determining the MCLG, the EPA traditionally added a risk management factor of from 1 to 10 to the denominator of the MCLG equation [equation (1)]; thereby adjusting the MCLG. At present the need for a risk management factor for a category II chemical is determined on a case-by-case basis. *Category III chemicals* are those with no evidence of carcinogenicity in animals via the oral route.

Maximum Contaminant Level

An MCL, not an MCLG, is the enforceable NPDWR. For chemicals, an MCL is generally set at the lowest feasible level that can be achieved technologically. Prior to 1996, costs and benefits were considered but to a more limited extent than required by the 1996 SDWA amendments. In most cases, where it was not possible to achieve an MCLG technologically, an MCL was set at the PQL (i.e., the concentration that can be measured in water by most analytical testing laboratories). In a few cases the MCL was established based on the level that can be achieved through treatment using the BAT rather than the PQL. The cancer risk associated with the resulting MCL is generally no greater than one person in a population of 10,000. The MCL was set based on technology for all carcinogens with a zero MCLG and for two noncarcinogens, 1,1,2-trichloroethane and thallium. For these two noncarcinogens, it was not possible to achieve the MCLG due to PQL limitations, leading to an MCL that is higher than the MCLG.

As noted earlier, the 1996 SDWA amendments require the EPA to determine whether or not the benefits of an MCL justify the costs based on a health risk reduction and cost analysis. Depending on the outcome of that assessment, the EPA can in certain cases set an MCL that is less stringent than the feasible level if the benefits do not justify the costs. The agency has used its discretionary authority for a limited number of drinking water regulations, including the final drinking water standard for uranium (2000 radionuclides rule), the final regulation for arsenic (2001), and the stage 2 disinfection by-product rule (2005). The MCLs covered under these regulations reflect concentrations in drinking water that are technologically achievable at costs that have a favorable relationship to the benefits that are expected once the regulation is in place.

Treatment Technique Regulations

When it is not economically or technically feasible to set an MCL, or when there is no reliable or economically feasible method to detect contaminants in the water, SDWA Section 1401(1)(C)(ii) directs the EPA to set a treatment technique (TT) requirement in lieu of an MCL. A TT specifies a type of treatment (e.g., filtration, disinfection, other methods of control to limit contamination in drinking water) and means for ensuring adequate treatment performance (e.g., monitoring of water quality to ensure treatment performance). The regulations for acrylamide and epichlorohydrin specify TT rather than MCL values because analytical methods for monitoring at levels within the risk range desired are not available. Accordingly, the best method of control was regulation of the treatment materials that are the major contributors of these contaminants to drinking water: the polyacrylamide and epichlorohydrin/dimethylamine coagulation aids. The regulations for these two contaminants restrict the amount of monomer in the polymer products and the maximum use level for the polymer coagulants, thereby controlling the concentrations of the contaminant in water.

Two other chemical contaminants, lead and copper, are treatment technique regulations. Since the major source of both of these contaminants is corrosion of the distribution system, the EPA regulation requires at-the-tap monitoring in a specified number of high-risk homes. If the concentrations in more than 10% of the homes tested exceed the action level value specified, the utility is required to implement measures that will reduce the corrosiveness of the water and decrease the leaching of lead and copper. The action level for lead (15 μg/L) is a concentration that will minimize the risk for neurodevelopmental effects in children and kidney problems in adults. The action level for copper (1.3 mg/L) protects against the acute gastrointestinal effect of copper and limits the opportunities for copper to accumulate in the livers of persons genetically sensitive to liver damage.

NONREGULATORY OPTIONS

Health Advisories

The development of *health advisory* (HA) values is one option that the SDWA provides for chemicals that occur in drinking water but not at a concentration or risk level of national concern. The EPA health advisory (HA) program was developed to assist local officials and utilities by establishing concentrations of concern for unregulated contaminants or when there are short-term excursions above the MCL for a regulated contaminant. HA values describe nonregulatory concentrations of drinking water contaminants at which adverse health effects are not anticipated to occur over specific exposure durations. They serve as technical guidance to federal, state, and local officials responsible for protecting public health when emergency spills or contamination situations occur. They are not legally enforceable federal standards and are subject to change as new information becomes available (U.S. EPA, 1989).

HA values are developed for 1-day, 10-day, longer-term, and lifetime exposures. They apply only to noncancer endpoints. In some but not all cases, chemicals that are classified as known or probable carcinogens lack a lifetime HA. The 1-day and 10-day values are established for a 10-kg (22-lb) child based on the premise that this group is most sensitive to acute toxicants. Longer-term exposures (estimated to be seven year or one-tenth of an average lifetime), are calculated for both a 10-kg child and for adults (70 kg). Each of these HA calculations assumes that drinking water is the only source of exposure to the chemical. Each HA is an estimate of the concentration of a chemical in drinking water that is not expected to cause adverse noncarcinogenic effects in a young child or adult for the duration specified. The values are developed from a study in humans or animals that provides a dose–response relationship for a comprehensive suite of endpoints for the appropriate duration (U.S. EPA, 1989). Data from long-term studies are not utilized for the derivation of short-term HA values. The lifetime HA is established only for the adult; it is adjusted to allow for nondrinking water sources of exposure.

The dose–response data are used to identify a BMDL, NOAEL, or LOAEL for the critical effect associated with the duration of interest. The LOAEL is used for calculation if a BMDL or NOAEL has not been identified, but only if the effect observed is an early marker of toxicity rather than a frank (severe) effect. Less-than-lifetime HA values are derived using the following equations:

$$\text{less-than-lifetime HA (child)} = \frac{\text{NOAEL or LOAEL} \times 10\ \text{kg}}{\text{UF} \times 1\ \text{L/day}} \tag{2}$$

$$\text{longer-term HA (adult)} = \frac{\text{NOAEL or LOAEL} \times 70\ \text{kg}}{\text{UF} \times 2\ \text{L/day}} \tag{3}$$

In the derivation of less-than-lifetime HAs, uncertainty factors (UFs) are most often employed for intraspecies adjustment, interspecies adjustment, and use of an LOAEL in place of an NOAEL. UFs are generally in multiples of three or 10, paralleling the UFs used in RfD development but not including a duration adjustment. The lifetime HA is the most conservative of the suite of HA values and is the equivalent of the MCLG for regulated noncarcinogens. In some cases the lifetime HA includes a risk management factor to account for potential carcinogenicity.

Secondary Maximum Contaminants Levels

The EPA also establishes *secondary maximum contaminant* levels (SMCLs) for drinking water contaminants. The SMCLs apply to factors such as color, taste, and odor that influence consumer acceptability of the water but are unrelated to human health. SMCLs have been established for aluminum, chloride, copper, fluoride, iron, manganese, silver, sulfate, and zinc, as well as for such characteristics of the water as color, pH, odor, and corrosivity. The SMCLs for fluoride and silver are related to cosmetic effects in humans that influence appearance but not physiological function. The cosmetic effect for fluoride is dental fluorosis, a form of discoloration of tooth enamel, and that for silver is argeria, a change in skin color caused by silver deposits. The SMCLs for the other chemical contaminants are based on their taste or color properties. SMCLs are not enforceable, but some may be adopted as regulations by individual states.

Drinking Water Advisories

In some cases (methyl tertiary butyl ether, sodium, and sulfate), the EPA has developed a drinking water advisory based on aesthetic effects (taste and/or odor) as the point of departure. Drinking water advisories provide a nonregulatory concentration of a contaminant in water that is likely to be without adverse effects on health and aesthetics. Human experimental data on the ability of subjects to detect the taste and/or odor of a contaminant in drinking water are used to set he drinking water advisory value. Drinking water advisories are similar to secondary standards but lack the regulatory status of SMCLs.

Disclaimer

The opinions expressed in this chapter are those of the authors and not necessaily those of the U.S. EPA.

REFERENCES

AWWA (American water works Association). 2005. *Brief History of Drinking Water*. AWWA, Denver, co. Accessed at: www.awwa.org/Advocacy/learn/info/HistoryofDrinkingWater.cfm.

Knotts, J. 1999. A brief history of drinking water regulations. *On Tap: Drinking Water News for America's Small Communities* 8(4):17–19. Accessed at: www.nesc.wvu.edu/ndwc/pdf/OT/OTw99.pdf.

U.S. EPA (Environmental Protection Agency). 1985. National primary drinking water regulations; volatile synthetic organic chemicals; proposed rulemaking. *Fed Reg* 50(219):46906, November 13.

——. 1986a. Guidelines for carcinogenic risk assessment. *Fed Reg* 51(185): 33992–34003, September 24, Accessed at: http://www.epa.gov/ncea/raf/car2sab/guidelines_1986.pdf.

______. 1986b. U.S. EPA history: president signs Safe Drinking Water Act amendments. EPA press release. U.S. EPA, washington, DC. June 20. Accessed December 30, 2005 at: http://www.epa.gov/history/topics/sdwa/04.htm.

______. 1989. *Guidelines for Authors of EPA Office of Water Health Advisories for Drinking Water Contaminants*. Office of Drinking Water, Office of Water, U.S. EPA, washington, DC, March.

______. 1991. National primary drinking water regulations; final rule. *Fed Reg* 56l(20): 3531–3535, January 30,

______. 1993. *Reference Dose (RfD): Description and Use in Health Risk Assessments*. U.S. EPA, Washington, DC. Accessed at: http://www.epa.gov/iris/rfd.htm.

______. 1996. U.S. EPA history: President Clinton signs legislation to ensure Americans safe drinking water. EPA press release. U.S. EPA, Washington, DC. August 6, Accessed December 30, 2005 at: http://www.epa.gov/history/topics/sdwa/05.htm#press.

______. 1999. *25 Years of the Safe Drinking Water Act: History and Trends*. EPA/816/R-99/007. U.S. EPA Washington, DC, December. Accessed at: http://permanent.access.gpo.gov/websites/epagov/www.epa.gov/safewater/sdwa/trends.html.

______. 2004. *Requirements of the Public Health Security and Bioterrorism Preparedness and Response Act of 2002 (Bioterrorism Act)*. U.S. EPA, Washington, DC. Accessed at: http://cfpub.epa.gov/safewater/watersecurity/bioterrorism.cfm.

______. 2005. *Guidelines for Carcinogen Risk Assessment*. Office of Research and Development, U.S. EPA, Washington, DC. Accessed at: http://www.epa.gov/iriswebp/iris/cancer032505.pdf.

______. 2006a. Small drinking water systems variances: – revision of existing national-level affordability methodology and methodology to identify variance technologies that are protective of public health. *Fed Reg* 71(41):19671, March 2.

______. 2006b. Questions and answers—Small drinking water systems variances: revision of existing national-level affordability methodology and methodology to identify variance technology that is protective of public health. U.S. EPA, Washington, DC. February 28. Accessed at: http://www.epa.gov/OGWDW/smallsys/affordability.html.

3

INTERPRETATION OF TOXICOLOGIC DATA FOR DRINKING WATER RISK ASSESSMENT

ROBERT A. HOWD AND ANNA M. FAN
California Environmental Protection Agency, Oakland, California

Most of the chemicals of concern in drinking water are widely used or widely distributed in the environment. Therefore, most have been studied extensively, and the data available on these chemicals tends to be quite voluminous. However, the quality of the toxicity data is highly variable, which provides a problem for risk assessment. An interesting analysis of this was provided by Festing (1994), who described three representative toxicity studies in experimental animals, detailing some common problems. More recently Festing and Altman (2005) discussed good experimental design and data analysis in greater detail. Every risk assessor must similarly evaluate and interpret the available data not only for individual studies but also considering consistency across all relevant studies. This can be a formidable challenge because of the complexity of different test methods. In this chapter we describe many of the available methods and their application to risk assessment of chemicals in drinking water.

Toxicity studies in animals provide the greatest source of data for risk assessment. Such data have become more important over the last few decades as exposure to toxic chemicals in the workplace has declined, thereby reducing the availability of human data. In the past, obvious toxic effects from occupational exposures provided the most compelling data on human responses, although actual doses were difficult to determine. Effects demonstrated through careful epidemiological studies have led to the institution of workplace standards. In many cases, these have been made more stringent over the years as sophisticated

Risk Assessment for Chemicals in Drinking Water, Edited by Robert A. Howd and Anna M. Fan

methods for data analysis were developed that documented effects at lower and lower levels. The corollary of this excellent progress is that risk assessment based on human exposures has become much more difficult. Most of the "low-hanging fruit" has been picked. At the same time, development of more stringent ethical guidelines for human studies has made purposeful exposures of people to low levels of toxic substances more restrictive. The quest for information on toxic effects of chemicals is therefore largely dependent on animal studies.

Studies in intact experimental animals can be divided into acute, subchronic, and chronic studies. These terms refer to administration of one dose, a few weeks of dosing (usually, 5 or 7 days per week for 90 or 120 days), or dosing for a major fraction of the expected life span, respectively. Other study types are also important, such as developmental toxicity, in which daily doses are administered to pregnant animals through the time of major organ development of their fetuses. These and other important study designs are summarized below. It should be kept in mind that every study is conducted according to a predetermined protocol, such as those required by the U.S. Environmental Protection Agency (EPA) for submission of data on pesticides and toxic substances (U.S. EPA, 2006a). Factors to be considered include, but are not limited to:

- Purpose of the study, including duration of exposures
- Species, strain, sex, age, source, and health of the animals
- Number of animals required for collection of useful data
- Animal husbandry conditions, including method of housing, facility maintenance, food and water supplied, room temperature, and day–night light cycle
- Chemical formulation and administration method
- Dose levels and dosing frequency
- Data to be collected during the experiment (e.g., body weights, food and water consumption, blood samples, urine or feces analyses, physiological parameters, functional tests)
- Animal sacrifice method
- Data to be collected at termination (e.g., organs taken, histological and biochemical analyses to be conducted)
- Statistical methods used for data analysis
- Physiological significance of the effects measured
- Quality control measures, including staff credentials and experience, record keeping, animal identification method

A more comprehensive list of parameters to be considered under good laboratory practices (GLP) guidelines is presented in 40 CFR part 792, subpart J; 40 CFR part 160; the principles of GLP prepared by the Organization for Economic Cooperation and Development (OECD, 1981); and the EPA's carcinogenicity testing guideline 870.4200 (U.S. EPA, 2006a).

Better-quality studies use enough animals in each dose group, including controls (zero dose), to allow a good statistical analysis, and provide enough details of the methods to facilitate an understanding of study design and conduct. The doses administered and examinations conducted are appropriate to evaluate the potential effects under consideration. Studies conducted according to GLPs are usually well designed and conducted, although not all good studies follow these protocols, which were developed for regulatory submissions on chemicals to the U.S. Food and Drug Administration (FDA) or EPA. Studies conducted for the National Toxicology Program (NTP), for example, do not necessarily follow the GLP guidelines, because they are not conducted for regulatory submission. GLP and NTP protocols include extensive peer review, so the reports of these studies are considered as reputable as are any published in peer-reviewed journals.

The risk assessor evaluates all of the factors tabulated above as well as the stated results in determining whether the study provides toxicity data that will be useful for risk assessment. Because many studies fall short in one or more of these factors, they may be of limited usefulness. The descriptions of study types below provide some details on data quality and requirements that risk assessors look for to make most use of the results in a risk assessment. However, there are too many parameters of study design to allow a comprehensive description of these variables here. Good sources of more in depth information on design and interpretation of toxicology studies include books by Williams and Hottendorf (1997), Hayes (2001), and Jacobson-Kram and Keller (2001), as well as the health effects test guidelines of the U.S. EPA (2006a).

The promise of in vitro studies for providing the basic data needed for risk assessment is still mostly unrealized. Such studies have provided excellent insights on acute cellular toxicity mechanisms (Gad, 2000), but are difficult to extrapolate to chronic effects in the intact organism. As better models are developed for quantitation of the processes of absorption, distribution, metabolism, and excretion (ADME), it becomes easier to apply the results of in vitro studies to risk assessment. However, tissue-specific effects—an effect on the thyroid, for example—are unlikely to be discovered using the more common in vitro preparations (e.g., fibroblast cell cultures or liver slices). In addition, critical information on cellular, tissue, and systemic adaptive mechanisms is not provided by the in vitro studies.

Studies involving oral administration are most relevant for risk assessment of chemicals in drinking water. However, studies involving other administration routes must also be considered. Systemic effects such as hormone disruption are likely to be independent of route of administration, whereas apparent direct, point-of-contact effects are more difficult to evaluate for cross-route extrapolation. Differences in absorption, distribution, and metabolism associated with the route of exposure must be factored into the evaluation. A major consideration in evaluation of inhalation studies, for example, is that absorption of a chemical may be more efficient by inhalation than after oral administration. Gastrointestinal (GI) availability may be limited by metabolism of a chemical within the GI tract, incomplete absorption, and first-pass metabolism in the liver. Chemicals absorbed from the GI tract pass through the liver before being delivered to the rest of the body, whereas chemicals absorbed through the lung are distributed to the entire body. Thus information about the distribution and metabolism of chemicals is critical to the interpretation of the available toxicity studies.

If the toxicity data appear relevant and reliable, the dose–response relationship is evaluated, with exposure calculated in terms of the systemic dose. This means more than counting the number of animals with such effects as liver necrosis or tumors, compared to controls, at each dose. For many measures, both incidence and severity of effects in each dose group can be assessed. The statistical significance as well as the probable toxicological significance of the effects must be considered. This information can be used to make conclusions about the progression of effects from perturbation of homeostasis to frank toxicity and/or carcinogenesis.

When multiple studies are being evaluated, the risk assessor considers the different effects observed, the doses causing each effect, and the potential relevance to humans to decide what the *critical effects* are. These are the effects used in extrapolating to acceptable exposure levels for humans. For risk assessment of chemicals in drinking water, effects worthy of concern may or may not be considered "adverse" in the traditional toxicological sense (i.e., death, malignancies, tissue damage, or clear functional impairment). For example, delayed development, even when fully reversible, is generally considered unacceptable. Reversible hormone changes are more controversial, and dependent on the physiological context, as are transient biochemical and behavioral effects. The risk assessor must attempt to put all such effects into the perspective of the dose–response continuum when evaluating the significance of the spectrum of effects observed in the available studies.

Although a single person cannot be expected to understand the details and limitations of all available toxicity study types, it is hoped that the following discussion will help provide useful perspectives on study evaluation for drinking water risk assessments.

ANIMAL TOXICITY STUDIES

Acute Toxicity

The best known acute toxicity study is probably determination of LD_{50}, the dose estimated to kill 50% of a group of animals tested. This crude measure of gross toxicity is not particularly useful for risk assessment and is considered a measure of last resort. Acute data of more interest are studies of short-acting chemicals that do not accumulate in the body or cause cumulative tissue damage. The short-term studies tend to have simple designs and statistical analyses. Standard protocols are available (U.S. EPA 2006a) for studies involving oral, inhalation, or dermal exposures. Twenty animals per sex per dose group are commonly used for such studies in rodents. Details of particular interest are whether the doses used were adequate to see the intended effects, whether the animals were tested at a time of maximum effect, and whether the analytical methods were appropriate. In a study involving oral administration, if the effect is quickly reversible, greater potency will be observed using gastric intubation (gavage) than with ad libidum water consumption, which usually occurs over an entire day. For that reason, gavage administration is usually considered most relevant, because bolus dosing is similar to the effect of a thirsty person consuming a large amount of water in a short period of time.

Evaluation of the organophosphate (OP) and carbamate pesticides, for which the predominant toxic effect is inhibition of the enzyme acetylcholinesterase (AChE), provides good examples of this study type for risk assessment. Percent inhibition of brain and muscle AChE after a single oral dose, in an aqueous vehicle, is the critical effect of most relevance. The rate of recovery of inhibited AChE after OP administration is sometimes slow enough that inhibition after two or three daily doses (often called *subacute*) is also relevant. This is not true for carbamates, which have very fast AChE recovery.

There are several other chemicals for which the acute effects are of most concern, or are observable at lower doses than effects after longer-term (repeated) administrations. For instance, gastric distress caused by oral administration of copper, and decrease of the oxygen-carrying capacity of the blood by nitrite and cyanide, are sensitive acute effects relevant to drinking water risk assessment.

Chemicals may also have effects, such as liver or kidney damage, which take some time to develop after an acute exposure, particularly when toxicity results from the effect of a metabolite. In such cases, maximum tissue damage might occur after 24 hours or more. In some studies, animals may be allowed to survive for a prolonged period, in order to assess the potential for further progression or reversibility of effects. Risk assessments based on acute administration often incorporate a 10-fold uncertainty factor (UF) to extrapolate to chronic exposures (of humans). This may not always be scientifically justifiable, especially when chronic studies are available showing higher no-observed-effect levels (NOELs) than after acute exposures. Logically, one should either choose the higher NOEL in the chronic study (if protection from chronic effects is the major concern) or use the lower NOEL from the acute study without the extra UF. Data from pharmacokinetic and mechanism-of-action studies may help in making such a decision.

Subchronic Toxicity

Effects of many chemicals become more severe with repeated doses, either because the chemical accumulates in the body or because the effects of each dose are not reversed before the next dosing. For such chemicals, the lowest dose effect level may be observed in a subchronic study. Daily dosing for 90 days is the most common duration for this study design, as in EPA standard subchronic oral toxicity method 87.3100 (U.S. EPA, 2006a); 28-, 120-, or 180-day studies are also found in the literature, intended for slightly different purposes. The studies are often conducted with both mice and rats, but sometimes with dogs or other species (EPA method 870.3150). Twenty animals per sex per dose are customary, although fewer animals may be used for some study types. Rodents are commonly put on test shortly after weaning, so the exposures take place during a period of rapid development.

Animals are commonly weighed weekly, and food and water consumption may be monitored, during the course of the study. Effects may also be monitored during the study (e.g., behavioral observations, muscle strength tests, analysis of blood samples). Time dependent effects may lead to more complicated statistical analyses and interpretations. For example, when a chemical is administered to animals in their drinking water, an initial dose-dependent dip in water consumption, associated with a lag in weight gain, may be observed. The problem is judging whether this is caused by a bad taste of the chemical in the water or by some direct physical effect which seems to be reversed as the animal matures. Although more definitive adverse effects would alleviate the need to use this equivocal type of endpoint, such situations do occur and the relevant knowledge and capability to make an appropriate determination is required of the risk assessor.

Repeated daily exposures allow accumulation of lipophilic (fat-soluble) chemicals in fatty tissues within the body. Metals may slowly accumulate in bone or perhaps kidney, until structural damage is noticeable, usually later in life. Adaptations to the daily dosing may include induction of metabolism of the chemical in the liver, which may be associated with liver enlargement or just an increase in hepatocyte microsomal structures, often seen with halogenated hydrocarbons. Damage to cellular components such as DNA may occur, which could lead to tumors if the experiment were continued long enough. An assay called the *liver foci test* was developed to provide a shorter-duration test reflective of liver carcinogenic potential (Ito, 2000; Ito et al., 2003; Fukushima et al., 2005). Cytological changes in liver or other tissues may or may not be interpreted as adverse, depending on the severity, reversibility, and presumed physiological significance of the change.

It has been suggested that most toxic effects are observable in subchronic studies, so that longer-term studies (entire lifetime or multigeneration developmental studies) are less valuable and may be a waste of resources (Ashby, 1996; Knight et al., 2006a,b). Indeed it is true that *many* effects are observable in subchronic studies, but appropriate tests to reveal or quantitate these effects may or may not be carried out in the subchronic studies, and the toxicological significance may not be appreciated (e.g., a change in kidney function that would result in a shortened life span may not be thought of as serious if the study is stopped before the early deaths are observed). When no adequate chronic studies are available, the risk assessor must make use of the data from shorter-term studies. A 10-fold UF is commonly applied by a risk assessor for using short-term studies in the absence of chronic data or when a subchronic study has a lower NOEL. As discussed above for acute studies, whether this UF is truly needed (i.e., represents real uncertainty about chronic effects) should be based on examination of all of the data.

Neurotoxicity

Many specialized study types have been developed to address the effects of chemicals on the central or peripheral nervous system. Exposures in such studies range from acute to chronic duration, and the tests may involve a wide variety of approaches. Assessment of peripheral neuropathy, for example, might be addressed by observation (the animal can no longer stand up; Battershill et al., 2004), by measurement of functional parameters such as muscle strength or nerve conduction velocity (Nichols et al., 2005), by chemical assay (Tonkin et al., 2004), or by histopathology (Tonkin et al., 2003). Only a few neurotoxicological studies have standardized procedures (see U.S. EPA, 2006a). Protocols vary in such aspects as species, strain, sex, number of doses, and number of animals per dose, as well as evaluation of the neural-related changes. In addition, the definition of neurotoxicity is not unanimous, with some risk assessors considering any change as neurotoxicity, whereas others consider only persistent changes as evidence of neurotoxicity. Despite such limitations, neurotoxicity tests in animals can be extremely valuable for predicting chemical effects in humans and have grown in significance over the last few years (Bushnell et al., 2006).

Neurobehavioral testing is a very complex specialty, with dozens of study types and designs; many involve training animals to perform special tasks to be observed and/or measured. Training may occur before chemical exposures or during exposure, and testing may be during exposure or long afterward. Behavioral tests may assess the ability to learn a task, to perform it while being exposed to a chemical, retention of memory, or many other factors. The effects may be acute and transient (reversing as the drug leaves the brain), somewhat persistent (involving resynthesis of an inhibited enzyme or receptor, for example), or virtually permanent (structural damage). Unfortunately, many of the behavioral test protocols do not distinguish among these types of effects, so interpretation of the results for risk assessment can be extremely difficult.

Nonbehavioral, functional neurological tests, such as visual evoked potentials (Boyes et al., 2005, EPA 870.6855) are often easier to interpret. A demonstration of impairment of a physiological response is self-evident as an adverse effect. In addition, a progression of effects can sometimes be obtained in this type of study, because the tests can be performed repeatedly in the same animal. Well-planned studies may reveal a spectrum of effects over time, from biochemical through functional to histological effects (Desai and Nagymajtenyi, 1999; Tanaka et al., 2006). A further benefit of many of the neurological test methods, such as auditory or visual evoked potentials or measurement of nerve conduction velocities, is that they can be conducted as easily in humans as in animals, providing an opportunity for direct cross-species comparisons.

Neurochemical parameters evaluated commonly include assays of neurotransmitters, metabolites, enzymes, receptors, and structural proteins. These are often non-GLP studies, carried out in university laboratories. Fewer animals may be used per group (five to 10 is common), and study details are often poorly reported. However, the greatest challenge for risk assessment with these studies is often interpreting the biological and toxicological significance of the changes.

Inhibition of brain AChE is one of the more common markers of a neurochemical effect, and the physiological and toxicological significance of this effect is relatively clear because of the extensive study of this effect for organophosphate pesticides. The effect at moderate AChE inhibition levels is considered to be fully reversible and may not correlate with any observable behavioral changes, while a much higher degree of inhibition will lead to death. A statistically significant inhibition of AChE (or sometimes, more than 20% inhibition) is generally deemed to be an adverse effect, worthy of regulatory concern. In this case, risk assessors have reached general agreement as to how the transient biochemical changes should be utilized in risk assessment, which sets a strong precedent by which other biochemical effects could be addressed. Clear neurochemical changes should generally be considered effects worthy of concern—although in many cases a transient, highly significant effect may not cause death or severe functional impairment, or the function of a biochemical component may be incompletely understood. The relevant principle is that people should not have to worry about such effects from chemicals in their drinking water.

Histology of peripheral and central nerves is a highly specialized technique, involving fixation of tissues, slicing tissues and preparing slides, chemical and immunostaining, and finally, evaluation of tissue changes on the slides. Interpretation of these data is often a great challenge for the risk assessor. Although cytological alterations have traditionally been the gold standard for neuropathological effects, tissue preservation artifacts or variations in brain slicing (U.S. EPA, 2002) may lead to equivocal results. In addition, many of the newer specific histological staining techniques provide a level of detail about neural histological and morphological changes that is quite difficult to interpret. For example, if a reorganization of a particular brain region appears to have occurred, does that reflect an actual functional change, and is it an adaptive response or an adverse effect? One good discussion and comparison of techniques is provided in a recent article about neurohistochemical biomarkers of domoic acid toxicity by Scallet et al. (2005).

Reproductive and Developmental Toxicity

Reproductive toxicity includes adverse effects on any aspects of male and female reproductive function, whereas *developmental toxicity* refers to adverse effects on offspring, occurring both pre- and postnatally. Effects on offspring may occur through maternal exposure prior to conception through weaning, or exposure of offspring from birth through sexual maturity. The standard *reproductive toxicity* test is a two-generation study in rats (EPA 870.3800). Both males and females of the parental generation are exposed daily, starting prior to breeding, using sufficient mating pairs to result in 20 pregnant females in each dose group. Offspring are exposed through maturation and production of a second generation of offspring. Parameters related to reproductive success and health of the offspring are evaluated in both offspring generations, including number and size of offspring, structural abnormalities, and growth. For this study type, two litters may be produced and evaluated in each generation (U.S. FDA, 2006a).

The basic *developmental toxicity test* involves treatment of females during pregnancy, which may be from 1 day prior to implantation through parturition, or for a more limited period such as the during organogenesis (EPA 870.3700). Usually, just before expected delivery, the dams are killed, the uterus and all its contents are evaluated, and the fetuses are examined for visceral and structural abnormalities.

A third study type involves treatment of both females and males for several weeks starting before conception, to evaluate effects of a substance on reproductive performance as well as on fetal development (EPA 870.3550). Males are treated for about four weeks including two weeks postmating, and then killed for testicular histological examination. Females are treated through a few days postparturition, then they and their offspring are killed and examined for a variety of reproductive and developmental parameters.

Another relevant procedure is the *developmental neurotoxicity study* [EPA 870.6300 or OECD TG 426 (draft 1999); Slotkin, 2004]. These studies, also called *behavioral teratology*, involve administering chemicals during pregnancy and lactation, followed by evaluation of postnatal indices of maturation, behavioral tests, and finally, determination of brain weights and a neurohistological examination.

These types of studies are critical for understanding the fundamental toxicological properties of chemicals. Production of birth defects and reproductive impairments are among the effects most feared by the general public. However, effects on these parameters are not necessarily readily extrapolated across species. This is partly because of the variations in hormonal regulatory systems among species and partly due to differences in the metabolism of chemicals. Different species may produce more or less of a teratogenic metabolite, for instance. Any accompanying metabolism studies may help clarify the mechanisms.

The standard design of the two-generation study does not involve lifetime (chronic) exposures to any of the generations, so it does not provide a backup for chronic toxicity and carcinogenicity studies. Nevertheless, longer-term administration does allow accumulation of chemicals in the body, including transfer of lipophilic chemicals to offspring in mother's milk. Bioaccumulation may result in more potent effects in the second offspring generation (Luebker et al., 2005; McKee et al., 2006).

A major statistical consideration in the analysis of reproductive and developmental studies is the *within-litter effect*. Offspring are more similar within a litter than between litters, so it is considered that within-litter abnormalities are not independent observations. Therefore, the proper n value for an experiment is the number of litters evaluated, not the number of fetuses or neonates. The statistical evaluations provided in some study reports do not follow this rule, so the risk assessor should redo the statistical analysis if enough individual-animal information is provided to make this possible.

Another consideration is the difference between a variation and a structural abnormality. For example, wavy ribs or a small extra rib are usually called *variations* and are not considered evidence of teratogenicity (Khera, 1981; Kast,

1994). Finally, judgment may be appropriate to distinguish between an effect secondary to maternal toxicity (e.g., slightly lower fetal body weight or length, or slightly delayed fetal development) and a direct effect on the fetus. The distinction is frequently made based on dose; if the (mild) fetal changes occur only at doses showing adverse maternal effects, the fetal effects are likely to be secondary. Although such a distinction may not have any effect on establishing a protective level, it may aid in elucidating the nature of the effects.

Immunotoxicity

Immune function can be impaired by exposure to environmental chemicals such as dioxin (Kerkvliet, 2002). Effects are frequently more severe when exposure occurs to the developing immune system (Luebke et al., 2006a) and can result in impaired responses to infectious organisms as well as enhanced sensitivity to some forms of cancer. Therefore, it is critical to assess the potential for immunotoxicity. However, the mammalian immune system is complex, with both antibody-dependent and cell-mediated responses. Natural killer cells, T-lymphocytes, and macrophages carry out the cell-mediated responses. B-lymphocytes produce the immunoglobin antibodies, and both T-cells and macrophages are involved in their activation or presentation of antibodies. Immune toxicity can be expressed either as impairment of responses or as an inappropriate stimulation of responses of any of the elements of the immune system. Therefore, no single test can assess all aspects of the immune system, and a battery of tests or tiered testing is necessary to evaluate the major elements of the immune system.

The only EPA standard guideline for an immunotoxicity test is for a sheep red blood cell (SRBC) response assay in rats or mice (EPA 870.7800). Effects of the test substance are evaluated on the ability of rodents to respond to the presentation of the antigen by a splenic anti-SRBC (IgM) response or increased serum anti-SRBC IgM levels. This assay tests only for inhibition of immune responsiveness, not for enhanced sensitization. The FDA has had immunological test guidelines for some time (Hinton, 2000), and recently released new proposed immunotoxicity testing guidelines which cover both immunosuppression and immunoenhancement (U.S. FDA, 2006b). International immunological testing guidelines for pharmaceuticals, which would also be applicable to testing of food and industrial chemicals, have been finalized and are now subject to implementation by participating countries (Schulte and Ruehl-Fehlert, 2006).

Inorganic mercury appears to be the only drinking water contaminant that is regulated based on an immunological response; an autoimmune response affecting the kidney is the critical endpoint (U.S. EPA, 2006b). However, there are important immune effects for several other contaminants, including beryllium, nickel, chromium, toluene, and 2,3,7,8-tetrachlorodibenzo-*p*-dioxin (TCDD). As immunological testing becomes more extensive and standardized, with better methods to relate effects in animals to those in humans, more emphasis on immunological effects is likely. This may become especially important in consideration of sensitive subpopulations.

Genotoxicity

Genotoxicity studies may be either in vivo or in vitro, involving evaluation of effects of chemicals on genes and chromosomes. Genetic alterations in somatic cells are associated with cancer, while mutations in reproductive cells will lead to birth defects. Dozens of assays have been developed to evaluate various aspects of chemical effects on genes, including changes on gene expression as well as induction of mutations. Mutations may involve chromosomal rearrangement or a change in a single nucleotide (point mutations). The battery of 18 test protocols listed by the EPA (U.S. EPA, 2006a) covers most common procedures, including Ames assays (method 870.5100), the *Drosophila* sex-linked recessive lethal test (EPA 870.5275), unscheduled DNA synthesis in cultured mammalian cells (EPA 870.5550), and the in vivo sister chromatid exchange assay (EPA 870.5915).

Although all of these procedures can provide important perspectives on the ability of chemicals to react with and alter genetic materials, interpretation of the results for risk assessment is not simple, with a high potential for false negative or false positive results (Kirkland et al., 2005). High concentrations of chemicals in the in vitro assays may cause cytotoxicity, and genetic changes may occur because cells are dying. Alternatively, the genetic changes caused by the chemical may have led to the cell death. One way to distinguish between these mechanisms is to determine whether the chemical can react with DNA, directly or through reactive metabolites. Another consideration is whether the chemical can enter the cell nucleus.

Chemicals that interact directly with DNA by such mechanisms as alkylation of nucleotides are considered to be capable of causing cancer by a nonthreshold mechanism. That is, even at very low, environmental doses (usually, 10^{15} or more molecules), the rate of interaction of the chemical with DNA is presumed to be linear. This is coincident with the ongoing natural process of DNA damage and repair, in which many thousands of individual DNA changes may occur before one of them happens to yield a tumor cell line. The relationship of concentration (or dose) to tumor incidence may be far below the level detectable in an animal study. Therefore, in vitro genotoxicity tests are extremely important in assessing whether a chemical is a potential carcinogen.

On the other hand, the in vitro tests are very difficult to interpret quantitatively. Concentrations of chemicals tested may be several orders of magnitude higher than is achievable in vivo. The availability of reactive metabolites is likely to be very different than it is in vivo. For the Ames assays, genotoxicity to the bacterial test cells is evaluated with and without a rat liver enzyme fraction added to produce metabolites of the chemicals tested. But in vivo, the reactive metabolites may be so short-lived that they never reach a sensitive cell population. For volatile chemicals, the in vitro system may not achieve a stable concentration because of volatilization. For low-solubility chemicals, biologically relevant concentrations may not be achieved because of the lack of carrier molecules, or unrealistically high concentrations may result from the use of solubilizers. Thus, the in vitro results are most relevant for qualitative judgments about potential genotoxicity.

The in vivo genotoxicity tests avoid many of the problems of interpretation of the in vitro assays. Administering a chemical to animals and assaying for chromosomal damage later, such as in the bone marrow chromosomal aberration test (EPA 870.5385) or sister chromatid exchange assay (EPA 870.5915), ensures that the results can be considered in the standard dose–response continuum. These tests use only one or a few daily treatments, and usually only five animals per sex per dose, so they are not as sensitive as many other test methods or as statistically robust. Therefore, although the results are certainly useful, risk assessments are virtually never based on potency in a genotoxicity test.

Chronic Toxicity

Chronic toxicity tests generally involve treatment of animals starting after weaning and continuing for a substantial portion of their expected lifetime, usually for one year or longer. Chemicals may be given in feed, in the drinking water, by inhalation, by gavage, or by dermal application. Administration of chemicals 7 days a week is recommended, although inhalation, gavage, or dermal dosing may be applied only 5 days a week. For risk assessment, doses would usually be recalculated for a 7-day/week equivalent (i.e., the daily dose for a 5-day/week schedule is multiplied by 5/7). The major factor in successful completion of such studies is good animal husbandry: maintaining a healthy colony with good quality control. Many studies have been compromised severely by intercurrent disease, wide temperature swings in the animal rooms when air conditioning failed, periods when the drinking water systems did not work, or times when dosages were prepared improperly (either too high or too low). Examination of the weekly animal weights helps reveal when conditions have varied. Several different study types are common, as follows.

Chronic Noncancer Studies Chronic studies in two species are recommended for regulatory submissions. For rodent studies, at least 20 animals per sex per dose are generally used. The most common second species for chronic toxicity studies is the dog; 8 animals per sex per dose are recommended by the EPA (test 870.4100). At least three dose levels, plus concurrent controls, are recommended, where the highest level elicits clear signs of toxicity without shortening the life span, and the lowest dose shows no toxic effects. However, this test protocol does not maintain animals for their complete expected life span, so only effects resulting in a severe problem with early mortality would normally be discovered.

The parameters assessed during exposures may include body weight, food consumption, water consumption (especially when the chemical is administered in water), behavioral observations, and clinical pathology (i.e., clinical chemistry, urinalysis, hematology). Ophthalmological examinations are recommended before the start and at the end of treatment. Gross pathology and histopathology are carried out at necropsy. A combined chronic toxicity–carcinogenicity study may also be conducted (EPA 870.4300), which expands the chronic study in duration of treatment, number of animals per test group, and the extent of the pathological examinations.

Carcinogenicity Studies Rodents are commonly used in cancer bioassay; studies in both rats and mice are usual (EPA 870.4200). At least 50 animals per sex per dose are recommended, with at least three dose groups plus concurrent controls, with daily or 5-day/week exposures from just past weaning to 18 months in mice and 24 months in rats. Chemicals may be administered in the food or water, by gavage, by inhalation, or dermally. The highest dose tested should produce visible toxic signs, and the lowest dose tested should produce no toxicity. Shortening of life span due to tumors is considered acceptable as long as survival is adequate for acceptable statistical analysis. Parameters examined are similar to those in the noncancer studies, with the addition of a much more detailed histopathological examination for tumors. Evaluation of the incidence of tumors in a wide variety of tissues is the ultimate intent, with classification of tumor type (e.g., dysplasia, benign, malignant), probable tissue of origin, and sometimes, a severity grade. Other pathological lesions and changes are also noted.

Risk assessors should keep in mind that slides are commonly not read blind; histopathologists commonly know the exposure group of every slide and use this information to help them make fine distinctions in gradation of effects observed. After evaluating the lesions at the highest dose versus the controls, and identifying the tissues where specific compound-related lesions occur, they can more profitably direct their time and attention to the tissues of most interest. This can be good, because an in-depth evaluation of a slide is very labor-intensive. On the other hand, this may lead to the pathologist's preconceptions influencing the ranking of every lesion. For this or other reasons, histopathologists can vary widely in their classification of tumors and other tissue changes. There are contradictory reports from different pathologists or even pathology panels for some of the most contentious chemicals (e.g., Keenan et al., 1991; Goodman and Sauer, 1992). The prudent path for a risk assessor is to avoid taking sides, and report on the range of opinions and the resulting tumor potencies.

HUMAN TOXICITY STUDIES

Application of data from human exposures is very desirable in risk assessment, to minimize uncertainty in extrapolations to toxic effects in people. However, good data are not easy to obtain. Limited data on acute incidental exposures (accidents, suicides, poisoning) is available for many chemicals, although the dose is rarely known accurately. Subchronic and chronic exposure and health-monitoring data are available for occupational exposures to industrially important chemicals. Human studies involving purposeful exposures are relatively rare, and generally only for acute and subchronic durations. Population studies involving exposure to chemicals in air or water are extremely important in risk assessment, although usually difficult to interpret because of confounding factors. Each of these sources of information will be summarized, with comments on their application to risk assessment of chemicals in drinking water.

Acute Toxicity

Accidental exposures and suicides or other poisoning incidents can provide important data on toxic effects of chemicals in humans, corroborating the effects observed in more systematic studies in animals. Often, the doses are very high in the acute exposures, so they do not contribute much to the determination of safe exposure levels. However, at times, risk assessments have been based on human acute exposure data. For example, gastric irritation caused by copper in adult women was the basis of the EPA risk assessment to derive the copper MCLG (U.S. EPA, 1991). The California OEHHA risk assessment for copper (1997) was based on similar effects in infants and children, who seem to be more sensitive (Spitalny et al., 1984). Follow-on studies involving repeated administration of solutions containing copper to both infants and adults up to the maximum level recommended by the World Health Organization (WHO) have confirmed and clarified the adverse effect level (Araya et al., 2001, 2004; Olivares et al., 2001).

When data obtained in humans are used in risk assessment, a 10-fold uncertainty factor for extrapolation from animal studies is eliminated. This increase in allowable exposures provides a powerful incentive to conduct human studies, especially for pesticide manufacturers. However, it also introduces the ethical issue of whether it is appropriate to expose humans to toxic substances in tests that offer no potential therapeutic benefit. Chemical manufacturers point to the societal benefit of more clearly delineating potential adverse effects in humans (McAllister, 2004; Tobia et al., 2004), as well as a great cost savings in cases where groundwater cleanup levels could be increased substantially without public health risk (CWQ, 2006).

A good example of designing studies to answer questions about human exposures, to influence regulatory decision making, is the series of small human studies carried out on the absorption of hexavalent chromium (Paustenbach et al., 1996; Corbett et al., 1997; Finley et al., 1997; Kerger et al., 1997). These studies showed that oral and dermal uptake of hexavalent chromium is relatively small, consistent with studies in animals. The EPA has been criticized for using human test data to determine acceptable exposure levels. In response, the EPA established a moratorium on the use of such data in evaluation of pesticide toxicity in 1998, subject to review by its science advisory board and science advisory panel. After extensive debate, the majority of the subcommittee charged with this task recommended that such studies should not be used to determine adverse effect levels, but were acceptable under strict guidelines to fill important data gaps (U.S. EPA, 2000). The EPA decided to seek further review of the issues (Goldman and Links, 2004), and commissioned the National Academy of Sciences (NAS) to convene a panel to address human testing, not only for pesticides but also for other chemical regulatory programs, including air and water contaminants. The NAS committee recommended that such studies be allowed, whether conducted by the EPA or by third parties with the intention of influencing a regulatory decision to either raise or lower a standard (NAS, 2001). This by no means settled the issue. Debate can be expected to continue indefinitely, based on very strong feelings in two opposing positions: (1) that experiments involving administration of toxic chemicals to human subjects in which no benefit to the subjects are intended are unethical and should never be allowed (Melnick and Huff, 2004; Needleman et al., 2005); and (2) that such studies provide important data relevant to protection of public health and should be allowed under stringent scientific and ethical standards (Charnley and Patterson, 2004; Resnick and Portier, 2005).

Dosimetry studies involving ongoing occupational exposures to chemicals, including pesticides, are not usually controversial, although this type of study is intended to evaluate exposure, not effects. Unfortunately, a similar long-planned study to determine children's exposure to pesticides under normal household use conditions became so controversial that it had to be canceled (U.S. EPA, 2005). On the other hand, the copper studies referred to above, which exposed infants to copper solutions up to or exceeding the nausea limit, do not seem to have attracted much attention or controversy (although they were conducted in Chile, not in the United States). The most important conclusion that should be derived from the foregoing studies and recommendations is that acute human testing is likely to continue to decrease.

Subchronic Toxicity

Subchronic human studies on environmentally relevant chemicals are relatively rare. Some studies on seasonal exposures to pesticides have been carried out, although they generally are designed more to assess exposure than effects, and many are unpublished. Repeated-dose human studies are required for registration of pharmaceuticals, and some of these may be of relevance to chemicals found in drinking water. In addition, some intermediate-duration occupational studies of solvents and other industrial chemicals exist, but these are not necessarily subchronic exposures; that is, the study covers a short period in the middle of an ongoing longer-term exposure.

Braverman et al. (2006) conducted a six-month study of perchlorate administered in drinking water to help establish safe concentrations of this drinking water contaminant. The study received such negative publicity that the researchers failed to sign up an adequate number of study subjects for good statistical inference, despite offering $1000 for participation. An earlier 14-day human exposure study (Greer et al., 2002) became the basis of several perchlorate risk assessments (California OEHHA, 2004; NAS, 2005; Massachusetts DEP, 2006). Problems with public perceptions have led researchers to conduct studies in humans outside the United States. Approval by institutional review boards, usually both in the foreign country and in the United States, is standard for these studies. Although some of the studies have been of high quality, differences in nutritional or demographic background may sometimes add complications to extrapolation to U.S.

populations. It should be noted, however, that the differences between humans in two countries are always less than the differences between humans and rodents.

Epidemiology Studies

The effects of chemicals on human health are also assessed using epidemiological studies, which correlate the health of a selected human population or group with exposure to various chemicals. Theoretically, this should be the best way to interpret potential health effects of chemicals in drinking water, because the studies are carried out with a population of potential concern, the doses are in the range of the doses found in the environment, and the exposed and control groups can often be very large. However, the doses to individuals are rarely known, and there are always confounding factors (variations within the exposed populations that could affect response to the chemical under study). The risk assessor must become familiar with the various types of studies, appropriate statistics used for each, and the limitations of interpretation that can be made for each.

There are six basic epidemiology study types. These are case–control studies, cross-sectional surveys, ecological studies, cohort studies, randomized controlled trials, and crossover design studies. The case series, a clinician's description of the characteristics of cases they see, can also be used to make inferences about effects of chemical exposures and causes of disease, but is not considered an epidemiology study for this discussion. The six study types are described briefly, including the advantages and disadvantages of each.

Case–Control Study In this study type, a group of people with a certain disease or condition is chosen, and a demographically similar group without that characteristic is identified and matched with the cases. The history of exposures to chemicals and other risk factors is correlated between the groups. This technique is the only feasible method for evaluating rare diseases or disorders, and is relatively quick and cheap. Fewer subjects are needed than with cross-sectional studies. However, selection of appropriate controls may be a problem, there may be many confounders (co-exposures or conditions that may increase or decrease the phenomenon being studied), and determination of the exposure history usually relies on recall, which may introduce recall bias or affect the control selection process. *Recall bias* is the tendency for people to think more about potential causes and precursor events when they get sick, so they are more likely to report something potentially bad that happened earlier (such as an exposure to pesticides or solvents) than people who are healthy. Because of the problems with confounders and bias, multiple regression analysis is commonly performed, and robust associations are needed to presume possible causation.

Cross-Sectional Survey Participants are sampled from within a population to answer questions or fill out a survey form. Their answers are used to make inferences about possible correlations between exposures and diseases (a "descriptive study"). This study type has few or no problems with ethical justification, since there is no manipulation of the population under study. This study method is also relatively quick and cheap, and much larger population sizes can be surveyed than for a case–control study. However, too few people with a particular disease or condition may be identified to make statistically valid inferences about that condition. Confounders may be unequally distributed among groups because of cross-correlations (i.e., a person who drinks bottled water rather than tap water may be more concerned about their health and may eat more healthfully or get better medical care). This means that such a study can only establish association, not causality. These studies are also subject to recall bias and survivor (Neyman) bias. *Survivor bias* occurs when a disease or exposure condition adversely affects mortality, thus truncating the population surveyed. The group sizes in a cross-sectional study are likely to be unequal, which complicates statistics.

Ecological Study Average exposures and effects in selected populations are assessed in this study type, which is another type of descriptive study. A question that might be asked, for example, is: "What is the incidence of osteoporosis in populations that are provided with fluoridated water compared to those who are not?" Data on individual behavior are not collected (i.e., "Who in the population drinks bottled water rather than tap water?"), although a survey may be conducted to determine group statistics ("What proportion of the people in the different age groups drink bottled water in each city, and what are the proportions in each subgroup who brush their teeth with fluoridated toothpaste?"). Differences among communities may not be obvious. For example, populations classified as Hispanic in various central California communities can be expected to differ greatly in their proportion of recent immigrants, depending on the type of crops grown in the area and the need for day labor. Their income levels, access to medical care, nutritional status, and exposure background will also differ. Multiple regression analysis may or may not be able to sort out the covariants.

As an example of the problems of covariants (confounders), in a study on exposure of pregnant women to drinking water containing various levels of perchlorate, a significant association of perchlorate with high thyroid-stimulating hormone (TSH) levels in the newborn offspring was noted (Brechner et al., 2003). However, TSH levels fall rapidly after birth, and most of the difference observed in the TSH levels was associated with a difference in the blood-sampling times between the cities the women lived in, which was related to a difference in medical practice in the hospitals serving the two communities (Goodman, 2000; Lamm, 2000). Whether the remaining small difference in TSH levels is statistically or biologically meaningful is difficult to determine. Because of such problems with confounders, ecological studies are generally considered as more hypothesis-generating than demonstrative of specific cause-and-effect relationships.

Cohort Study In a cohort study, a group of people is studied over time to assess the incidence of diseases that develop and to look for associations with other individual characteristics, exposures or lifestyle factors, compared to the other people in the group. This is called a *prospective study*, rather than a *retrospective study* as in the case–control study and cross-sectional survey. This design eliminates recall bias, and timing of events and exposures can be determined as the study proceeds. Subjects can be matched within the disease and nondisease groups for multiple characteristics, thus minimizing confounders. However, in general, only healthy people are chosen to participate in such studies, so potentially susceptible populations may be eliminated at the outset. This is a variant of the "healthy worker" effect (only healthy workers are employed to work in a chemical plant or factory), which is responsible for the common observation that occupational exposures to a chemical appear to be associated with a *decreased* incidence of disease compared to a randomly chosen control population. Careful subject matching may minimize this bias. However, subjects are not necessarily randomly distributed internally in the population, and they are not blind to the exposure conditions and study design. For example, knowing that a study is in progress concerning exposure to perchloroethylene in dry cleaning fluid might influence participants' behavior regarding perchloroethylene exposures or frequency of medical checkups. The major problem with this type of study is that it is very time consuming and expensive. A 10- to 20-year follow-up is required for the study of carcinogens, for example.

Randomized Controlled Trial This is the "gold standard" of epidemiological design, in which subjects are randomly assigned to different groups. Treatments are usually administered to subjects in a double-blind fashion, which means that neither the subjects nor those evaluating the health of the subjects know who was given the treatment(s). Sometimes the nature of the treatment makes blinding impossible. For example, if a chemical stains your skin, causes all your hair to fall out, or has a particularly strange taste, this may invalidate the blinding. This is the basic design of clinical trials for new drugs. The random assignment to treatment groups means that there is an unbiased distribution of confounders, and statistical treatment is more straightforward. Such studies usually have a moderate duration (weeks to months), but are relatively expensive. Also, in many cases there is no untreated control, because for dire diseases for which a treatment is established, it is unethical to withhold treatment from the control group. When treatments are compared only to each other, it is more difficult to achieve statistically significant differences of results. A *volunteer bias* is also possible, in which only those most likely to profit from the treatment, or least likely to profit from existing treatments, agree to participate in the study.

Crossover Design Study In this study type, subjects are randomly, sequentially assigned to treatment groups. Each subject receives one or more periods of daily treatments and one or more periods of daily treatment with a placebo, another effective chemical, or a different dose of the test chemical. The treatments are usually double-blind. Physiological responses are evaluated repeatedly during the treatment periods. In effect, each subject serves as his or her own control. Another variant of this design is a placebo period followed by treatment, followed by a placebo, with the daily dose randomized among individuals during the treatment period, but not the order in which treatment and placebo are given. The crossover study design minimizes interindividual variance, thus reducing the sample size needed to demonstrate effects. The study by Greer et al. (2002) used to estimate effect levels for perchlorate utilized this design. One advantage of this design for a drug treatment study is that all subjects are treated with the active compound at least some of the time. For evaluation of effect levels for a toxic chemical, the ability to estimate interindividual differences is maximized because all the volunteers are treated. However, long-acting chemicals cannot be evaluated in this way, and if the washout period is unknown, there may be questions of residual effects persisting through subsequent periods.

Quantitative Reporting of Epidemiology Results The statistical analysis methods utilized for epidemiology studies differ from those used for other evaluations of toxic responses. Effects are commonly reported in terms of an *odds ratio* (OR) (or *cross-product ratio*), which is a measure of the odds of exposure to a risk factor among cases versus the exposure among controls, or a *relative risk ratio* (RR), which is the ratio of the probability of an adverse outcome among those exposed to a risk factor, compared with the probability of developing the outcome if the risk factor is not present. For both these calculations, values larger than 1.0 represent increased risk. The 95% confidence limits of the ratios are also reported; a lower bound greater than 1.0 indicates statistical significance of an increase at the $p < 0.05$ level. A mean ratio of as little as 1.3 or so can be statistically significant with a large enough sample size. An RR greater than 2.0 could be interpreted as meaning that an adverse effect is more likely to be due to the exposure of concern than to competing causes, but it should be emphasized that this is only an association, not proof of cause and effect. The risk assessor should exercise great restraint in concluding that any particular case was a result of the exposure.

Meta-analysis Limited sample sizes and biases tend to result in small effects and conflicting results of epidemiology studies. Meta-analysis, which is a methodology for statistical analysis of a group of studies, taken together, can help overcome these limitations. However, the studies must be as comparable as possible, and must be selected and interpreted in a consistent manner. Essential steps include defining the problem and criteria for admission of studies, locating studies using a well-described literature search method, ranking the quality of the individual studies, converting the results of each to a common scale, then analyzing and interpreting the combined results (Thacker, 1988). However, the individual studies are inherently heterogeneous, which has led to much discussion of appropriate ranking and study selection methods (Engels et al., 2000). Overall, the most important criterion in evaluating a meta-analysis appears to be transparency—inclusion of enough detail so that its analysis is understandable and replicable (Stroup et al., 2001). Important contributions of meta-analysis to evaluation of the effects of contaminants in drinking water include those for chlorination by-products (Morris et al., 1993; Villanueva et al., 2003), lead (Pocock et al., 1994; Schwartz, 1994), and arsenic (Navas-Acien et al., 2006).

Epidemiology Summary and Conclusions Epidemiology studies have been very useful for helping estimate maximum acceptable levels for chemicals in drinking water, although such studies have been used more frequently as supporting evidence of an adverse effect rather than to calculate the acceptable level. This is partly because of the difficulty of calculating actual or average doses from epidemiological data, and partly because the many confounders in human studies make it difficult to establish cause-and-effect relationships. Drinking water contaminants for which the acceptable exposure level has been derived from epidemiology studies include arsenic, barium, benzene, cadmium, fluoride, lead, nitrite, selenium, and the radionuclides (especially radium and uranium).

Reproductive Toxicity

Effects of environmental chemicals on humans have frequently been investigated in epidemiological studies, because reproductive disorders and developmental problems are major public health issues. Effects of concern include altered menstrual cycles, increased time to pregnancy (related to effects in both males and females), decreased sperm counts, decreased libido, infertility, increased spontaneous abortions, birth defects, altered growth and development, and functional deficiencies such as mental retardation. Such effects may be caused by metabolic and hormonal alterations, genotoxicity, or adverse effects of chemicals on specific maternal or paternal reproductive-related tissues. Reproductive effects may be discovered through observation (case reports), which leads to more specific studies, or epidemiological studies may be instituted because of adverse effects reported in animal studies.

All states maintain records of fetal deaths and of live births and deaths. Some states have specific birth defect registries. California has, in addition, an ongoing birth defects monitoring program (see http://www.cbdmp.org/spd_overview.htm). The registries have provided a rich data source for investigations of the effects on reproduction of exposures to chemicals. An excellent example of this is a series of investigations on reproductive outcomes in California's Santa Clara Valley that was triggered by anecdotal reports of increased birth defects after groundwater was found to be contaminated with solvents. Adverse reproductive outcomes (higher rates of spontaneous abortions) were associated with trihalomethanes in drinking water (Swan et al., 1998; Waller et al., 1998). Increasingly sophisticated methods have been brought to bear on the problem of proper classification of exposures to better characterize the relationship of disinfection by-products to spontaneous abortion (Waller et al., 2001; Windham et al., 2003, Weinberg et al., 2006).

Effects of dibromochloropropane (DBCP), a fumigant, on fertility were suspected because of poor fertility in families of workers in DBCP manufacturing plants. Greatly decreased sperm production and depressed fertility were confirmed in follow-up investigations (Biava et al., 1978; Potashnik et al., 1979; Potashnik, 1983). This widely used fumigant permeated readily to groundwater; because of its long persistence, exceedances of the MCL by DBCP are still common, despite its cancellation in the continental United States in 1979.

Other investigations into reproductive toxicity issues have been instigated by concern about pesticides and environmental estrogens commonly found in groundwater (McLachlan et al., 2006; Swan, 2006). An excellent summary of techniques for investigation of reproductive outcomes is available in a report from the NAS (2001).

Neurotoxicity

A wide variety of procedures have been developed to evaluate neurological endpoints in humans, all of which are potentially applicable to evaluation of chemical exposures. Available human neurotoxicity tests include measurements of nerve conduction velocities, brain evoked potentials, electroencephalograms, reaction times, body sway (balance), muscle strength and fine motor control (for peripheral neuropathies), reflexes, and various psychological tests (IQ, mood, affect, recall). Developmental neurotoxicity testing in infants and young children is also relevant for environmental exposures (Hass, 2006). However, evaluation of human neurotoxicity studies is often a great challenge in risk assessment. Study reports often provide limited reporting of experimental and statistical details and inadequate control groups. Many of the reports are from occupational studies with inhalation exposures and limited air monitoring, which represents an additional problem in relating them to potential drinking water exposures.

Few human neurotoxicity studies are available involving a drinking water exposure route for either occupational or environmental exposures. Studies on lead exposure and effects are perhaps the most relevant, although drinking water has not been the major exposure source in recent times (Weitzman et al., 1993). Other chemicals that might be of interest regarding potential neurotoxicity after exposure in drinking water include aluminum (Caban-Holt et al., 2005; Halatek et al., 2005; Kawahara, 2005), the organophosphates (Slotkin, 2004), and chlorine dioxide plus chlorite (Toth et al., 1990; Gill et al., 2000). No such effects have yet been confirmed in human studies involving drinking water exposures. Many other chemicals of great interest to neurotoxicologists, such as methyl mercury (Davidson et al., 2004; van Wijngaarden et al., 2006) and organic solvents (van der Hoek et al., 2000; Bushnell et al., 2006), are found at such low concentrations in drinking water that this source is not expected to be a major exposure route for the toxicants.

Immunotoxicity

There are several important tests of immunological functions in humans, although results are not necessarily readily relatable to the effects of specific chemicals. The major human immunological tests include measurements of serum immunoglobins and blood T, B, and natural killer cells, skin allergen sensitivity tests, and respiratory sensitization or reactivity tests. Specific antigen and antibody reactions can also be evaluated in vitro. Immunostimulation from chemicals causing allergic sensitization, such as beryllium and nickel, might result in enhanced autoimmune reactions, leading to arthritis and other autoimmune diseases. Impaired immune response can leave humans vulnerable to infections with bacteria, viruses, and fungi, as well as leading to enhanced susceptibility to cancer (Penn, 1986; Moore and Chang, 2003; Gerlini et al., 2005). There is at present no established method to assess possible developmental immunotoxicological effects (Holsapple, 2002). Standardization of requirements for immunotoxicology testing, mostly in animals, has been a major goal of pharmaceutical manufacturers (Schulte and Ruehl-Fehlert, 2006; Spanhaak, 2006).

Allergic sensitization potential has been considered in toxicological evaluation and regulatory standards development for several chemicals, including beryllium, chromium, nickel, and mercury. The sensitization potency has been calculated for both respiratory and dermal exposure routes. Application of these techniques for drinking water risk assessment is not common, however. The most important reason for this is that concentrations of chemicals in drinking water are far lower than those encountered in occupational settings for which the respiratory and dermal exposure guidelines were developed. Another reason is that it is more difficult to estimate allergic sensitization potential and potency for oral exposures, although oral exposure pathways may be relevant to sensitization (Deubner et al., 2001). Drinking water risk assessments do not typically provide estimates of potential effects in the subpopulation of allergically sensitized individuals, because of a lack of adequate dose–response and incidence data. The putative toxicological response called *multiple chemical sensitivity*, which is usually described by its advocates as a heightened responsiveness to synthetic chemicals (not necessarily exclusively immune mediated), is also not reflected in risk assessments, because it cannot be related to specific chemicals and toxicological endpoints.

Improved understanding of the immune system and its responses has not yet led to a process for rating the immunotoxicity of chemicals. Because of the huge influence of both genetics and environment on individual responses, it is not clear whether such a process can be developed at all. However, recent advances in immunotoxicogenomics have provided hope that a chemical's immunotoxicologic potential may eventually be quantifiable (Luebke et al., 2006b).

Genotoxicity

Genotoxic effects of chemicals in humans have been assessed by several cytological methods. These include evaluations of chromosomes in lymphocytes or other readily available cells (Mark et al., 1994; Mahata et al., 2003; Osswald et al., 2003) and various methods to evaluate specific DNA changes. The presence of micronuclei or sister chromatid rearrangements are the usual endpoints of concern in cytological examinations. Effects on DNA as evaluated by the Comet assay (Giannotti et al., 2002), alkylated nucleotides (Ricicki et al., 2006), or hyper- or hypomethylation (Luczak and Jogodzinski, 2006) are increasingly being measured. Evaluation of sperm morphology is also relevant to potential genotoxic effects of environmental chemicals (Swan et al., 2005; Swan, 2006; Toft et al., 2006).

In vitro studies of the effects of chemicals in human cell cultures are useful for evaluating the potential of chemicals to damage DNA (Clare et al., 2006;

Lorge et al., 2006; Parry and Parry, 2006). However, such studies are difficult to relate to in vivo toxic potency and thus are used only as supporting evidence in risk assessment. For human genotoxicity testing, evaluation of the significance of chromosome changes in occupationally exposed populations is not simple, because of the difficulty of establishing a dose–response relationship and eliminating confounders. However, detailed chemical studies, such as those on arsenic (Andrew et al., 2006), asbestos (Zhao et al., 2006), and styrene (Godderis et al., 2004), have great potential for providing quantitative measures of genotoxic potential of chemicals as well as helping to identify susceptible populations.

CONCLUSIONS

Risk assessment is not conducted by strict application of a formula or guidelines to the results of the toxicity tests described above; it involves having a good knowledge of biological systems and multidisciplinary training in chemistry, biochemistry, and physiology as well as an understanding of toxicological responses. The basic principles of dose–response and statistical evaluation must be applied to interpret these results properly, and specialized expertise in certain fields such as neurotoxicity or reproductive and developmental toxicity is often required. A risk assessor needs an ability for critical evaluation and interpretation, remaining open-minded in response to new data and conflicting opinions. He or she must keep in mind that the more important a particular drinking water contaminant is, the more varied the opinions will be. Skills in interpreting the data are gained with time and practice, so consultation with other risk assessors is always useful. In addition, communication skills are critical. A risk assessor needs to be able to explain the basis for all the steps in a risk assessment and to summarize them succinctly to the general public or inquisitive reporters. The development of risk assessments for drinking water chemicals carries a significant professional, public health, and social responsibility and requires a career-long commitment to keeping abreast of the state-of-the-art science and methodology.

Disclaimer

The opinions expressed in this chapter are those of the authors and not necessarily those of the Office of Environmental Health Hazard Assessment or the California Environmental Protection Agency.

REFERENCES

Andrew AS, Burgess JL, Meza MM, Demidenko E, Waugh MG, Hamilton JW, Karagas MR. 2006. Arsenic exposure is associated with decreased DNA repair in vitro and in individuals exposed to drinking water arsenic. *Environ Health Perspect* 114(8): 1193–1198.

Araya M, McGoldrick M, Klevay L, Strain J, et al. 2001. Determination of an acute no-observed-adverse-effect level (NOAEL) for copper in water. *Regul Toxicol Pharmacol* 34: 137–145.

Araya M, Olivares M, Pizarro F, Llanos A, Figueroa G, Uauy R. 2004. Community-based randomized double-blind study of gastrointestinal effects and copper exposure in drinking water. *Environ Health Perspect* 112(10): 1068–1073.

Ashby J. 1996. Alternatives to the 2-species bioassay for the identification of potential human carcinogens. *Hum Exp Toxicol* 15(3): 183–202.

Battershill JM, Edwards PM, Johnson MK. 2004. Toxicological assessment of isomeric pesticides: a strategy for testing of chiral organophosphorus (OP) compounds for delayed polyneuropathy in a regulatory setting. *Food Chem Toxicol* 42(8): 1279–1285.

Biava C, Smuckler E, Whorton D. 1978. The testicular morphology of individuals exposed to dibromochloropropane. *Exp Mol Pathol* 29(3): 448–458.

Boyes WK, Bercegeay M, Krantz T, Evans M, Benignus V, Simmons JE. 2005. Momentary brain concentration of trichloroethylene predicts the effects on rat visual function. *Toxicol Sci* 87(1): 187–196.

Braverman LE, Pearce EN, He X, Pino S, Seeley M, Beck B, Magnani B, Blount BC, Firek A. 2006. Effects of six months of daily low-dose perchlorate exposure on thyroid function in healthy volunteers. *J Clin Endocrinol Metab* 91(7): 2721–2724.

Brechner RJ, Parkhurst GD, Humble WO, Brown MB, Herman WH. 2000. Ammonium perchlorate contamination of Colorado River drinking water is associated with abnormal thyroid function in newborns in Arizona. *J Occup Environ Med* 42(8): 777–782. Comments in: *J Occup Environ Med* 2001 Apr; 43(4): 305–309; *J Occup Environ Med* 2001 Apr; 43(4): 307–309; *J Occup Environ Med* 2003 Nov; 45(11): 1131–1132.

Bushnell PJ, Boyes WK, Shafer TJ, Bale AS, Benignus VA. 2006. Approaches to extrapolating animal toxicity data on organic solvents to public health. *Neurotoxicology* 28(2): 221–226.

Caban-Holt A, Mattingly M, Cooper G, Schmitt FA. 2005. Neurodegenerative memory disorders: a potential role of environmental toxins. *Neurol Clin* 23(2): 485–521.

California OEHHA (Office of Environmental Health Hazard Assessment). 1997. *Public Health Goal for Copper in Drinking Water*. California Environmental Protection Agency, Oakland, CA. Accessed at: http://www.oehha.ca.gov/water/phg/pdf/copper-c.pdf.

———. 2004. *Public Health Goal for Perchlorate in Drinking Water*. California Environmental Protection Agency, Oakland, CA. Accessed at: http://www.oehha.ca.gov/water/phg/pdf/finalperchlorate31204.pdf.

Charnley G, Patterson J. 2004. Ethical standards of studies involving human subjects. *Environ Health Perspect* 112: A152–A153.

Clare MG, Lorenzon G, Akhurst LC, Marzin D, van Delft J, Montero R, Botta A, Bertens A, Cinelli S, Thybaud V, Lorge E. 2006. SFTG international collaborative study on in vitro micronucleus test, II. Using human lymphocytes. *Mutat Res* 607(1): 37–60.

Corbett GE, Finley BL, Paustenbach DJ, Kerger BD. 1997. Systemic uptake of chromium in human volunteers following dermal contact with hexavalent chromium (22 mg/L). *J Expo Anal Environ Epidemiol* 7(2): 179–189.

CWQ (Council on Water Quality). 2006. Economic impacts: Overly strict standards would create unnecessarily high costs to consumers, society. CWQ (a perchlorate industry lobbying group), Sacramento, CA. Accessed at: http://www.councilonwaterquality.org/issue/economic_impact.html.

Davidson PW, Myers GJ, Weiss B. 2004. Mercury exposure and child development outcomes. *Pediatrics* 113(4 suppl): 1023–1029.

Desai I, Nagymajtenyi L. 1999. Electrophysiological biomarkers of an organophosphorous pesticide, dichlorvos. *Toxicol Lett* 107(3–1): 55–64.

Deubner DC, Lowney YW, Paustenbach DJ, Warmerdam J. 2001. Contribution of incidental exposure pathways to total beryllium exposures. *Appl Occup Environ Hyg* 16(5): 568–578.

Engels EA, Schmid CH, Terrin N, Olkin I, Lau J. 2000. Heterogeneity and statistical significance in meta-analysis: an empirical study of 125 meta-analyses. *Stat Med* 19(13): 1707–1728.

Festing MF. 1994. Reduction of animal use: experimental design and quality of experiments. *Lab Anim* 28(3): 212–221.

Festing MF, Altman DG. 2002. Guidelines for the design and statistical analysis of experiments using laboratory animals. *ILAR J* 43(4): 244–258. Erratum in: *ILAR J* 2005; 46(3):320.

Finley BL, Kerger BD, Katona MW, Gargas ML, Corbett GC, Paustenbach DJ. 1997. Human ingestion of chromium(VI) in drinking water: pharmacokinetics following repeated exposure. *Toxicol Appl Pharmacol* 142(1): 151–159.

Fukushima S, Morimura K, Wanibuchi H, Kinoshita A, Salim EI. 2005. Current and emerging challenges in toxicopathology: carcinogenic threshold of phenobarbital and proof of arsenic carcinogenicity using rat medium-term bioassays for carcinogens. *Toxicol Appl Pharmacol* 207(2 suppl): 225–229.

Gad SC. 2000. *In Vitro Toxicology*, 2nd ed. Taylor & Francis, New York.

Gerlini G, Romagnoli P, Pimpinelli N. 2005. Skin cancer and immunosuppression. *Crit Rev Oncol Hematol* 56(1): 127–136.

Giannotti E, Vandin L, Repeto P, Comelli R. 2002. A comparison of the in vitro Comet assay with the in vitro chromosome aberration assay using whole human blood or Chinese hamster lung cells: validation study using a range of novel pharmaceuticals. *Mutagenesis* 17(2): 163–170.

Gill MW, Swanson MS, Murphy SR, Bailey GP. 2000. Two-generation reproduction and developmental neurotoxicity study with sodium chlorite in the rat. *J Appl Toxicol* 20(4): 291–303.

Godderis L, De Boeck M, Haufroid V, Emmery M, Mateuca R, Gardinal S, Kirsch-Volders M, Veulemans H, Lison D. 2004. Influence of genetic polymorphisms on biomarkers of exposure and genotoxic effects in styrene-exposed workers. *Environ Mol Mutagen* 44(4): 293–303.

Goldman LR, Links JM. 2004. Testing toxic compounds in human subjects: ethical standards and good science. *Environ Health Perspect* 112(8):A458-A459. Comment on: *Environ Health Perspect* 2004 Mar; 112(3):A150–A151; author reply, A151–A152; discussion, A152–A156.

Goodman G. 2001. The conclusions of the Arizona perchlorate study require reexamination. *J Occup Environ Med* 43(4): 305–309. Comment on: *J Occup Environ Med* 2000 Aug; 42(8): 777–782.

Goodman DG, Sauer RM. 1992. Hepatotoxicity and carcinogenicity in female Sprague–Dawley rats treated with 2,3,7,8-tetrachlorodibenzo-p-dioxin (TCDD): a pathology working group reevaluation. *Regul Toxicol Pharmacol* 15(3): 245–252.

Greer MA, Goodman G, Pleus RC, Greer SE. 2002. Health effects assessment for environmental perchlorate contamination: the dose response for inhibition of thyroidal radioiodine uptake in humans. *Environ Health Perspect* 110(9): 927–937. Erratum in: *Environ Health Perspect* 2005 Nov; 113(11):A732.

Halatek T, Sinczuk-Walczak H, Rydzynski K. 2005. Prognostic significance of low serum levels of Clara cell phospholipid-binding protein in occupational aluminium neurotoxicity. *J Inorg Biochem* 99(9): 1904–1911.

Hass U. 2006. The need for developmental neurotoxicity studies in risk assessment for developmental toxicity. *Reprod Toxicol* 22(2): 148–156.

Hayes W. 2001. *Principles and Methods of Toxicology*, 4th ed. Taylor & Francis, Philadelphia, PA.

Hinton DM. 2000. U.S. FDA "Redbook II" immunotoxicity testing guidelines and research in immunotoxicity evaluations of food chemicals and new food proteins. *Toxicol Pathol* 28(3): 467–478.

Holsapple MP. 2002. Developmental immunotoxicology and risk assessment: a workshop summary. *Hum Exp Toxicol* 21(9–10): 473–478.

Ito N. 2000. Marriage of surrogate markers with medium term tests for carcinogenicity and modification potential. *Asian Pac J Cancer Prev* 1(4): 347–353.

Ito N, Tamano S, Shirai T. 2003. A medium-term rat liver bioassay for rapid in vivo detection of carcinogenic potential of chemicals. *Cancer Sci* 94(1): 3–8.

Jacobson-Kram D, Keller KA. 2001. *Toxicology Testing Handbook: Principles, Applications, and Data Interpretation*. Marcel Dekker, New York.

Kast A. 1994. "Wavy ribs": a reversible pathologic finding in rat fetuses. *Exp Toxicol Pathol* 46(3): 203–210.

Kawahara M. 2005. Effects of aluminum on the nervous system and its possible link with neurodegenerative diseases. *J Alzheimer's Dis* 8(2): 171–182; discussion, 209–215.

Keenan RE, Paustenbach DJ, Wenning. RJ, Parsons AHJ. 1991. Pathology reevaluation of the Kociba et al. (1978) bioassay of 2,3,7,8-TCDD: implications for risk assessment. *Toxicol Environ Health* 34(3): 279–296.

Kerger BD, Finley BL, Corbett GE, Dodge DG, Paustenbach DJ. 1997. Ingestion of chromium(VI) in drinking water by human volunteers: absorption, distribution, and excretion of single and repeated doses. *J Toxicol Environ Health* 50(1): 67–95.

Kerkvliet NI. 2002. Recent advances in understanding the mechanisms of TCDD immunotoxicity. *Int Immunopharmacol* 2(2–3): 277–291.

Khera KS. 1981. Common fetal aberrations and their teratologic significance: a review. *Fundam Appl Toxicol* 1(1): 13–18.

Kirkland D, Aardema M, Henderson L, Muller L. 2005. Evaluation of the ability of a battery of three in vitro genotoxicity tests to discriminate rodent carcinogens and non-carcinogens, I: Sensitivity, specificity and relative predictivity. *Mutat Res* 584(1–2): 1–256. Erratum in: *Mutat Res* 2005 Dec 7; 588(1):70.

Knight A, Bailey J, Balcombe J. 2006a. Animal carcinogenicity studies, 3: Alternatives to the bioassay. *Altern Lab Anim* 34(1): 39–48.

_____. 2006b. Animal carcinogenicity studies: implications for the REACH system. *Altern Lab Anim* 34 (suppl 1): 139–147.

Lamm SH. 2003. Perchlorate exposure does not explain differences in neonatal thyroid function between Yuma and Flagstaff. *J Occup Environ Med* 45(11): 1131–1132. Comment on: *J Occup Environ Med* 2000 Aug; 42(8): 777–782.

Lorge E, Thybaud V, Aardema MJ, Oliver J, Wakata A, Lorenzon G, Marzin D. 2006. SFTG international collaborative study on in vitro micronucleus test, I: General conditions and overall conclusions of the study. *Mutat Res* 607(1): 13–36.

Luczak MW, Jagodzinski PP. 2006. The role of DNA methylation in cancer development. *Folia Histochem Cytobiol* 44(3): 143–154.

Luebke RW, Chen DH, Dietert R, Yang Y, King M, Luster MI. 2006a. The comparative immunotoxicity of five selected compounds following developmental or adult exposure. *J Toxicol Environ Health B Crit Rev* 9(1): 1–26.

Luebke RW, Holsapple MP, Ladics GS, Luster MI, Selgrade M, Smialowicz RJ, Woolhiser MR, Germolec DR. 2006b. Immunotoxicogenomics: the potential of genomics technology in the immunotoxicity risk assessment process. *Toxicol Sci* 94(1): 22–27.

Luebker DJ, Case MT, York RG, Moore JA, Hansen KJ, Butenhoff JL. 2005. Two-generation reproduction and cross-foster studies of perfluorooctanesulfonate (PFOS) in rats. *Toxicology* 215(1–2): 126–148.

Mahata J, Basu A, Ghoshal S, Sarkar JN, Roy AK, Poddar G, Nandy AK, Banerjee A, Ray K, Natarajan AT, Nilsson R, Giri AK. 2003. Chromosomal aberrations and sister chromatid exchanges in individuals exposed to arsenic through drinking water in West Bengal, India. *Mutat Res* 534(1–2): 133–143.

Mark HF, Naram R, Pham T, Shah K, Cousens LP, Wiersch C, Airall E, Samy M, Zolnierz K, Mark R Jr, et al. 1994. A practical cytogenetic protocol for in vitro cytotoxicity and genotoxicity testing. *Ann Clin Lab Sci* 24(5): 387–395.

Massachusetts DEP (Department of Environmental Protection). 2006. Update to "Perchlorate Toxicological Profile and Health Assessment." In support of: Perchlorate Maximum Contaminant Level (310 CMR 22.06) Perchlorate Cleanup Standards (310 CMR 40.0000). June. Office of Research and Standards, Massachusetts. DEP, Boston, MA. Accessed at: http://mass.gov/dep/toxics/perchlorate-toxicity-061206.doc#_Toc1270-93376.

McAllister RS. 2004. Statement of CropLife America on pesticide testing involving human subjects. *Environ Health Perspect* 112:A 154–155.

McKee RH, Pavkov KL, Trimmer GW, Keller LH, Stump DG. 2006. An assessment of the potential developmental and reproductive toxicity of di-isoheptyl phthalate in rodents. *Reprod Toxicol* 21(3): 241–252.

McLachlan JA, Simpson E, Martin M. 2006. Endocrine disrupters and female reproductive health. *Best Pract Res Clin Endocrinol Metab* 20(1): 63–75.

Melnick RL, Huff J. 2004. Testing toxic pesticides in humans: health risks with no health benefits. *Environ Health Perspect* 112(8): A459–A461. Comment on: *Environ Health Perspect* 2004 Mar; 112(3): A150–A151; author reply, A151–A152; discussion, A152–A156.

Moore PS, Chang Y. 2003. Kaposi's sarcoma–associated herpesvirus immunoevasion and tumorigenesis: Two sides of the same coin? *Annu Rev Microbiol* 57: 609–639.

Morris RD, Audet AM, Angelillo IF, Chalmers TC, Mosteller F. 1992. Chlorination, chlorination by-products, and cancer: a meta-analysis. *Am J Public Health* 82(7): 955–963.

Erratum in: *Am J Public Health* 1993 Sep; 83(9): 1257. Comment in: *Am J Public Health* 1993 Sep; 83(9): 1347–1348.

NAS (National Academy of Sciences). 2001. *Evaluating Chemical and Other Agent Exposures for Reproductive and Developmental Toxicity*. Commission on Life Sciences, National Research Council. National Academies Press, Washington, DC.

———. 2004. *Intentional Human Dosing Studies for EPA Regulatory Purposes: Scientific and Ethical Issues*. National Research Council. National Academies Press, Washington, DC.

———. 2005. *Health Implications of Perchlorate Ingestion*. National Research Council. National Academies Press, Washington, DC.

Navas-Acien A, Sharrett AR, Silbergeld EK, Schwartz BS, Nachman KE, Burke TA, Guallar E. 2005. Arsenic exposure and cardiovascular disease: a systematic review of the epidemiologic evidence. *Am J Epidemiol* 162(11): 1037–1049. Comment in: *Am J Epidemiol* 2006 Jul 15; 164(2): 194–195; author reply, 195–196.

Needleman HL, Reigart JR, Landrigan P, Sass J, Bearer C. 2005. Benefits and risks of pesticide testing on humans. *Environ Health Perspect* 113(12): A804–A805; author reply, A805. Comment on: *Environ Health Perspect* 2005 Jul; 113(7): 813–817.

Nichols CM, Myckatyn TM, Rickman SR, Fox IK, Hadlock T, Mackinnon SE. 2005. Choosing the correct functional assay: a comprehensive assessment of functional tests in the rat. *Behav Brain Res* 163(2): 143–158.

OECD (Organization for Economic Cooperation and Development). 1981. *Guidelines for Testing of Chemicals*, Section 4, Health Effects, Part 451, Carcinogenicity Studies. OECD, Paris.

Olivares M, Araya M, Pizarro F, Uauy R. 2001. Nausea threshold in apparently healthy individuals who drink fluids containing graded concentrations of copper. *Reg Toxicol Pharmacol* 33: 271–275.

Osswald K, Mittas A, Glei M, Pool-Zobel BL. 2003. New revival of an old biomarker: characterization of buccal cells and determination of genetic damage in the isolated fraction of viable leucocytes. *Mutat Res* 544(2–3): 321–329.

Parry JM, Parry EM. 2006. The use of the in vitro micronucleus assay to detect and assess the aneugenic activity of chemicals. *Mutat Res* 607(1): 5–8.

Paustenbach DJ, Hays SM, Brien BA, Dodge DG, Kerger BD. 1996. Observation of steady state in blood and urine following human ingestion of hexavalent chromium in drinking water. *J Toxicol Environ Health* 49(5): 453–461.

Penn I. 1986. Cancer is a complication of severe immunosuppression. *Surg Gynecol Obstet* 162(6): 603–610.

Pocock SJ, Smith M, Baghurst P. 1994. Environmental lead and children's intelligence: a systematic review of the epidemiological evidence. *BMJ* 309(6963): 1189–1197. Comment in: *BMJ* 1995 Feb 11; 310(6976):397; *BMJ* 1995 Feb 11; 310(6976):397; *BMJ* 1995 Feb 11; 310(6976):397; *BMJ* 1995 Feb 11; 310(6976): 397–398; *BMJ* 1995 May 27; 310(6991): 1408–1409; *BMJ* 1995 May 27; 310(6991):1408; author reply, 1409.

Potashnik G. 1983. A four-year assessment of workers with dibromochloropropane-induced testicular dysfunction. *Andrologia* 15(2): 164–170.

Potashnik G, Yanai-Inbar I, Sacks M, Israeli R. 1979. Effect of dibromochloropropane on human testicular function. *Isr J Med Sci* 15(5): 438–442.

Resnik DB, Portier C. 2005. Pesticide testing on human subjects: weighing benefits and risks. *Environ Health Perspect* 113(7): 813–817. Comment in: *Environ Health Perspect* 2005 Dec; 113(12): A804–A805; author reply, A805.

Riicki EM, Luo W, Fan W, Zhao LP, Zarbl H, Vouros P. 2006. Quantification of *N*-(deoxyguanosin-8-yl)-4-aminobiphenyl adducts in human lymphoblastoid TK6 cells dosed with *N*-hydroxy-4-acetylaminobiphenyl and their relationship to mutation, toxicity, and gene expression profiling. *Anal Chem* 78(18): 6422–6432.

Scallet AC, Schmued LC, Johannessen JN. 2005. Neurohistochemical biomarkers of the marine neurotoxicant, domoic acid. *Neurotoxicol Teratol* 27(5): 745–752.

Schulte A, Ruehl-Fehlert C. 2006. Regulatory aspects of immunotoxicology. *Exp Toxicol Pathol* 57(5–6): 385–389.

Schwartz J. 1994. Low-level lead exposure and children's IQ: a meta-analysis and search for a threshold. *Environ Res* 65(1): 42–55.

Slotkin TA. 2004. Guidelines for developmental neurotoxicity and their impact on organophosphate pesticides: a personal view from an academic perspective. *Neurotoxicology* 25(4): 631–640.

Spanhaak S. 2006. The ICH S8 immunotoxicity guidance. Immune function assessment and toxicological pathology: autonomous or synergistic methods to predict immunotoxicity? *Exp Toxicol Pathol* 57(5–6): 373–376.

Spitalny KC, Brondum J, Vogf RL, Sargent HE, Kappel S. 1984. Drinking water induced copper intoxication in a Vermont family. *Pediatrics* 74: 1103–1106.

Stroup DF, Thacker SB, Olson CM, Glass RM, Hutwagner L. 2001. Characteristics of meta-analyses related to acceptance for publication in a medical journal. *J Clin Epidemiol* 54(7): 655–660.

Swan SH. 2006. Semen quality in fertile US men in relation to geographical area and pesticide exposure. *Int J Androl* 29(1): 62–68; discussion, 105–108.

Swan SH, Waller K, Hopkins B, Windham G, Fenster L, Schaefer C, Neutra RR. 1998. A prospective study of spontaneous abortion: relation to amount and source of drinking water consumed in early pregnancy. *Epidemiology* 9(2): 126–133. Comment in: *Epidemiology* 1999 Mar; 10(2): 203–204.

Swan SH, Kruse RL, Liu F, Barr DB, Drobnis EZ, Redmon JB, Wang C, Brazil C, Overstreet JW. 2003. Semen quality in relation to biomarkers of pesticide exposure. *Environ Health Perspect* 111(12): 1478–1484. Comment in: *Environ Health Perspect* 2005 Oct; 113(10):A652; author reply, A652–A653.

Tanaka H, Ikenaka K, Isa T. 2006. Electrophysiological abnormalities precede apparent histological demyelination in the central nervous system of mice overexpressing proteolipid protein. *J Neurosci Res* 84(6): 1206–1216.

Thacker SB. 1988. Meta-analysis. a quantitative approach to research integration. *JAMA* 259(11): 1685–1689.

Tobia A, Ayers A, Blacker A, Hodges L, Carmichael N. 2004. Aldicarb study misrepresented in human testing debate. *Environ Health Perspect* 112: A155–A156.

Toft G, Rignell-Hydbom A, Tyrkiel E, Shvets M, Giwercman A, Lindh CH, Pedersen HS, Ludwicki JK, Lesovoy V, Hagmar L, Spano M, Manicardi GC, Bonefeld-Jorgensen EC, Thulstrup AM, Bonde JP. 2006. Semen quality and exposure to persistent organochlorine pollutants. *Epidemiology* 17(4): 450–458.

Tonkin EG, Valentine HL, Zimmerman LJ, Valentine WM. 2003. Parenteral *N,N*-diethyldithiocarbamate produces segmental demyelination in the rat that is not dependent on cysteine carbamylation. *Toxicol Appl Pharmacol* 189(2): 139–150.

Tonkin EG, Valentine HL, Milatovic DM, Valentine WM. 2004. *N,N*-Diethyldithiocarbamate produces copper accumulation, lipid peroxidation, and myelin injury in rat peripheral nerve. *Toxicol Sci* 81(1): 160–171.

Toth GP, Long RE, Mills TS, Smith MK. 1990. Effects of chlorine dioxide on the developing rat brain. *J Toxicol Environ Health* 31(1): 29–44.

U.S. EPA (Environmental Protection Agency). 1991. Maximum contaminant level goals and national primary drinking water regulation for lead and copper; final rule. Fed Reg 56(110): 26460–26464, June 7.

———. 2000. *Comments on the Use of Data from the Testing of Human Subjects: A Report by the Science Advisory Board and the FIFRA Scientific Advisory Panel*. EPA/SAB/EC-00/017. U.S. EPA, Washington, DC. Accessed at: http://www.epa.gov/sab/pdf/ec0017.pdf.

———. 2002. *Report on the Peer Review of the U.S. Environmental Protection Agency's Draft External Review Document "Perchlorate Environmental Contamination: Toxicological Review and Risk Characterization*." EPA/635/R-02/003, June. National Center for Environmental Assessment, U.S. EPA, Washington, DC. Accessed at: http://www.epa.gov/ncea/pdfs/perchlorate/final_rpt.pdf.

———. 2004. *An Examination of EPA Risk Assessment Principles and Practices*. Staff paper prepared for the U.S. Environmental Protection Agency by members of the Risk Assessment Task Force. EPA/100/B-04/001. Office of the Science Advisor, U.S. EPA, Washington, DC. Accessed at: http://www.epa.gov/osa/pdfs/ratf-final.pdf.

———. 2005. Children's Environmental Exposure Research Study. This study was cancelled. Statement by Stephen L. Johnson, Acting Administrator, U.S. EPA, Washington, DC. Accessed at: http://www.epa.gov/cheers/.

———. 2006a. *OPPTS Harmonized Test Guidelines, Series 870*. Office of Prevention, Pesticides and Toxic Substances, U.S. EPA, Washington, DC. Accessed at: http://www.epa.gov/opptsfrs/publications/OPPTS_Harmonized/870_Health_Effects_Test_Guidelines/Series/.

———. 2006b. *Mercuric Chloride* ($HgCl_2$) (CASRN 7487-94-7). Integrated Risk Information System, U.S. EPA, Washington, DC. Accessed at: http://www.epa.gov/iris/subst/0692.htm.

U.S. FDA (Food and Drug Administration). 2006a. *Guideline for Toxicological Testing*. Center for Veterinary Medicine, U.S. FDA, Rockville, MD. Accessed at: http://www.fda.gov/cvm/Guidance/guideline3pt2.html.

———. 2006b. Notice of availability: Guidance on S8 immunotoxicity studies for human pharmaceuticals. International Conference on Harmonisation, U.S. FDA. *Fed Reg* Apr 13; 71(71): 19193–19194.

van der Hoek JA, Verberk MM, Hageman G. 2000. Criteria for solvent-induced chronic toxic encephalopathy: a systematic review. *Int Arch Occup Environ Health* 73(6): 362–368. Comment in: *Int Arch Occup Environ Health* 2000 Aug; 73(6):361.

van Wijngaarden E, Beck C, Shamlaye CF, Cernichiari E, Davidson PW, Myers GJ, Clarkson TW. 2006. Benchmark concentrations for methyl mercury obtained from the 9-year follow-up of the Seychelles Child Development Study. *Neurotoxicology* 27(5): 702–709.

Villanueva CM, Fernandez F, Malats N, Grimalt JO, Kogevinas M. 2003. Meta-analysis of studies on individual consumption of chlorinated drinking water and bladder cancer. *J Epidemiol Community Health* 57(3): 166–173. Erratum in: *J Epidemiol Community Health* 2005 Jan; 59(1):87.

Waller K, Swan SH, DeLorenze G, Hopkins B. 1998. Trihalomethanes in drinking water and spontaneous abortion. *Epidemiology* 9(2): 134–140. Comment in: *Epidemiology* 1999 Mar; 10(2): 203–204.

Waller K, Swan SH, Windham GC, Fenster L. 2001. Influence of exposure assessment methods on risk estimates in an epidemiologic study of total trihalomethane exposure and spontaneous abortion. *J Expo Anal Environ Epidemiol* 11(6): 522–531.

Weinberg HS, Pereira VR, Singer PC, Savitz DA. 2006. Considerations for improving the accuracy of exposure to disinfection by-products by ingestion in epidemiologic studies. *Sci Total Environ* 354(1): 35–42.

Weitzman M, Aschengrau A, Bellinger D, Jones R, Hamlin JS, Beiser A. 1993. Lead-contaminated soil abatement and urban children's blood lead levels. *JAMA* 269(13): 1647–1654. Comments in: *JAMA* 1993 Apr 7; 269(13): 1679–1681; *JAMA* 1993 Aug 18; 270(7): 829–30; *JAMA* 1993 Nov 3; 270(17): 2054–2055.

Williams PD, Hottendorf GH. 1997. *Comprehensive Toxicology*, Vol 2, *Toxicological Testing and Evaluation*. Elsevier Science, New York.

Windham GC, Waller K, Anderson M, Fenster L, Mendola P, Swan S. 2003. Chlorination by-products in drinking water and menstrual cycle function. *Environ Health Perspect* 111(7): 935–941; discussion, A409.

Zhao XH, Jia G, Liu YQ, Liu SW, Yan L, Jin Y, Liu N. 2006. Association between polymorphisms of DNA repair gene XRCC1 and DNA damage in asbestos-exposed workers. *Biomed Environ Sci* 19(3): 232–238.

4

EXPOSURE SOURCE AND MULTIROUTE EXPOSURE CONSIDERATIONS FOR RISK ASSESSMENT OF DRINKING WATER CONTAMINANTS

KANNAN KRISHNAN
Université de Montréal, Montréal, Québec, Canada

RICHARD CARRIER
Health Canada, Ottawa, Ontario, Canada

Health risks associated with human exposure to drinking water contaminants (DWCs) are determined by their intrinsic toxicity and extent of exposure. Exposure in the present context refers to the contact between the individual or population and the DWCs. Assessment of exposure should account for the intensity, duration, and frequency of contact for each exposure route and source. Three approaches are often used for conducting quantitative exposure assessment: (1) direct measurement, (2) the exposure scenario, and (3) biomonitoring (U.S. EPA, 1992a; IPCS, 2000; Paustenbach, 2002). The *direct measurement approach* involves measurement of the chemical at the point of contact and integrating the measures over a period of time. The *exposure scenario approach* involves estimation or prediction of chemical concentrations in the exposure medium and integration of this information with other data relevant to the individual or population (activity pattern, duration and extent of contact, and body weight) for estimating exposure. The *biomonitoring* or *biological monitoring approach* involves measurement of the chemical in parental form or in other forms (metabolites or adducts) in biological matrices (urine, exhaled air, blood,

Risk Assessment for Chemicals in Drinking Water, Edited by Robert A. Howd and Anna M. Fan

TABLE 1. Definitions of Dose-Related Terms Employed in Exposure Assessment

Terminology	Definition
Potential dose	Amount of a substance in the environmental medium or material ingested, breathed, or in contact with the skin. Synonymous with *administered dose*.
Internal dose	Amount of a substance penetrating across the absorption barriers of an organism. Synonymous with *absorbed dose*.
Biologically effective dose	Amount of a substance reaching the cells or target site where an adverse effect occurs. The amount available for interaction with a cell (or tissue) is also termed the *delivered dose* for that cell (or tissue).

Source: Based on U.S. EPA (1992a).

etc.), and use of that information to infer the exposure that has occurred. These methods allow exposure dose calculations, depending on the data availability and needs of the risk assessment. Such calculations relate to the potential dose, internal dose, or biologically effective dose (Table 1). The temporal nature and magnitude of the dose received might vary according to the route and sources of exposure. Further, drinking water may not be the sole source of exposure, and in fact, multiple exposure sources may contribute to the entry of DWCs into the human body. In this chapter we describe current approaches for consideration of sources and routes of exposure in the process of health risk assessment of DWCs.

EXPOSURE SOURCE CONSIDERATIONS IN RISK ASSESSMENT

The sources of exposure to DWCs may be sole or multiple. In establishing the guideline values for noncarcinogenic DWCs, the proportion of the daily dose resulting from drinking water consumption is accounted for by the use of a relative source contribution (RSC) or source allocation factor (SAF) as follows (U.S. EPA, 1993; Health Canada, 1995; Howd et al, 2000; WHO, 2004):

$$\text{guideline value} = \frac{\text{tolerable daily intake (mg/kg·day)} \times \text{body weight (kg)} \times \text{RSC}}{\text{water consumption rate (L/day)}} \tag{1}$$

Frequently, the sources of exposure are diverse and multiple, consisting not only of drinking water but also of food, soil, air, and consumer products. Accordingly, the RSC represents the fraction of the daily dose resulting from the drinking water source. If drinking water is the sole source of exposure to a chemical, it is relatively straightforward to consider this aspect in the risk assessment process (RSC=1), else the relative contribution of drinking water to the total dose needs to be known or assumed. The U.S. EPA (1993, 2000) uses a default RSC value of 0.2 in the absence of adequate data. This default RSC factor implies that exposure via sources other than drinking water (i.e., food, soil, air, and consumer products) is likely to account for most of the daily dose (80%). Conversely, when the drinking water has been identified as the major exposure source, the EPA uses 0.8 as the RSC (Figure 1). The use of 0.8 represents a ceiling value that is low enough to provide protection to people for whom the dose received via exposure to other sources is probably not negligible (U.S. EPA, 2000). Similarly, a floor value of 0.2 has been used when exposure via drinking water is thought to contribute from 0 to 20% of the daily dose (Figure 1). The default RSC value of 20% used in DWC risk assessment appears to have arisen from historical use rather than from any standard scientific approaches (Ritter et al., 2005).

Existing compilations and comparisons of RSC factors used by various regulatory agencies suggest that the default value is used frequently (Howd et al., 2004; Ritter et al., 2005). The default RSCs used by some regulatory agencies for inorganic and organic substances in drinking water are provided in Table 2. Of course, there have been instances where a RSC value different from the default has been used on the basis of exposure data or knowledge of exposure sources. For example, a RSC value of 0.4 for antimony has been used on the basis of indications that approximately 60% of antimony exposure is likely to be associated with sources other than drinking water (i.e., food and air) (California OEHHA, 1997; Ritter et al., 2005). A RSC value of 0.8 has been used by Health Canada and WHO in the case of microcystin: LR, a blue-green algal toxin, for which domestic and recreational water use together represent the major source of exposure (Health Canada, 2002; WHO, 2004). In the case of barium and fluoride, an RSC of 1 has been used, reflecting the fact that the risk assessments were based on exposures through drinking water, without corrections for other sources of exposure to these chemicals (California OEHHA, 2003; Howd et al., 2004; WHO, 2004).

On the contrary, for chemicals for which drinking water does not represent a significant source of exposure, a RSC value much lower than the default value has been used [e.g., 0.01 in the case of EDTA and di(2-ethylhexyl)phthalate;

TABLE 2. Summary of Default Values Used by Various Agencies in the Risk Assessment of Drinking Water Contaminants

Jurisdiction	Body Weight (kg)		Drinking Water Intake (L/day)		Source Allocation Factor (%)
	Adult	Child	Adult	Child	
Canada	70	13	1.5	0.8	20
United States	70	10	2	1	20
WHO	60	10	2	1	10
Australia	70	13	2	1	10–20
European Union	60–70	10	2	1	10

Source: Ritter et al. (2005).

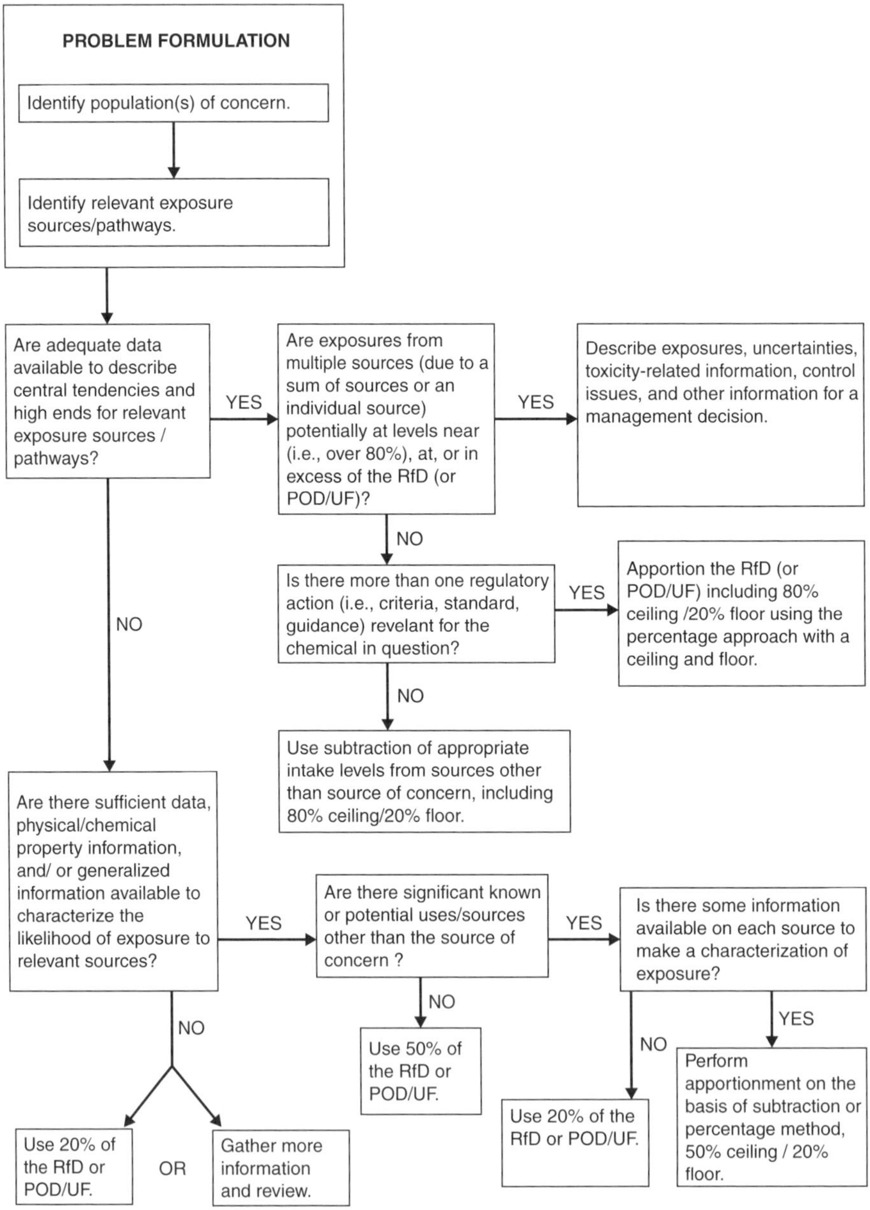

Figure 1. Exposure decision tree for defining apportionment of the reference dose. (Redrawn from U.S. EPA, 2000.)

WHO, 2004]. Food is often the major source of exposure to pesticides and micronutrients. In the case of highly persistent and bioaccumulating chemicals particularly, food is likely to be much more important as a source of exposure than drinking water such that the daily dose received from drinking water should be very small. Accordingly, in the case of such chemicals (e.g., aldrin, dieldrin, chlordane, heptachlor, lindane) a RSC of 0.01 has been used (WHO, 2004). However, for a number of inorganic chemicals and pesticides, the default RSC of 20% has been used either due to lack of exposure data or on the basis of the assumption that drinking water is not the sole or significant source of exposure to these contaminants.

Ideally, for establishing RSC for DWCs, media-specific intake (mg/kg·day) should be estimated for each subpopulation and then the proportion of the dose associated with each medium of exposure can be computed. Table 3 shows results of such an exercise conducted for methyl *tert*-butyl ether (MTBE), for which the data on the daily dose received from air and water were used to derive population-specific RSCs (California OEHHA, 1999). When data on actual concentrations of a DWC in the various media are unavailable or inadequate, mathematical models may be used to predict plausible concentrations on the basis of the usage pattern and physicochemical properties. One example is the

TABLE 3. Relative Source Contribution (RSC) Estimates for Various Combinations of Air and Drinking Water Exposures to Methyl *tert*-butyl Ether (MTBE)[a]

Air Exposure Estimate (mg/kg·day)	Air Exposure Scenario	RSC (%)			
		0.36 ppb	2 ppb	12 ppb	70 ppb
0.0085	Combined U.S. population grand average	0.6	3	16	52
0.002	Los Angeles basin at 4 ppbv ambient	0.5	2.8	15	50
6.7×10^{-5}	Milwaukee, Wisconsin air	13	46	84	97
0.37	MTBE distribution of fuel mixture time-weighted average (TWA) for workers	0.003	0.02	0.09	27
1.3×10^{-4}	Albany, New York air	7	30	72	94

Source: Based on California OEHHA (1999).

[a] $RSC = (I_{water} \times 100)/(I_{water} + I_{air})$. Food and soil sources are considered negligible for MTBE. I_{water} is the uptake by ingestion of tap water containing MTBE at the concentrations noted assuming 2 L/day and 100% intestinal absorption. I_{air} is the uptake by inhalation of airborne MTBE assuming 20 m^3 air inhaled per day and 50% absorption. I_{water} and I_{air} are both assumed for a 70-kg person.

fugacity model for predicting distribution and concentration of DWCs in various environmental media (Mackay, 1991). The model predictions of environmental concentrations can then be used along with contact rates and body weight information to compute the dose received from each source, for eventually calculating the RSC. No effort has so far been taken to use a systemic or internal dose in calculating RSCs, and this may be extremely important in some cases, given that the bioavailability and fraction of dose absorbed by the various routes is not 100%, as assumed in the calculations. Such an internal dose-based assessment of RSCs is likely to augment the scientific basis of current approaches in setting the guideline values for DWCs.

ROUTES OF EXPOSURE AND DOSE CALCULATIONS

The oral route is often the primary route of human exposure to DWCs. Contaminants entering by the oral route are absorbed in the various sections of the gastrointestinal tract to different degrees, resulting in their entry into systemic circulation and internal organs. A typical time-course profile of the blood concentration of a DWC following its oral absorption is presented in Figure 2. In this case, the maximal blood concentration of the parent chemical (C_{max}) is determined primarily by the rate of absorption and the extent of the first-pass effect in the liver. The *first-pass effect* refers to the extent of removal (usually by metabolism) by the port of entry before or immediately after systemic absorption. In the case of orally absorbed chemicals, the first-pass effect occurs in the liver, which is rich in phase I and phase II enzymes (Klaassen et al., 1996).

Some DWCs may also be absorbed by the inhalation and dermal routes. The dermal route of absorption is relevant for DWCs that are absorbed to a significant extent during use conditions that result in dermal contact (e.g., showering, bathing). The pulmonary route of exposure is particularly important for those DWCs that volatilize into the atmosphere under normal use conditions. The extent of absorption during exposure by the inhalation route is determined primarily by the breathing rate, cardiac output, rate of metabolism, and the blood–air partition coefficient (Medinsky, 1990; Krishnan and Andersen, 2001). As determined on the basis of Henry's law, the blood–air partition coefficient is the ratio of solubility of a DWC in blood and air (Poulin and Krishnan, 1996). The pulmonary absorption of a DWC with a relatively high blood–air partition coefficient is limited by the alveolar ventilation rate, whereas that of a DWC with low blood–air partition coefficient is limited by cardiac output (Klaassen et al., 1996). For lipophilic volatile DWCs, airways such as the nasal passages, larynx, trachea, bronchi, and bronchioles can be considered as inert tubes that carry the chemical to the alveolar region where the diffusion occurs. However, for highly polar chemicals, the adsorption along the respiratory tree during inhalation and desorption during exhalation should also be accounted for (Johanson, 1991). The skin absorption of airborne DWCs is unlikely to be a major process of concern (e.g., Brooke et al., 1998). However, dermal absorption is likely to be more important in the case of waterborne chemicals.

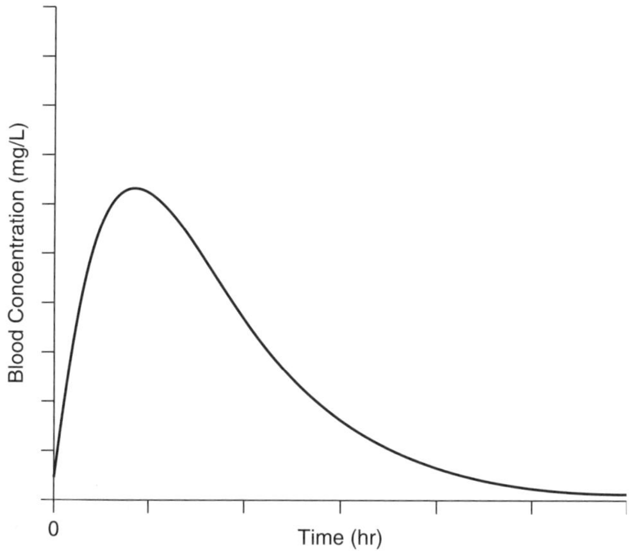

Figure 2. Profile of the blood concentration versus time course of a hypothetical drinking water contaminant following oral absorption.

Contaminants in water may penetrate the stratum corneum and the underlying epithelial cells, thus gaining entry into the bloodstream, resulting in systemic exposure. Human dermal exposure needs to be considered for showering and bathing scenarios, given the extent of the skin area exposed. For a given exposure scenario, the extent of dermal absorption of a DWC is determined by the duration of contact, skin area exposed, and the skin permeability coefficient, which is the result of the diffusion coefficient, skin–water partition coefficient, and the path length of diffusion. The path length of diffusion, in turn, is determined by the skin lipid and water content and the lipophilicity of the chemicals (U.S. EPA, 1992b; Poulin and Krishnan, 2001). A number of algorithms and experimental systems have been used for estimation of the skin permeability coefficient (K_p) (Mattie et al., 1994; Poulin et al., 1999; Prah et al., 2002). The simplest of the existing approaches relates K_p to molecular weight and/or the n-octanol/water partition coefficient of the nonionized DWCs (U.S. EPA, 1992b; Durkin et al., 1995). With knowledge of the chemical- and exposure-specific parameters, dose calculations can be performed for the various routes.

The average daily potential dose received by a person following oral exposure is calculated as follows:

$$\text{dose (mg/kg·day)} = \frac{\text{water concentration of DWC} \times \text{water consumption rate}}{\text{BW}} \tag{2}$$

TABLE 4. Summary of Tap Water Intake by Various Age Groups

	Intake (mL/day)		Intake (mL/kg·day)	
Age Group	Mean	10th–90th Percentiles	Mean	10th–90th Percentiles
Infants (<1 yr)	302	0–649	43.5	0–100
Children (1–10 yr)	736	286–1294	35.5	12.5–64.4
Teens (11–19 yr)	965	353–1701	18.2	6.5–32.3
Adults (20–64 yr)	1366	559–2268	19.9	8.0–33.7
Adults (65+yr)	1459	751–2287	21.8	10.9–34.7
All ages	1193	423–2092	22.6	8.2–39.8

Source: Paustenbach (2002), based on Ershow and Cantor (1989).

Equation (2) implies or assumes that 100% of the DWC is absorbed. On the contrary, as supported by data, the appropriate chemical-specific value representing the fraction absorbed can be used in the dose calculation process. The water consumption rate used in equation (2) typically corresponds to the daily rate. Table 4 summarizes the range of daily water intake values for infants, children, and adults. The default values used by various regulatory agencies are given in Table 2.

Similarly, the inhaled dose for an airborne DWC is calculated as follows:

$$\text{dose (mg/kg/}\cdot\text{day)} = \frac{C_{\text{air}} Q_{\text{alv}} t F_{\text{in,abs}}}{\text{BW}} \tag{3}$$

where

C_{air} = atmospheric concentration
Q_{alv} = alveolar ventilation rate
t = duration of exposure
$F_{\text{in,abs}}$ = fraction of the dose absorbed by the inhalation route
BW = body weight

The potential dose received through skin contact can be calculated as follows:

$$\text{dose (mg/kg}\cdot\text{day)} = \frac{C_{\text{water}} K_{\text{p}} A t F_{\text{sk,abs}}}{\text{BW}} \tag{4}$$

where

K_p = skin permeability coefficient
$F_{\text{sk,abs}}$ = fraction of the dose absorbed by the dermal route
A = area of skin exposed

For calculating the total dose received by all three routes during a single day, equations (2) to (4) can be combined as follows:

$$\text{total dose} = \text{dose}_{\text{ingestion}} + \text{dose}_{\text{inhalation}} + \text{dose}_{\text{dermal}} \tag{5}$$

where the route-specific doses are calculated according to equations (2) to (4). Over the past several years, researchers and regulators have contributed to the development of data and knowledge regarding the evaluation of multiroute exposure to DWCs and its role in setting the guideline values for DWCs.

Consideration of Multiroute Exposures in Deriving Guideline Values for DWCs

In setting the acceptable exposure levels for DWCs, it is important to account for the exposure and dose received by multiple exposure routes. It is obvious that multiroute exposures could result in a greater dose than the assumed exposure by a single route (i.e., oral route). This has become a particularly important risk assessment issue in the case of several DWCs, for which additional exposure by dermal and inhalation routes should be accounted for (Jo et al., 1990a, 1990b; Blancato and Chiu, 1993; Weisel and Jo, 1996; Lévesque et al., 1994, 2000, 2002; Lindstrom and Pleil, 1996; Rao and Ginsberg, 1997; Nuckols et al., 2005).

Considering the multiroute exposure to DWCs, the total potential dose received can be calculated as follows:

$$\text{total dose} = \frac{C_{\text{water}} V_{\text{water}} + C_{\text{air}} Q_{\text{alv}} t + C_{\text{water}} K_p A t}{\text{BW}} \qquad (6)$$

Since C_{air} resulting from C_{water} of a DWC is related by some factor ($F_{\text{a:w}}$), equation (6) can be rewritten as

$$\text{total dose} = \frac{C_{\text{water}} V_{\text{water}} + C_{\text{water}} F_{\text{a:w}} Q_{\text{alv}} t + C_{\text{water}} K_p A t}{\text{BW}} \qquad (7)$$

Taking C_{water} out of equation (7) gives us

$$\text{total dose} = \frac{C_{\text{water}} [V_{\text{water}} + F_{\text{a:w}} Q_{\text{alv}} t + K_p A t]}{\text{BW}} \qquad (8)$$

In equation (8) the second term in the numerator represents the *L-equivalent* associated with the multiroute exposure to a DWC. Here, there are three components. The first is V_{water}, which simply refers to the volume of water ingested per day. The second component, $F_{\text{a:w}}\ Q_{\text{alv}} t$, represents the L-equivalent resulting from the inhalation route. In other words, this term facilitates conversion of the potential dose received by the inhalation route in terms of the volume of drinking water. Finally, the term $K_p A t$ gives the L-equivalent associated with the dermal route. The L-equivalents, compared to the daily water consumption level used in regulatory assessments [e.g., 2 L (U.S. EPA), 1.5 L (Health Canada)], would give an idea of the relative importance of the inhalation and dermal pathways. Considering the percent absorption, the L-equivalents associated with dermal and inhalation routes can be computed as follows:

$$\text{L-equivalent, dermal exposure} = K_p A t F_{\text{sk,abs}} \qquad (9)$$

$$\text{L-equivalent, inhalation exposure} = F_{a:w} Q_{alv} t F_{in,abs} \tag{10}$$

In the following paragraphs we focus on two case studies relating to the consideration of multiroute exposure in risk assessment. These studies relate to trichloroethylene and chloroform. For trichloroethylene, several literary sources indicate the importance of inhalation and dermal routes of exposure. Bogan et al. (1988) on the basis of extensive indoor modeling and older data on skin permeability coefficients, reported L-equivalent values of 2.9 and 2, respectively, for inhalation and dermal exposures to trichloroethylene associated with household uses of tap water. Weisel and Jo (1996) evaluated the relative importance of inhalation and dermal routes of exposure compared with oral ingestion of trichloroethylene in an experimental study. These authors concluded that dermal and inhalation routes contribute similar internal doses, and that the dermal + inhalation dose is greater than that received from ingestion of tap water. Lindstrom and Pleil (1996) calculated the trichloroethylene dose received by the ingestion, dermal, and inhalation routes, which indicated that the dose ingested is more important than the dose inhaled, which in turn is greater than the dermal dose for trichloroethylene. Health Canada (2004a) recently used the results of a physiologically based pharmacokinetic (PBPK) modeling effort to determine the systemic exposures resulting from a dermal dose and inhalation of trichloroethylene following a 10-minute shower and a 30-minute bath. PBPK model–generated data onanabsorbed fraction (Tables 5 and 6) were used to calculate the L-equivalents for the risk assessment process according to equations (9) and (10). The L-equivalent values calculated for an adult taking a 30-minute bath were 1.7 and 0.71, respectively, for inhalation and dermal exposures to trichloroethylene. The multiroute exposure considerations might become more relevant particularly for children of various age groups, because of the fact that

TABLE 5. PBPK Model Simulations of Dose Absorbed by Adults and Children Through Inhalation Exposure During Showering (10-Minute Exposure) and Bathing (30-Minute Exposure) with Water Contaminated with 5 μg/L Trichloroethylene

Age Group	Exposure Duration (min)	Dose Absorbed ($\times 10^{-2}$ mg)	Fraction Absorbed[a]
6-year-old child	10	0.0531	0.58
	30	0.16	0.68
10-year-old child	10	0.0875	0.64
	30	0.2638	0.64
14-year-old teenager	10	0.1185	0.66
	30	0.3572	0.66
Adult	10	0.1530	0.66
	30	0.461	0.66

Source: Krishnan (2003a).

[a]Calculated as the ratio of dose absorbed to potential dose associated with inhalation exposure.

TABLE 6. PBPK Model Simulations of Dose Absorbed by Adults and Children Through Dermal Exposure During Showering (10-Minute Exposure) and Bathing (30-Minute Exposure) with Water Contaminated with 5 μg/L Trichloroethylene

Age Group	Exposure Duration (min)	Dose Absorbed ($\times 10^{-2}$ mg)	Fraction Absorbed[a]
6-year-old child	10	0.05	0.57
	30	0.15	0.57
10-year-old child	10	0.07	0.64
	30	0.22	0.64
14-year-old teenager	10	0.10	0.65
	10	0.10	0.65
Adult	10	0.12	0.66
	30	0.36	0.67

Source: Krishnan (2003a).

[a]Calculated by dividing the dose absorbed by the potential dose associated with dermal exposure.

TABLE 7. L-Equivalents of Trichloroethylene for the Various Exposure Scenarios and Age Groups

Age Group	Exposure Scenario	Exposure Route		Total L-Equivalents
		Inhalation	Dermal	
Adult	Showering	0.56	0.24	0.80
	Bathing	1.70	0.71	2.41
14-year-old teenager	Showering	0.54	0.21	0.75
	Bathing	1.63	0.62	2.25
10-year-old child	Showering	0.48	0.15	0.63
	Bathing	1.45	0.45	1.90

Source: Krishnan (2003a).

the rate of water consumption and area of skin exposed per kilogram of body weight are greater in children, invoking the possibility of greater internal dose in exposed children. Table 7 gives the L-equivalents for the dermal and inhalation route for children of various age groups in comparison to adults.

Several studies have evaluated the relative importance of inhalation and dermal routes for chloroform on the basis of experimental data or mechanistic models (e.g., Jo et al., 1990a; Corley et al., 2000; Lévesque et al., 2000, 2002; Xu 2002). Xu (2002) concluded that inhalation exposure during a 10-to15-minute shower would be equivalent to ingesting 0.5 to 1.4 L of drinking water. Jo et al. (1990a) have reported that inhalation exposure during a 10-minute shower would be equal to ingestion of 0.7 L of water, and the dermal exposure during a 10-minute shower would be equivalent to the ingestion of 0.66 L of tap water. Corley et al. (2000) reported that dermal exposure during a 30-minute bath is equivalent to ingesting 0.49 L of tap water. Lévesque et al. (2002) used a multiroute PBPK model to simulate the amount of chloroform metabolized and bound to renal and hepatic macromolecules over a 24-hour period. This study indicated that, following a 10-min shower, the metabolite concentration would be about 0.01 μg $CHCl_3$-equivalent per kilogram of tissue in kidneys and liver, whereas multiroute exposures (shower + drinking water ingestion + indoor air inhalation) would result in 5 to 10 times greater exposure of the tissues. In a recent health risk assessment for chloroform, Health Canada (2004b) accounted for the contribution of multiroute exposures as indicated in equations (9) and (10), with the use of data on percent absorption obtained with PBPK models (Tables 8 and 9). The L-equivalent values derived in this study (Table 10) are comparable with those of previous studies (Jo et al., 1990a; Corley et al., 2000; Lévesque et al., 2002; Xu, 2002), once the differences in exposed skin area and duration of dermal and inhalation exposure are accounted for. However, it is difficult to evaluate the relative importance of the various exposure routes and compute the L-equivalents for each DWC. In this context, use of a two-tier approach may be useful. The objective of tier I should be to evaluate qualitatively whether or not a particular route of exposure is of significance to the risk assessment of a DWC. Tier II can then facilitate the derivation of chemical-specific L-equivalents on the basis of known physicochemical characteristics.

In the two-tier approach developed for Health Canada by Krishnan (2004), the dermal and inhalation routes are evaluated separately. For the dermal route, tier I focuses on an evaluation of whether or not this route of exposure would contribute at least 10% of the drinking water consumption (i.e., 10% of 1.5 L = 0.15 L). The Scientific Committee Group on Occupational Exposure Limits (SCOEL) of the European Commission has recommended that dermal absorption be considered important if it contributes to at least 10% of the daily uptake from the principal exposure route (e.g., respiratory uptake in case of occupational

TABLE 8. PBPK Model Simulations of Dose Absorbed by Adults and Children Through Inhalation Exposure During Showering (10-Minute Exposure) and Bathing (30-Minute Exposure) with Water Contaminated with 5 μg/L Chloroform

Age Group	Exposure Duration (min)	Dose Absorbed ($\times 10^{-2}$ mg)	Fraction Absorbed[a]
6-year-old-child	10	0.048	0.51
	30	0.141	0.52
10-year-old-child	10	0.08	0.57
	30	0.24	0.59
14-year-old teenager	10	0.11	0.61
	30	0.33	0.61
Adult	10	0.14	0.64
	30	0.43	0.66

Source: Krishnan (2003b).

[a]Calculated as the ratio of absorbed dose to potential dose associated with inhalation exposure.

TABLE 9. PBPK Model Simulations of Dose Absorbed by Adults and Children Through Dermal Exposure During Showering (10-Minute Exposure) and Bathing (30-Minute Exposure) with Water Contaminated with 5 μg/L Chloroform

Age Group	Exposure Duration (min)	Dose Absorbed ($\times 10^{-2}$ mg)	Fraction Absorbed[a]
6-year-old child	10	0.047	0.43
	30	0.14	0.41
10-year-old child	10	0.305	0.47
	30	0.22	0.47
14-year-old teenager	10	0.10	0.48
	30	0.31	0.48
Adult	10	0.146	0.63
	30	0.457	0.63

Source: Krishnan (2003b).

[a]Calculated by dividing the absorbed dose by the potential dose associated with dermal exposure.

TABLE 10. L-Equivalents of Chloroform for the Various Exposure Scenarios and Age Groups

Age Group	Exposure Scenario	Exposure Route		Total L-Equivalents
		Inhalation	Dermal	
Adult	Showering	0.55	0.30	0.85
	Bathing	1.7	0.91	2.61
14-year-old teenager	Showering	0.51	0.20	0.71
	Bathing	1.53	0.61	2.14
10-year-old child	Showering	0.43	0.15	0.68
	Bathing	1.35	0.44	1.79
6-year-old child	Showering	0.40	0.10	0.50
	Bathing	1.17	0.28	1.45

Source: Krishnan (2003b).

exposures) (SCOEL, 1999). Similarly, the EPA in its guidance for Superfund assessments EPA, (U.S. 1998) has suggested that the dermal pathway be considered for water contaminants only when the dermal dose exceeds 10% of the dose ingested. Accordingly, the focus of the tier 1 evaluation was to determine whether a chemical is likely to contribute at least 0.15 L-equivalent (i.e., 10% of the daily dose from drinking water ingestion). Based on equation (9), the tier I goal of 0.15 L-equivalent was associated with a K_p value of 0.024 cm/h. It implies that, for DWCs with K_p values below 0.024 cm/h, dermal absorption is unlikely to contribute significantly to the dose received during showering and bathing. For chemicals with a K_p value of 0.025 cm/h or greater, however, dermal notation is implied, and as such, tier II calculation is appropriate (Figure 3).

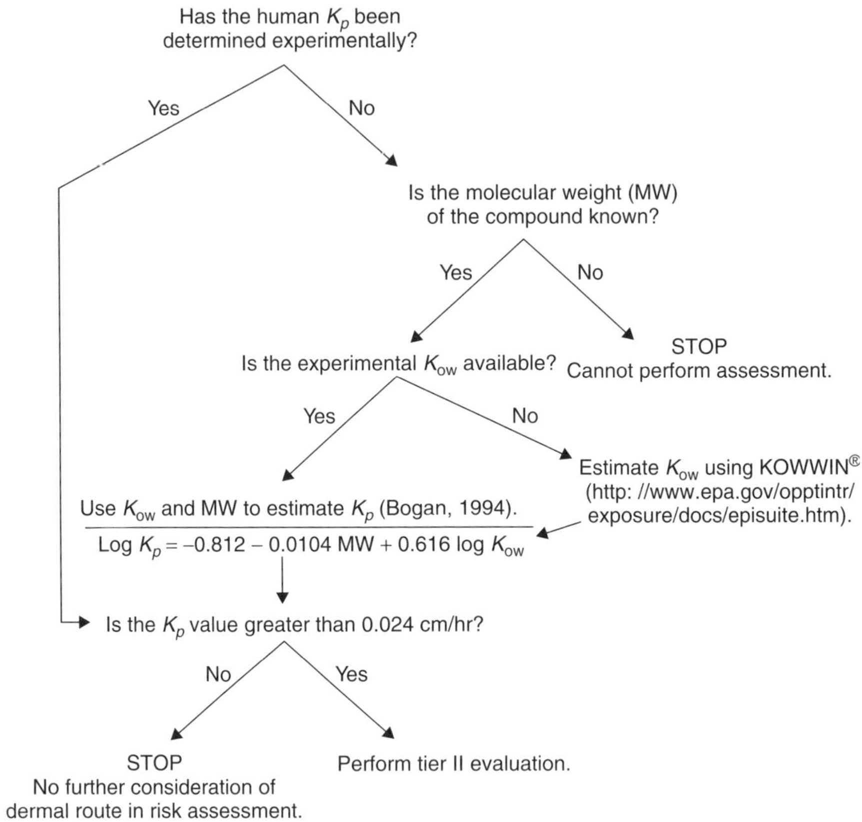

Figure 3. Tier I evaluation for dermal exposure to drinking water contaminants. (From Krishnan, 2004.)

Tier 2 addresses the question of what value of L-equivalent should be used for a DWC to account for dermal exposure. Again, substituting the exposure-specific values in equation (9), the L-eq for the dermal route can be calculated as a function of K_p using the following formula:

$$\text{L-equivalent, dermal exposure} = 6.3K_p \text{ (cm/h)} \tag{11}$$

If the experimentally determined K_p value is not available for a DWC, its molecular weight and n-octanol/water partition coefficient can be used to predict K_p (U.S. EPA, 1992b). Table 11 presents the tier I and II results for a few DWCs. For example, the K_p value of 0.0267 cm/h (molecular weight: 78.5, n-octanol/water partition coefficient: 0.09) for chloroacetaldehyde would indicate the marginal importance of the dermal route during showering and bathing (e.g., L-equivalent of 0.17).

TABLE 11. Two-Tier Evaluation of the Dermal Exposure to Drinking Water Contaminants Based on Molecular Weight (MW) and Log *n*-Octanol/Water Partition Coefficient (log K_{ow})

Chemical	MW	Log K_{ow}	K_p	Tier I Result	Tier 2 Result
Dibromoacetic acid	218.4	0.7	0.00223	No	NA[a]
Dichloroacetonitrile	110.9	0.29	0.0163	No	NA
Dibromoacetonitrile	199.8	0.47	0.0025	No	NA
Chloracetaldehyde	78.5	0.09	0.0267	Yes	0.17 L-eq

Source: Krishnan (2004).
[a]NA, not applicable.

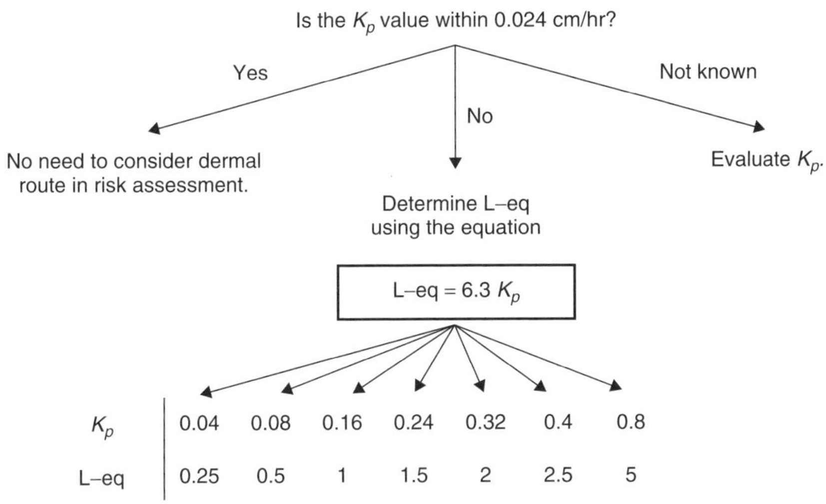

Figure 4. Tier II evaluation for dermal exposure to drinking water contaminants. (From Krishnan, 2004.)

Figure 4 summarizes the two-tier approach for evaluating the relevance of the dermal route for DWCs. Several "bins" representing the range of K_p values and the associated L-equivalent values are also presented. Accordingly, a DWC with a K_p value lower than 0.024 would not require the consideration of the dermal route in setting the drinking water goals, whereas DWCs with K_p values > 0.024 cm/h would require the use of an appropriate L-equivalent value to account for the dermal route.

Using a similar rationale, a two-tier evaluation of the inhalation route has been proposed (Krishnan, 2004). Accordingly, tier I evaluation focuses on determining whether a DWC is likely to contribute at least 0.15 L-equivalent. Setting $t = 0.5$, $Q_{alv} = 675$ L/h (adults) and $F_{abs} = 0.7$ in equation (10), we get a cutoff $F_{air/water}$ value of 0.00063 (Figure 5). The calculation above implies that for a

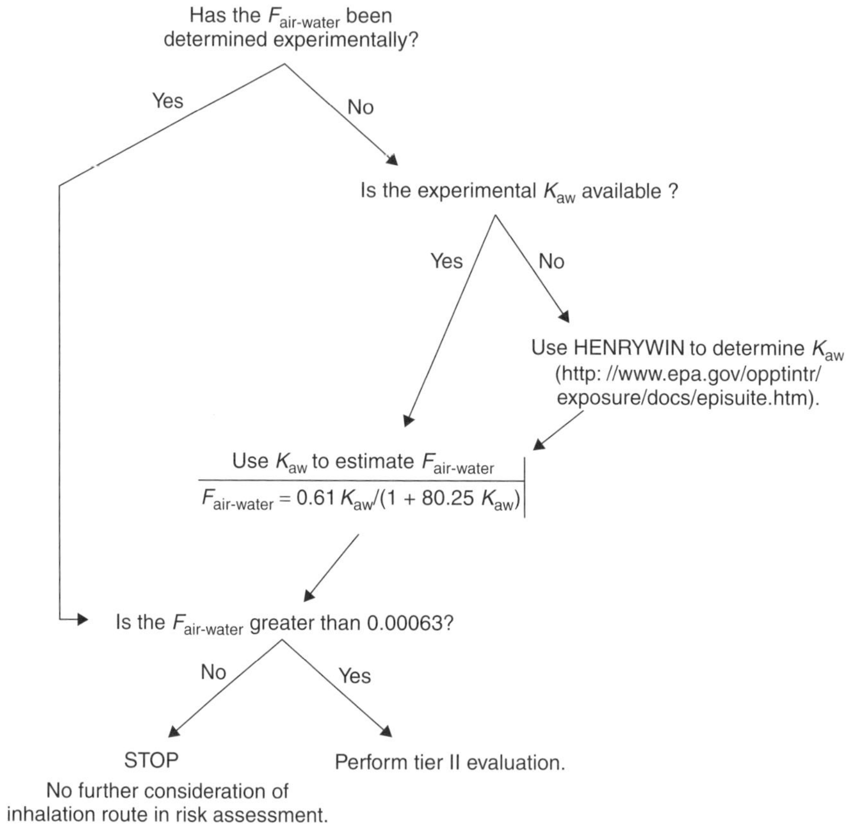

Figure 5. Tier I evaluation for inhalation exposure to drinking water contaminants. (From Krishnan, 2004.)

DWC with a $F_{air/water}$ value equal to or less than 0.00063, inhalation exposure related to showering and bathing would be negligible (i.e., it would not contribute significantly to the daily dose of DWC). For DWCs with $F_{air/water}$ values greater than 0.00063, however, inhalation notation is implied, and as such, tier II evaluation should be pursued. Table 12 presents examples of tier I evaluation for some DWCs. In this case, several chemicals are identified as being ones for which the inhalation route would be significant, for which tier II evaluation is indicated.

Tier II addresses the question of what value of L-equivalent should be used for a DWC to account for inhalation exposure. The L-equivalent for the inhalation route is calculated as a function of $F_{air/water}$ using the formula

$$\text{L-equivalent, inhalation exposure} = 236.25 F_{air/water} \tag{12}$$

TABLE 12. Two-Tier Evaluation of the Inhalation Exposure to Drinking Water Contaminants Based on Air/Water Concentration Ratio ($F_{air/water}$) and Henry's Law Constant (K_{aw})

Chemical	K_{aw}	$F_{air/water}$	Tier I Result	Tier 2 Result
Methanol	0.000174	0.000105	No; stop	NA[a]
Methyl ethyl ketone	0.00269	0.00135	Yes; tier II	0.32 L-eq
Toluene	0.243	0.00723	Yes; tier II	1.71 L-eq
Carbon tetrachloride	1.04	0.007511	Yes; tier II	1.77 L-eq
n-Hexane	6.98	0.007588	Yes; tier II	1.79 L-eq

Source: Krishnan (2004).

[a]NA, not applicable.

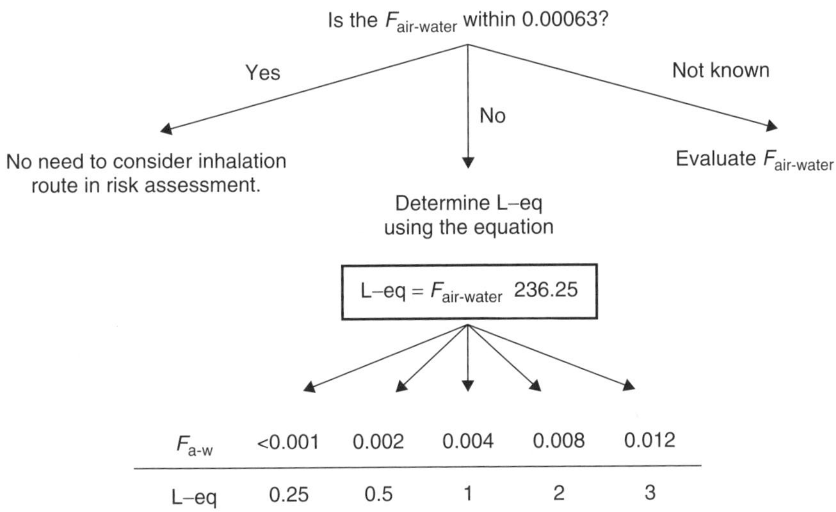

Figure 6. Tier II evaluation for inhalation exposure to drinking water contaminants. (From Krishnan, 2004.)

The value of 236.25 was derived from equation (10) using the following values: Q_{alv} = alveolar ventilation rate (675 L/h), t = time duration of exposure (0.5 h), and F_{abs} = fraction absorbed [0.7, based on Krishnan (2003a, 2003b, 2004)].

Figure 6 summarizes the procedure and outcome of tier II evaluation for the inhalation route. Several "bins" of the range of $F_{air/water}$ values and the associated L-equivalent value are also presented. Accordingly, a DWC with a $F_{air/water}$ value lower than 0.00063 would not require consideration of the inhalation route in setting the drinking water goals, whereas those that possess $F_{air/water}$ values > 0.00063 would require use of an appropriate value to account for the inhalation route.

TABLE 13. Representative Exposure Rates as a Function of Exposure Source, Route, and Age Class Based on U.S. EPA's Exposure Factors Handbook

Exposure Source/Media	Route	Age Class <1	1–5	6–19	Units
Air	Inhalation	4.5	6	10	m^3/day
Water	Oral	300	694	904	mL/day
Food	Oral				
Fruit		135.6	145.1	161.6	g/day
Vegetables		61.9	108.0	209.1	g/day
Grains		155.1	143.7	227.7	g/day
Meat		26.1	59.1	113.8	g/day
Fish			4	6	g/day
Dairy products		570.9	339.6	438.7	g/day
Mother's milk		775	0	0	g/day
Soil					
Intake	Oral		0.11	0.087	g/day
Contact area	Dermal		0.43	0.62	m^2

Source: Adapted from Mackay (1991).

A two-tier approach such as the one developed for Health Canada by Krishnan (2004) allows qualitative and quantitative evaluations of the relevance of dermal and inhalation routes for developing guideline values for DWCs. Accordingly, a specific route of exposure (dermal or inhalation) can be neglected in the process of establishing drinking water goals for a chemical if it does not contribute to 0.15 L-equivalent by that route. It is important to realize that 0.15 L-equivalent is only a cutoff value and not a number to be used in the risk assessment per se. The key parameters for the two-tier approach include the physicochemical parameters (K_p, $F_{\text{air/water}}$, K_{ow}, and K_{aw}). Although a number of experimental systems and algorithms are available for determining K_p, $F_{\text{air/water}}$, K_{ow}, and K_{aw} values, the relevance of these data for risk assessment should be evaluated on a case-by-case basis. The calculations performed by Krishnan (2004) were for an adult consuming 1.5 L of water per day. Such calculations can be made for children of various age groups, using the appropriate exposure and contact rates (Table 13). These age-specific parameters may also be used to derive the L-equivalent values for evaluating the potential age dependency of the importance of dermal and inhalation exposure routes in developing guidelines for DWCs.

Exposure Source and Multiroute Considerations in DWC Risk Assessment

The consideration of exposure source and multiroute exposures in risk assessment of DWCs is facilitated by the use of a relative source factor (RSC) and

L-equivalents:

$$\text{guideline value} = \frac{\text{Tolerable daily intake (mg/kg·day)} \times \text{body weight (kg)} \times \text{RSC}}{\text{water use–related exposure rate (L-eq/day)}} \tag{13}$$

The RSC represents the relative contribution of drinking water to the total dose, whereas the L-eq/day represents the sum total of the rate of contact via oral, dermal, and inhalation routes. Allocating one-fifth of the total exposure through the oral route, a default RSC of 20% has been used in setting guideline values for DWCs. The dose associated with drinking water should then be equal to the total amount received through all significant uses of drinking water. In other words, guideline value times L-eq/day should be equal to the daily dose received by the various exposure routes (oral + dermal + inhalation). Therefore, the guideline values for DWCs will be lower with increasing values of L-equivalents (i.e., with increasing importance of the dermal and inhalation exposure routes). The relevance of these routes may potentially be accounted for by altering either the L-equivalent values (increasing them) or the RSC values (decreasing them). In the presence of multiple exposure routes, the RSC associated with the *oral route* of a DWC will be lower than the default value. However, this reduction in RSC should be accompanied by a simultaneous reassessment of the RSC associated with the inhalation (ambient air) risk assessment. In such an approach, however, the contribution of the dermal pathway cannot readily be included without the calculation of a L-equivalent. Therefore, it is logical and pragmatic to use L-equivalent values in the risk assessment of DWCs to account for the contribution of multiple exposure routes rather than alter the RSC.

The L-equivalent for DWCs might vary depending on the relative consumption by a person of tap water and bottled water. Whereas chlorine is commonly used to disinfect public water systems, ozone is used as a disinfectant for most bottled water. In the United States the testing requirements, monitoring requirements and regulatory considerations are different for tap water and bottled water, resulting in potentially different profiles and levels of contaminants (Olson, 1999). Table 14 lists the contaminants that are usually monitored in tap water but not in bottled water. As a result, bottled water is not necessarily cleaner or safer than tap water, but its contribution to human exposure might vary from one DWC to another (Olson, 1999).

Variability in the extent of consumption of tap water and bottled water along with the variability of other exposure parameters may be taken into consideration in the context of a probabilistic exposure assessment, which commonly makes use of Monte Carlo simulation techniques (Paustenbach, 2002). Such approaches allow computation of the median and percentile values of the DWC dose received by an exposed population. Further, conducting a sensitivity analysis facilitates evaluation of the relative importance of specific exposure sources and routes for each DWC. Such an analysis is critical to direct the limited resources to identify areas of uncertainty that serve to enhance the credibility and scientific basis of exposure and risk assessments for DWCs.

TABLE 14. Contaminants Required to Be Monitored in U.S. Tap Water but Not in Bottled Water

Class	Contaminants
Regulated contaminants[a]	Asbestos
	Bromate (big cities in the past, soon all systems)
	Di(2-ethylhexyl)phthalate
	Haloacetic acids (big cities in the past, soon all systems)
Unregulated contaminants[b]	Dibromomethane
	m-Dichlorobenzene
	1,1-Dichloropropene
	1,1-Dichloroethane
	1,1,2,2-Tetrachloroethane
	1,3-Dichloropropane
	Chloromethane Bromomethane
	1,2,3-Trichloropropane
	1,1,1,2-Tetrachloroethane
	Chloroethane
	2,2-Dichloropropane
	o-Chlorotoluene
	p-Chlorotoluene
	Bromomenzene
	1,3-Dichloropropene

Source: Olson (1999).

[a]Unregulated contaminants are contaminants not subject to enforceable maximum contaminant levels or treatment requirements, but still required to be monitored for in tap water.
[b]Regulated contaminants are those subject to enforceable regulations.

Disclaimer

The opinions expressed in this chapter are those of the authors and not necessarily those of the office of Environment Health Hazard Assessment or the California Environment Protection agency.

REFERENCES

Blancato JN, Chiu N. 1993. Predictive modeling for uptake and tissue distribution from human exposures. In: *Safety of Water Disinfection: Balancing Chemical and Microbial Risks*. Craun G, ed. ILSI Press, Washington, DC.

Bogan KT, Hall LC, Perry L, Fish R, McKone T, Dowd P, Patton SE, Mallon B. 1988. *Health Risk Assessment of Trichloroethylene (TCE) in California Drinking Water*. Environmental Sciences Division, Lawrence Livermore National Laboratory, University of California, Livermore, CA.

Brooke I, Cocker J, Delic JI, Payne M, Jones K, Gregg NC, Dyne D. 1998. Dermal uptake of solvents from the vapour phase: an experimental study in humans. *Ann Occup Hyg* 42: 531–540.

California OEHHA (Office of Environmental Health Hazard Assessment). 1999. *Public Health Goal for Methyl Tertiary Butyl Ether (MTBE) in Drinking Water*. California Environmental Protection Agency, Sacramento, CA.

______. 1997. *Public Health Goal for Antimony in Drinking Water*. California Environmental Protection Agency, Sacramento, CA.

______. 2003. *Public Health Goal for Barium in Drinking Water*. California Environmental Protection Agency, Sacramento, CA.

Corley RA, Gordon SM, Wallace LA. 2000. Physiologically based pharmacokinetic modeling of the temperature dependent dermal absorption of chloroform by humans following bath water exposures. *Toxicol Sci* 53: 13–23.

Durkin PR, Rubin L, Withey J, Meylan W. 1995. Methods of assessing dermal absorption with emphasis on uptake from contaminated vegetation. *Toxicol Ind Health* 11: 63–77.

Ershow AG, Cantor KP. 1989. *Total Tap Water Intake in the United States: Population-Based Estimates of Quantities and Sources*. Life Sciences Research Office, Federation of the American Society of Experimental Biologists, Bethesda, MD.

Health Canada. 1995. *Guidelines for Canadian Drinking Water Quality*. Supporting documentation, part I: Approach to the derivation of drinking water guidelines. Environmental Health Directorate, Ottawa, Ontario, Canada.

______. 2002. *Guidelines for Canadian Drinking Water Quality*. Supporting document. Cyanobacterial toxins. Health Canada, Ottawa, Ontario, Canada.

______. 2004a. Trichloroethylene in drinking water. Document for public comment. Federal-Provincial-Territorial Committee on Drinking Water, Ottawa, Ontario, Canada. Accessed at: www.hc-sc.gc.ca/hecs-sesc/water/pdf/TCE-Consultation.pdf.

______. 2004b. Trihalomethanes in drinking water. Document for public comment. Federal-Provincial-Territorial Committee on Drinking Water, Ottawa, Ontario, Canada. Accessed at: www.hc-sc.gc.ca/hecs-sesc/water/pdf/trihalomethanes_drinking_water.pdf.

Howd RA, Brown JP, Morry DW, Wang YY, Bankowska J, Budroe JD, Campbell M, DiBartolomeis MJ, Faust J, Jowa L, Lewis D, Parker T, Polakoff J, Rice DW, Salmon AG, Tomar RS, Fan AM. 2000. Development of California Public Health Goals (PHGs) for chemicals in drinking water. *J Appl Toxicol* 20: 365–380.

Howd RA, Brown JP, Fan AM. 2004. Risk assessment for chemicals in drinking water: estimation of relative source contribution. *Toxicologist* 78.

IPCS (International Programme on Chemical Safety). 2000. *Environmental Health Criteria 214: Human Exposure Assessment*. World Health Organization, Geneva, Switzerland.

Jo WK, Weisel CP, Lioy PJ. 1990a. Chloroform exposure and the health risk associated with multiple uses of chlorinated tap water. *Risk Anal* 10: 581–585.

______. 1990b. Routes of chloroform exposure and body burden from shower with chlorinated tap water. *Risk Anal* 10: 575–580.

Johanson G. 1991. Modelling of respiratory exchange of polar solvents. *Ann Occup Hyg* 35: 323339.

Klaassen C, Amdur MO, Doull J, eds. 1996. *Casarett and Doull's Toxicology: The Basic Science of Poisons*, 5th ed. McGraw-Hill, New York.

Krishnan K. 2003a. Evaluation of the relative importance of dermal and inhalation routes of exposure for trichloroethylene. Final report submitted to the Water Quality and Health Bureau, Safe Environment Programs, Health Canada, Ottava, Ontario, Canada.

———. 2003b. Evaluation of the relative importance of dermal and inhalation routes of exposure for trihalomethanes. Final report submitted to the Water Quality and Health Bureau, Safe Environment Programs, Health Canada, Ottava, Ontaria, Canada.

———. 2004. Development of a two tier approach for evaluating the relevance of multiroute exposures in establishing drinking water goals for volatile organic chemicals. Final report submitted to the Water Quality and Health Bureau, Safe Environment Programs, Health Canada, Ottava, Ontaria, Canada.

Krishnan K, Andersen ME. 2001. Physiologically based pharmacokinetic modeling in toxicology. In: *Principles and Methods of Toxicology*. Hayes AW,ed. Taylor & Francis, Philadelphia, PA, pp. 193–241.

Lévesque B, Ayotte P, LeBlanc A, Dewailly E, Prud'Homme D, Lavoie R, Allaire S, Levallois P. 1994. Evaluation of dermal and respiratory chloroform exposure in humans. *Environ Health Perspect* 102: 10821087.

Lévesque B, Ayotte P, Tardif R, Charest-Tardif G, Dewailly E, Prud'Homme D, Gingras G, Allaire S, Lavoire R. 2000. Evaluation of the health risk associated with exposure to chloroform in indoor swimming pools. *Toxicol Environ Health A* 61: 225–243.

Lévesque B, Ayotte P, Tardif R, Ferron L, Gingras S, Schlouch E, Gingras G, Levallois P, Dewailly E. 2002. Cancer risk associated with household exposure to chloroform. *J Toxicol Environ Health A* 65: 489–502.

Lindstrom AB, Pleil JD. 1996. A methodological approach for exposure assessment studies in residence using volatile organic compound-contaminated water. *J Air Waste Manag Assoc* 46:- 1058–1066.

Mackay D. 1991. *The Fugacity Approach*. CRC Press, Boca Raton, FL.

Mattie DR, Grabau JH, McDougal JN. 1994. Significance of the dermal route of exposure to risk assessment. *Risk Anal* 14: 277–284.

Medinsky MA. 1990. Critical determinants in the systemic availability and dosimetry of volatile organic chemicals. In: *Principles of Route-to-Route Extrapolation for Risk Assessment*. Gerrity TR. Henry CJ, eds. Elsevier Science, New York, pp. 155–171.

Nuckols JR, Ashley DL, Lyu C, Gordon SM, Hinckley AF, Singer P. 2005. Influence of tap water quality and household water use activities on indoor air and internal dose levels of trihalomethanes. *Environ Health Perspect* 113: 863–870.

Olson ED. 1999. *Bottled Water: Pure Drink or Pure Hype?* Natural Resources Defense Council, New York.

Paustenbach DJ. 2002. Exposure assessment. In: *Human and Ecological Risk Assessment: Theory and Practice*. Paustenbach DJ, ed. Wiley, Hoboken, NJ, pp. 189–292.

Poulin P, Krishnan K. 1996. A mechanistic algorithm for predicting blood:air partition coefficients of organic chemicals with the consideration of reversible binding in haemoglobin. *Toxicol Appl Pharmacol* 136: 131–137.

———. 2001. Molecular structure-based prediction of human abdominal skin permeability coefficients for several organic compounds. *J Toxicol Environ Health* 62: 143–159.

Poulin P, Beliveau M, Krishnan K. 1999. Mechanistic animal-replacement approaches for predicting pharmacokinetics of organic chemicals. In: *Toxicity Assessment Alternatives: Methods, Issues, Opportunities*. Salem H, Katz SA, eds. Humana Press. Totowa, NJ, pp. 115–139.

Prah JD, Blount B, Cardinali FL, Ashley DL, Leavens T, Case MW. 2002. The development and testing of a dermal exposure system for pharmacokinetic studies of administered and ambient water contaminants. *J Pharmacol Toxicol Methods* 47: 89195.

Rao HV, Ginsberg GL. 1997. A physiologically-based pharmacokinetic model assessment of methyl t-butyl ether in groundwater for a bathing and showering determination. *Risk Anal* 17: 583–598.

Ritter L, Totman C, Watson T. 2005. Evaluation of assumptions for chemical drinking water guidelines. Final report submitted to the Water Quality and Health Bureau, Health Canada, Ottawa, Ontario, Canada.

SCOEL (Scientific Committee Group on Occupational Exposure Limits). 1999. Methodology for the Derivation of Occupational Exposure Limits: Key Documentation. European Commission, Luxembourg, p. 28.

U.S. EPA, (Environmental Protection Agency). 1992a. Guidelines for exposure assessment; notice. *Fed Reg* 57(104): 22888–22938.

______. 1992b. *Dermal Exposure Assessment: Principles and Applications*. EPA/600/8-91/011B. Office of Health and Environmental Assessment, U.S. EPA, Washington, DC.

______. 1993. *Health Advisories for Drinking Water Contaminants*. Lewis Publishers (CRC Press), Boca Raton, FL.

______. 1998. *Risk assessment guidance for superfund (RAGS)*, Vol. I, *Human Health Evaluation Manual* (pt E, Supplemental guidance for dermal risk assessment). Interim document. U.S. EPA, Washington, DC; p A-31.

______. 2000. *Methodology for Deriving Ambient Water Quality Criteria for the Protection of Human Health*. EPA/822/B-00/004. Office of Science and Technology, Office of Water, U.S. EPA, Washington, DC.

Weisel CP, Jo WK. 1996. Ingestion, inhalation and dermal exposures to chloroform and trichloroethene from tap water. *Environ Health Perspect* 104: 48–51.

WHO (World Health Organization). 2004. *Guidelines for Drinking Water Quality*, 3rd ed. WHO, Geneva, Switzerland. Accessed at: www.who.int/water_sanitation_health/dwq/gdwq3/en/.

Xu X. 2002. *Dermal and Inhalational Exposure to Disinfection By-products in Drinking Water*. State University of New Jersey, New Brunswick, NJ.

5

TOXICOKINETICS FOR DRINKING WATER RISK ASSESSMENT

JOHN C. LIPSCOMB
National Center for Environmental Assessment, U.S. Environmental Protection Agency, Cincinnati, Ohio

Risk assessment often involves the extrapolation of findings from animals to humans and consideration of human interindividual variability. The ultimate determinants of the dose response are (1) the inherent sensitivity of the target tissue to the toxicant, and (2) exposure of the target tissue to the toxicant. The former represents toxicodynamics (TD) and the latter, toxicokinetics (TK). These areas are sometimes described as what a chemical does to the body and how the body handles a chemical, respectively. Thus, the ultimate description of *dose* is concentration of the toxicant at the target.

Dose–response evaluation is the primary concept of toxicology and is one of the four components of the risk assessment paradigm (NAS, 1983). With respect to this paradigm, TK is useful in both the refinement of dose–response evaluation and the exposure assessment. The first step in dose–response evaluation is to describe the relationship between dose and response. This should be done for each of the adverse effects noted in the study; the effect occurring at the lowest of increasing doses is identified as the critical effect. Following selection of the critical effect, the data are evaluated to identify a point of departure (POD) for extrapolation in the remaining steps of the risk assessment process. The POD is the dose–response point that marks the beginning of a low-dose extrapolation. This point can be the lower bound on dose for an estimated incidence or a change in response level from a dose–response model [benchmark dose (BMD)], a no-observed-adverse-effect level (NOAEL), a lowest-observed-adverse-effect level (LOAEL) for an observed incidence, or a change in level of response

Risk Assessment for Chemicals in Drinking Water, Edited by Robert A. Howd and Anna M. Fan

(U.S. EPA, 2006a). Various levels of response may be selected as the POD. For noncarcinogens, these may be the NOAEL, the LOAEL, or a benchmark dose level (BMDL). Once the POD is determined, the reference value can be determined. A reference value may be the reference concentration for inhaled substances or the reference dose for orally ingested substances. The relationship between the POD and the reference value is demonstrated by

$$\text{reference value} = \frac{\text{POD}}{\text{UF}} \tag{1}$$

where UF is the uncertainty factor, the product of five individual factors: UF_H representing human interindividual variability, UF_A representing interspecies extrapolation, UF_L representing extrapolation from a LOAEL to a NOAEL, UF_S representing extrapolation from subchronic to chronic duration, and UF_D representing the overall completeness of the database for the chemical (U.S. EPA, 2006a). The potential for TK to affect the values for UF_A and UF_H quantitatively is being recognized increasingly, but it may also influence the choice of value for UF_S. In this chapter we focus on the quantitative application of TK data in the context of UF_A and UF_H. This dose–response extrapolation process is illustrated in Figure 1.

Uncertainty factors are used to extrapolate findings and most commonly have default values. These default values can and should be considered as placeholder values to be replaced when and if sufficient data exist. The Agency's Reference Concentration methodology (U.S. EPA, 1994) advanced a hierarchy of approaches, in which categorical default values were developed for UF_A based on knowledge of chemical reactivity and site of insult within the respiratory tract. The ultimate method of developing uncertainty factors is through the application

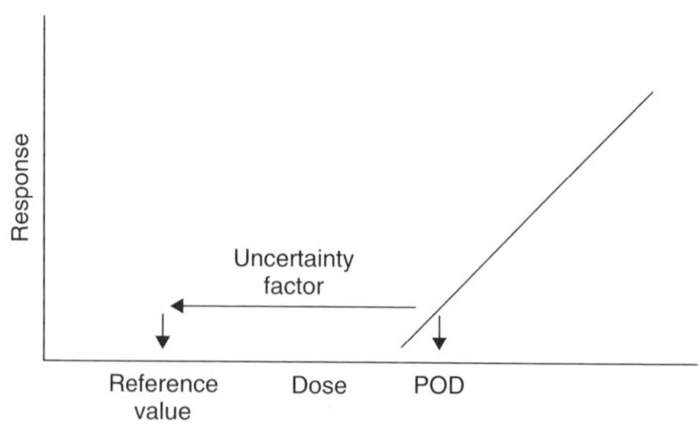

Figure 1. Uncertainty factors define the relationship between the POD and the reference value. Once the POD has been defined from animal studies or studies with human subjects, uncertainty factors are applied to extrapolate this effect level to be protective of human health. For noncancer risk assessment, the resulting reference value may be an inhalation RfC or an oral RfD value.

of physiologically based pharmacokinetic (PBPK) modeling (U.S. EPA, 1994). A refined approach to developing default uncertainty values for UF_A for orally encountered and systemically active toxicants has recently been proposed (U.S. EPA, 2006b).

Contaminants found in drinking water can be encountered via the oral, dermal, and inhalation routes. Routinely, exposure assessments for each route have not been available. The approach to drinking water risk assessment, however, differs between the U.S. EPA (EPA) and Canada, in that Health Canada has developed a methodology to estimate inhalation exposures. Under this methodology, several assumptions about inhalation exposure (as well as oral ingestion) are made and drinking water standards are developed based on the combined exposure anticipated. The EPA considers oral ingestion only in establishing drinking water contaminant levels. This restriction of the exposure assessment to the ingestion route may be a concern for some contaminants (U.S. EPA, 2006c), based on volatilization of the compound into indoor air and subsequent inhalation.

The application of TK data to the risk assessment of drinking water contaminants is a data-intensive endeavor. Not only must the toxicity data identify a target tissue, but information on the contribution of metabolism to toxicity is required. Finally, the development of reliable TK models for the test species and humans are required. Once these individual pieces of information and models are available, TK can be used to refine the dose–response evaluation and refine estimates of human exposure. In the remaining sections of this chapter we (1) present some considerations when evaluating toxicity data for quantitative use in risk assessment, (2) discuss the value of toxicokinetics in refining the dose–response evaluation and discuss PBPK modeling for risk assessment, (3) address some of the guiding principles for TK and PBPK analyses, and (4) present some examples of the quantitative application of TK data in risk assessments.

EVALUATION OF TOXICITY DATA

Results from toxicity tests with animals can be valuable qualitatively (e.g., hazard identification) and/or quantitatively (e.g., dose–response evaluation). In this section we describe some points for data evaluation pertinent to dose–response evaluation.

Recently, guidance has been developed to aid risk assessors as they evaluate toxicology studies (U.S. EPA, 2002). The following are some considerations that should be given to animal toxicity studies:

- Study purpose, design, protocol, hypothesis, analysis, and results should be made clear.
- Endpoints and the techniques for their evaluation should be appropriate; the size of groups should provide adequate statistical power, the study should demonstrate a dose–response relationship, and a lower response level (BMD, NOAEL, or LOAEL) should be established.

- The dose–response relationship should be consistent with any TK information and alterations of metabolism at higher doses (if observed) should be evaluated; the doses, durations, and routes of exposure, and the rationale for their selection, should be presented.
- The effects should agree with any mode of action data available, and the results should be biologically plausible.
- The study should involve an appropriate species, the stability of the test compound should be demonstrated, and the study should be documented adequately.

In addition to these points, this document stresses the importance of a multidisciplinary expert-based approach to support judgments of mode of action and whether an effect is adverse. In determining whether an event is adverse, both the biological and statistical significance should be weighed. *Biological significance* is "the determination that the effect is likely to impair the performance or reduce the ability of an individual to function or to respond to additional challenge from the agent. ... Precedence is given to biological significance, and a statistically significant change that lacks biological significance is not considered an adverse response" (U.S. EPA, 2002).

After determining that the study report reflects a study of adequate purpose, design, control, conduct, and reporting, a dose–response analysis should be made. For each of the adverse effects reported, data should be evaluated for a pattern. Data displaying no clear dose–response trend, but demonstrating doses for which the response differs significantly from the control, may be more amenable to the traditional NOAEL-based approach than to the benchmark dose (BMD) modeling approach. [Benchmark dose modeling software and tutorial materials are available at www.epa.gov/NCEA (Risk Tools).] When the data demonstrate a trend of increasing response with multiple doses, the benchmark dose modeling approach can be used; this models the entire data set and can extrapolate the dose–response relationship below the range of doses evaluated. In this manner, a dose (or confidence bound on dose) can be estimated for a given response level, and this can be used in place of a NOAEL value (when justified) to serve as the POD for reference value determination.

Every toxicologist understands that the dose defines the poison; the dose–response relationship is the fundamental principle of toxicology. This point is not lost on risk assessors. The dose–response relationships of the various adverse effects are used to determine which will serve as the critical effect—that effect observed at the lowest of ascending doses. Once this effect has been selected, the next step is to determine the POD. This is the dose that will be extrapolated through division by the total uncertainty factor to develop the human reference value. The POD may be an animal or human LOAEL or NOAEL value or a value identified through benchmark dose modeling. Benchmark dose modeling offers a superior alternative to the traditional NOAEL approach in that it makes use of the full range of dose–response information rather than a single point.

Recently increasing attention has been paid to mode-of-action (MOA) information. MOA data can be useful qualitatively in that they may substantiate or refute the extrapolation of toxicologically active pathways from animals to humans. An example of the latter is provided by α 2u-globulin associated nephropathy observed in male rats, its biochemical underpinnings, and species differences therein (U.S. EPA, 1991). Developing an understanding of the MOA and archiving this understanding via the application of available guidance (e.g., U.S. EPA, 2005) can support the choice of the dose metric. However, although developing an MOA description can be valuable, it is not necessary to identify a dose metric or a target tissue.

TOXICOKINETICS: PBPK MODELING

Whether evaluating the dose–response relationship or refining the exposure assessment, concentration of the toxicant in the target tissue is the most technically accurate description of dose or exposure. Andersen et al. (1992, 1995) addressed this point in coining the term *exposure–dose–response*. Toxicokinetics (TK) is the descriptive science that characterizes the absorption, distribution, metabolism, and elimination of chemicals in the body. Classical descriptions of TK (pharmacokinetics) date back many decades, although a different, more rigorous approach has been developed in the past two decades.

The chief advantage of determining target tissue exposure is the reduction of uncertainty in exposure. There are several ways that this information can be obtained, not the least of which is empirical observation in test animals. However, this approach cannot be applied to humans, so alternative methods must be developed. One of these approaches to predicting target tissue dosimetry is PBPK modeling. The seminal publication in this area occurred more than two decades ago (Ramsey and Andersen, 1984). The method predicted tissue dosimetry by combining ordinary differential equations solved simultaneously that contained parameters for tissue compartment volumes, blood flows, chemical partitioning, and metabolism, whose values could be determined experimentally. Groups of organs or tissues with similar characteristics were grouped into compartments whose volumes and blood flows were determined from the physiological literature. The complete organism was represented in physiological terms as necessary. Since that time, the concepts, terms, and model structure defined by Ramsey and Andersen 1984 have served as the basis for the development of dozens of PBPK models. While both classical (compartmental) and PBPK-based approaches have their advantages (Dixit and Ward, 2007), PBPK models offer specific advantages in health risk assessment. These advantages include physiological, anatomical, and biochemical representations of the target tissue of interest, the opportunity for flexible model structure in that models can be as simple as practical and as complex as necessary, and the ability of PBPK models to extrapolate across duration of exposure, route of exposure, dose, and species. When designing a PBPK model, adding compartments and additional parameters may seem like a good idea. However, adding unnecessary details to the model structure increases the number of uncertainties and decreases the reliability of model predictions; with the addition of more parameters, small adjustments made in their values can readily be accomplished to better "fit" model predictions to data observed. This will result in a false sense of confidence in the model and can lead to inappropriate applications of the model.

PBPK models can be developed for humans, as well as for the test species. When the models will be used to address interspecies differences in dosimetry, they should be constructed so that they predict the exposure of the average animal (of the sensitive species and strain) and human. When this approach is extended to address human variability (intraspecies differences), it is not critical to prove that the model used to infer animal–human differences represents human dosimetry at the median (or other defined measure of the central tendency of the population). However, it is important that predictions from this model be used as a starting point when examining human interindividual variability. This is done to avoid "gaps" or "double-counting" in extrapolating dosimetry from animals to sensitive humans.

There have been several treatments of the subject of pharmacokinetic studies for quantitative application in health risk assessment. The reader may also wish to consult some of these works (e.g., Reddy et al., 2005; Lipscomb and Ohanian, 2007; U.S. EPA, 2006b, 2006d).

PBPK Model Development and Evaluation

The development of PBPK models requires several sets of data. These data must convey the physiological and anatomical attributes of the test species as well as the chemically specific biochemical characterizations of the interactions between a toxicant and the tissue and organism. The former data, considered physiological parameters, include tissue compartment sizes, blood flows, and ventilation rates—and their values are independent of the chemical examined. Many of these parameter values can be found in compilations of data (e.g., Arms and Travis, 1988; Brown et al., 1997; Gentry et al., 2005). These values (e.g., fraction of cardiac output delivered to the liver) may differ among species and even among strains of the same species. The latter data, considered chemical-specific parameters, include tissue partition coefficients, metabolic rate constants, and elimination constants. Their values are determined by the biochemical and physicochemical characteristics of the toxicant. Partition coefficients (PCs) for volatile and nonvolatile compounds can be measured experimentally in vitro following published methods (Gargas et al., 1989; Jepson et al., 1994). Alternatively, their values can be predicted using information describing protein binding, water solubility, lipid solubility and tissue protein, and water and lipid content (see Krishnan et al., 2007). Metabolic rate constants for volatile compounds can be developed from gas uptake data using a preliminary form of the PBPK model to "optimize" the values to best fit the data demonstrating a loss of concentration in a closed chamber containing a living test animal (Gargas et al., 1986). Once these data have been collected, the PBPK model is compiled and evaluated. In addition to the points cited below, additional areas for consideration are detailed in a recent report (U.S. EPA, 2006d).

Model Evaluation Although PBPK models produce quantitative predictions of dosimetry, they should be evaluated critically before their predictions can be used in quantitative risk assessment. Clark et al. 2004 archived a set of considerations for model evaluation that can lead to a justification for reliability. These include (1) assessment of model purpose, (2) assessment of model structure and biological characterizations, (3) assessment of mathematical descriptions, (4) assessment of computer implementation, (5) parameter analysis and assessment of model fit, and (6) assessment of specialized analyses. Developing a uniform guide to model evaluation will enforce consistency on the inherently creative field of model development and application.

Useful Outcomes The goal of TK evaluations should be to identify exposure of the target organ to the toxic chemical moiety. This may be the parent chemical or a metabolite. PBPK modeling offers the flexibility of developing descriptions of target tissues and incorporating them into the model structure. With confidence in the values for compartment-specific parameters (i.e., flows, partition coefficients, metabolic capacity, etc.), it will be possible to develop an acceptable level of confidence in model predictions of target tissue exposure. This may not always be the case, often for the reason that measurements of toxicant in the target tissue are not available to use in evaluating the accuracy of model predictions for that tissue. In that instance, some risk assessors have relied on measures of the toxicologically active chemical in the blood (central) compartment as a means of refining exposure and dose–response evaluations. In this case, interspecies comparisons of exposure may be developed, assuming that partitioning from blood into target tissue will be roughly the same in the mammalian test species as in the human. This assumption is based largely on results presented by Thomas 1975 which demonstrated that species differences in the blood–air partitioning of chemicals was greater than species differences in the tissue–air partitioning for the same chemicals. This rationale has also been used as the basis of combining human blood–air PC measures with animal tissue–air PC values to determine tissue–blood PC estimates for inclusion in human PBPK models. This is because high-quality human blood is much more readily available than high-quality human tissues (i.e., liver, muscle, fat, etc.).

Dose Metrics Dose metrics demonstrate the exposure of the tissue of interest to the toxicant and have unit values that reflect the determinant of response. The choice of the unit of measurement ($C_{\max}$ or AUC) for the dose metric for simulation and comparison across species will influence the numerical outcome. This choice of dose metric unit should be based on available toxicity information. Available evidence suggests that for acute exposures, $C_{\max}$ (the maximally attained tissue concentration; in units of concentration, mass/volume) may be an appropriate choice. However, unless mode of action data dictate a contrary choice, for longer-duration exposures to parent compounds or stable and toxicologically active metabolites, AUC measurements (integrated exposure over time; in units of mass/volume × time) represent the best choice (Collins, 1987; Voisin et al., 1990); Clewell et al., 2002; IPCS, 2005; U.S. EPA, 2006d).

Uncertainty and Variability

EPA *reference values* are not intended to be "bright lines," regardless of their application. However, it is important that all of the underlying data be as specific and reliable as possible. When model predictions do not fit observed data adequately, uncertainty (a diminished level of confidence) in model reliability and predictions follows. The lack of fit may be based on the variability of parameter values which is not adequately captured by the model, or because of uncertainty in parameter values which were not determined (measured) empirically. Estimates of uncertainty and variability can be used in probabilistic modeling approaches, and predictions can be restated in terms of confidence limits instead of point estimates. Variability (e.g., of parameter values) can be determined, but uncertainty is best determined mathematically. Because these models are used to extrapolate predictions below the range of observations, the fit between observations and predictions is critical. There seem to be no established guidelines for acceptability of fit; expert judgment seems to rule the day. Evaluations of fit should include a comparison of simulations for multiple dose levels and durations, paying attention to the relative magnitude of maximally attained concentrations, times to attain maximal concentrations, and elimination profiles. Given that in vivo toxicokinetic measurements may be taken from animals of different ages, sexes, weights, species, strains, and suppliers, under different conditions of handling, dosing, lighting, housing, and diet, may have been dosed with different grades of compound at different times of the day using different dosing solutions, that differences in accuracy and precision of analytical techniques relative to parent compound and metabolites may exist, and that studies may have been carried out in different climates and even decades apart, it should not be surprising that a single set of model parameters may fail to provide the same degree of fit to multiple sets of toxicokinetic measurements.

For PBPK models constructed for animal species, it should be considered that the "general" animal is the starting point for species extrapolation. In contrast to the attention devoted to mathematical comparisons of predictions to observations, the fundamental requirement to define the physiological, anatomical, and biochemical characteristics of the animal in which the point of departure is defined should not be overlooked. Differences among test animals should be accounted for as completely as possible in PBPK models used for interspecies extrapolation. A Bayesian approach to uncertainty analysis for PBPK models is useful; that is, one should keep in mind that (1) the numerical values for all physiological parameters should be known or bounded by reason, (2) tissue (compartment) volumes should sum to a value at least slightly lower than body mass, (3) the sum of the blood flows to the compartments should be equal to cardiac output, (4) chemical mass balance should be maintained, and (5) when sampling values from distributions, the biological linkage (covariance) of parameters should be maintained. Once this has been accomplished, other sources and effects of uncertainty can be examined.

There are many sources for values of physiological parameters and often of chemical-specific parameters (e.g., partition coefficient values for well-studied chemicals). Commonly, values for parameters are adjusted in a given model application to improve model fit. These adjustments should be supported by a logical and justifiable reason for the adjustment. Adjustments that fall within the range of observed values seem reasonable. A sensitivity analysis can be conducted to determine whether the variability or uncertainty of parameter values can have an appreciable impact on model predictions (see Evans and Andersen, 2000); this analysis can identify the most deterministic (sensitive) parameters. First, the toxicologically relevant dose metric (i.e., the AUC of the parent chemical in venous blood) is selected; it may be considered under dynamic or static (steady-state) conditions, depending on the intent of the model. Then the physiological or biochemical parameter value selected is manipulated with all other parameter values held constant. The *sensitivity coefficient* is the ratio of a change in the dose metric to a change in the parameter value. It is common practice not to report parameters with a sensitivity coefficient below an absolute value of 0.1. Sensitivity coefficients may be positive or negative. Negative sensitivity coefficients indicate an inverse relationship between the values for the parameter and the dose metric. The results of a sensitivity analysis can indicate the parameters that have the greatest influence on the dose metric. This information could be used to guide additional research to refine estimates of sensitive parameter values.

Quantitative methods have been used to examine variability and uncertainty in model predictions of dosimetry. Some of these methods or approaches have been focused on a given parameter and are hypothesis driven, such as studies to examine whether enzymic variability may alter dosimetry (Lipscomb et al., 2003; Lipscomb, 2004). A recent publication by Clewell and colleagues (Covington et al., 2007) demonstrates the application of Monte Carlo simulations to develop confidence bounds on predictions of a dose metric important for tetrachloroethylene. Other, more general treatment of the subject of statistical analysis of PBPK models can be found in articles by Bernillon and Bois (2000) and Chiu (2007). At present, there is neither codified guidance nor consensus on methods for quantitative uncertainty analysis of PBPK models. However, it can be anticipated that this subject will be addressed in upcoming international workshops (e.g., sponsored by the International Programme on Chemical Safety).

Refining the Point of Departure

Traditionally, the POD has been expressed in terms of applied (external) dose. Two assessments on the EPA's IRIS database demonstrate how PBPK modeling may refine the POD. In the case of vinyl chloride, the POD was expressed as an external dose in rats, a NOAEL value. Information from toxicity studies identified a metabolite as the causative agent in liver pathology, and the dose metric was chosen to reflect the cumulative amount of metabolite formed per unit liver mass. PBPK modeling in the rat determined the level of this dose metric associated with exposure at the externally defined POD, and PBPK modeling simulating the vinyl chloride–exposed human identified the human external dose associated with the level of the dose metric associated with the animal POD. Under the traditional approach, animal–human differences in external dose would have been used to determine the value of the TK component of the interspecies uncertainty factor. Beyond the historical division of UF_A into TK and TD components in the EPA's inhalation reference concentration methodology (U.S. EPA, 1994), other approaches have been taken (IPCS, 2005). Under these approaches, the default uncertainty factor value of 10 for UF_A has been divided into equal parts of one-half order of magnitude (mathematically, 3.16) each, representing TK and TD contributions. With adequate data on either TK or TD, the default value can be replaced with the ratio of species differences. This approach has also been applied to UF_H, although not yet codified in guidance. Under the approach suggested recently (U.S. EPA, 2002), PBPK modeling would have simply identified the human equivalent dose, (HED), and uncertainty factors would be applied to the HED. When dose–response modeling is used to identify the POD, the optimal approach is for each of the applied doses to be expressed in units of the internal dose metric and BMD analysis performed on the dose metric to identify the POD (U.S. EPA, 2006d).

Route-to-Route Extrapolations

Often, toxicity data for a given chemical exist for a single route, and this is typically the oral route. However, exposures via the dermal and inhalation routes may be a concern. In this case, toxicity and tissue exposures need to be extrapolated across routes. For this to be valid, several considerations should be met (IGHRC, 2005):

- The toxicity of the compound should relate to a systemic target rather than a portal of entry tissue.
- The determinants of target organ dose (absorption, metabolism) and their relationship to route of administration should be characterized.
- The contribution of route-specific enzymes to the metabolism of the compound should be characterized.
- The contribution of the oral first-pass effect on systemic availability of the toxicologically active chemical moiety should be evaluated.
- Inter-route differences in the absorption and systemic availability of the compounds should be characterized.

A review by the European Centre for Ecotoxicology and Toxicology of Chemicals (ECETOC, 2003) of the use of assessment factors (uncertainty factors or adjustment factors) in human health risk assessment of chemicals developed similar conclusions:

- Route-to-route extrapolation is feasible only for substances with a systemic mode of action and should take dose rate and toxicokinetic data into account.
- If route-to-route extrapolation implies a lower systemic concentration due to slower absorption (which applies only to the dermal route), this can be considered to provide a built-in safety margin. In such cases, no assessment (adjustment) factor is needed (i.e., an assessment factor of 1 is considered to be appropriate).
- Extrapolation from oral to dermal data may be considered on a case-by-case basis, provided that appropriate information is available on dermal penetration. It is not appropriate to define a default assessment factor.

Guidance on route-to-route extrapolation is also available from the U.S. EPA (1988, 1994). This guidance is based on the principles communicated above and includes additional details [many of which can also be found in an article by Corley and Reitz (1990)] on the application of PBPK modeling for route-to-route extrapolation.

In addition to the above, route extrapolations may be most successful for water-soluble compounds with biological half-lives long enough to indicate that metabolism is low and that the first-pass effect does not contribute significantly to systemic availability. The route-to-route extrapolation of dosimetry for vinyl chloride (U.S. EPA, 2006a) is included in the case study at the end of the chapter, and the statistical considerations are also addressed by Chiu (2007). In addition, the EPA performed a route-to-route extrapolation in the IRIS assessment for benzene (U.S. EPA, 2006a). For benzene, the data supported an inhalation absorption factor of 50%; this was used in the method to determine the oral unit risk. An example of successful, quantitative route-to-route extrapolation is provided in the case study for vinyl chloride at the end of the chapter.

RISK ASSESSMENT

TK tools offer the opportunity to develop an understanding of tissue dosimetry for use in risk assessment. A recent report by the EPA has addressed the application of PBPK modeling in risk assessment. (U.S. EPA, 2006d). Although this document is aimed specifically at PBPK modeling, its principles apply to any type of TK analysis. The report reiterates three basic requirements for successful application of TK data *in risk* assessment:

1. The analysis must contain a compartment that represents or contains the target tissue or is a valid surrogate for the target tissue.
2. The analysis must contain values for parameters that are biologically plausible.

3. The analysis must have undergone an evaluation of its structure, implementation, and predictive capabilities.

The report goes on to indicate some criteria for acceptability:

1. The analysis must represent the species and life stage relevant for the assessment.
2. The analysis must have undergone a peer review for the adequacy of its structure and parameter values.
3. The results must adequately simulate the toxicologically active chemical species in the target tissue or in an acceptable surrogate for the appropriate route of administration and for the time course for which the chemical is present in the target tissue.

This report provides useful guidance in the evaluation of PBPK models intended for use in risk assessment and addresses many key points therein. Model structure, ideally guided, would contain the target tissue as well as compartments representing tissues of unique physiological and biochemical attributes relevant to the TK of the chemical. The mathematical representation of the structure should be evaluated for the adequacy of the number and forms of the equations used to represent the biological tissues and processes. Values for parameters in the model should be verified against reliable ranges of values; actual values can be measured in vivo or scaled from in vitro observations. The characterization of the mathematical implementation of the model via computer should make clear the integration algorithm and integration interval. Model evaluation (or validation) comes when predictions are compared to observations. Complications may arise when different sets of experimental observations fail to agree and professional judgment is required. A statistical comparison of model predictions to observations may best be accomplished by a multivariate analysis of variance (Haddad et al., 1995). However, a visual inspection seems to be the most common evaluation tool. With respect to the choice of dose metrics, when several should be considered, some have suggested that preference may be given to the metric that produces the most conservative estimate of risk (see Clewell et al., 2002). The POD should be demonstrated in units of the dose metric; in instances where the NOAEL approach is used, POD_{NOAEL} can be converted to units of dose metric and used for extrapolation. However, if BMD modeling is to be used, doses should be converted individually to dose metrics and BMD applied to the dose metric (see the EGBE case study later in the chapter). In addition to these points, this report addresses route, dose, duration, species extrapolation, and variability; it discusses methods of application to reference dose (RfD), reference concentration (RfC) and cancer assessments, mixture risk assessments, and PBPK model linkage to toxicodynamic models.

Uncertainty Factors

In this section we discuss how TK data can be used to develop nondefault values for uncertainty factors and how species-specific physiologic and anatomical data

	TK	TD	
Interspecies	(3.16)	(3.16)	10
Intraspecies	(3.16)	(3.16)	10
			100

Figure 2. Subdivision of inter- and intraspecies uncertainty factors. Over the past decade, the uncertainty factor for interspecies extrapolation has been divided into TK and TD components; this has recently been extended to the uncertainty factor for intraspecies variability. The default values for the TK and TD components are values at one-half order of magnitude each. Subdivision of uncertainty factors into these components and the establishment of their default values represents the framework through which quantitative TK data can be incorporated into noncancer risk assessments.

can be used to develop uncertainty factor values intermediate between traditional default values and data-derived values.

Once determined in animals, various means are available to extrapolate effect levels between and among species. For several decades, "standard" default values for uncertainty factors have been used. UF_S, UF_L, UF_A, and UF_H all have established default values of 10. The default values for UF_A and UF_H were described by the National Academy of Sciences (NAS, 1980) based on a composite safety factor of 100, described over a half-century ago (Lehman and Fitzhugh, 1954). In the relatively recent past, these factors (UF_A and UF_H) have been subdivided into TK and TD components, which better allows their replacement with chemical-specific data (Figure 2).

Dosimetric Adjustments For noncancer risk assessment, the EPA's guidance indicates that TK can be included quantitatively in inter- and intraspecies extrapolations. This is enabled by dividing the two respective uncertainty factors into TK and TD components as shown in Figure 2. Under the agency's reference concentration methodology (U.S. EPA, 1994), data describing the site of toxicity are used as the basis for selecting an approach to develop uncertainty factor values. Here, the uncertainty factor values for species differences in the exposure (TK) of inhaled systemic toxicants can be (1) the default value, (2) a categorical default value developed using species differences in blood–air partitioning, or (3) a data-derived value developed via PBPK modeling, the preferred method. For orally encountered agents, specific guidance is not available (see U.S. EPA, 1993), but a precedent has been established for subdivision and quantitative replacement of default uncertainty factors. Recently, the EPA has recommended the application of allometric scaling for orally encountered agents that are producing systemic toxicity. This approach is a default approach, and uncertainty factors are developed on the basis of quantitative TK data when possible (U.S. EPA,

2002). These uncertainty factors represent an intermediate approach between the full (traditional) default approach and that based on chemical-specific TK data, and as such, less confidence can be placed in them than in factors developed from quantitative TK data.

The EPA has relied on dosimetric principles to develop methods to extrapolate dosimetry from animals to humans based on knowledge of the sensitive tissue (U.S. EPA, 1994). This methodology subdivides UF_A into TK and TD components and comprises a hierarchy of default and chemical-specific methods that (1) provides separate methods for direct-acting respiratory tract toxicants (those that produce portal-of-entry effects; category 1 gases) and toxicants that produce systemic effects (category 3 gases), (2) guides the development of default values for the TK component of UF_A for category 1 gases based on the anatomical location of their effect, (3) guides the development of default values for the TK component of UF_A by considering blood–air partitioning of category 3 gases, and (4) recommends PBPK modeling for inhaled toxicants as the optimal approach to characterize inhalation dosimetry for interspecies extrapolation.

Based on recommendations (U.S. EPA, 2002) to further refine the approach to the RfD and RfC processes, a panel of EPA scientists recommended that the RfD methodology (U.S. EPA, 1993) be modified to subdivide UF_A into TK and TD components and address the development of UF values in a manner similar to that done for inhaled toxicants. One approach suggested was that of allometric scaling of orally ingested doses of toxicants (U.S. EPA, 2006b). Since the time of those recommendations, the cancer guidelines (U.S. EPA, 2005) have included a recommendation of body weight scaling for orally ingested nonmutagenic carcinogens. In this approach, a scaling factor is derived based on the ratio of body weights:

$$\mathrm{HED} = \frac{D_\mathrm{A}(\mathrm{SF})}{\mathrm{BW_H}} \tag{2}$$

$$\mathrm{SF} = \left(\frac{\mathrm{BW_H}}{\mathrm{BW_A}}\right)^{3/4} \tag{3}$$

where HED is the human equivalent dose in mg/kg, D_A is the animal dose in milligrams, SF is the body weight scaling factor, BW_H is the human body weight, and BW_A is the test animal body weight. This process results in scaling factor values that are different for different test species and are inversely proportional to test species body weight. This results in lower human equivalent doses when dose is extrapolated from smaller species (i.e., mice vs rats; Table 1). Because this approach is developed to be analogous to guidance for developing a human equivalent concentration (U.S. EPA, 1994), interspecies differences are not presented as a value for an uncertainty factor; rather, they are applied directly to the animal POD for extrapolation. Their impact is captured implicitly and subsumed in an extrapolation of the POD to the human.

TABLE 1. Oral Dose Scaling for a 10-mg/kg Dose Translated to Default Values for TK Components of Interspecies UF[a]

Species	Body Weight (kg)	Dose (mg)	Scaling Factor[b]	Scaled Dose	Human Equivalent Dose (mg/kg)	UF_A–TK[c]	UF_A[d]
Mouse	0.025	0.25	385	96.25	1.38	7.2	22.9
Rat	0.25	2.5	68	170	2.43	4.1	13.1
Dog	12	120	3.7	444	6.34	1.6	5.1

[a]Human body weight = 70 kg.
[b]Scaling factor = [human BW (kg)/animal BW (kg)]$^{3/4}$.
[c]This is provided to compare what the default value for the TK component would have been under an even subdivision of UF_A into TK and TD components (3.16), and is calculated by dividing the animal dose of 10 mg/kg by the HED.
[d]$UF_A = UF_{A\text{-}TK}$ (3.16).

Guidance In the *reference concentration* (RfC) methodology (U.S. EPA, 1994), the concentrations of chemicals inhaled that are directly toxic to the various different regions of the respiratory tract are extrapolated to humans based on the site of insult and species differences in surface area and minute ventilation. For compounds that are toxic to sites beyond the respiratory tract (i.e., those tissues for which chemical exposure is through delivery via the bloodstream), interspecies extrapolations are often based on differences in blood–air partitioning. (A Blood–air partition coefficient describes the concentration of chemical in blood relative to air at equilibrium; it is a unitless measure of solubility of a chemical in blood.)

However, the reference concentration methodology indicates that the preferred method for interspecies extrapolation is via physiologically based pharmacokinetic (PBPK) modeling. In this approach (discussed later), ordinary differential equations are developed and solved simultaneously through computer programs to describe the internal concentrations of chemicals and metabolites. Parameters for PBPK models describe the organism (e.g., body weights, tissue volumes, blood flows) and describe the behavior of the chemical in the organism (e.g., metabolic rate constants, partition coefficient values), and values for many of the parameters are determined experimentally. This method to species extrapolation offers the most technical approach. PBPK modeling has been used for interspecies extrapolation of inhaled and orally encountered substances on the EPA's integrated risk information system (U.S. EPA, 2006a).

In addition, the International Programme on Chemical Safety (IPCS) has finalized guidance for the replacement of default uncertainty factors with chemical-specific adjustment factors (IPCS, 2005). Under this approach (see Meek and Renwick, 2007), the factors for interspecies extrapolation and human interindividual variability are subdivided into TK and TD components (Figure 3). In this

Adjustment Factors	
AK_{AF} (any)	AD_{AF} (any)
HK_{AF} (any)	HD_{AF} (any)

Uncertainty Factors	
AK_{UF} (4.0)	AD_{UF} (2.5)
HK_{UF} (3.2)	HD_{UF} (3.2)

Figure 3. IPCS composite uncertainty factor. The IPCS has divided the uncertainty factors for interspecies extrapolation and intraspecies variability into TK and TD components. Under this scheme, the uncertainty factor for interspecies differences has been divided unequally in favor of TK. The default values in parentheses may be replaced with chemical-specific values of any magnitude. The composite factor (CF) is the product of four factors; uncertainty factors (UF) and adjustment factors (AF) for a given component (i.e., AK) are mutually exclusive. A, animal; H, human; K, toxicokinetic; D, dynamic. Subscripts: A, adjustment; U, uncertainty; F, factor.

document, those replacement values are called *chemical-specific adjustment factors*, indicating that they pertain to a given chemical and that they are adjustment rather than uncertainty factors—they have their basis in quantitative data rather than in default assumptions. This document guides several considerations and the quantitation of TK data, as summarized below.

Identification of the Active Chemical Moiety Without identification of toxicologically active species, default approaches should be undertaken. Various data sets, including those developed in vitro and in vivo using structurally similar chemicals, metabolites, and co-exposures to chemicals that induce or inhibit the metabolism of the test chemical, can be used to determine whether the parent chemical or a metabolite is responsible for toxicity.

Choice of Relevant Toxicokinetic Parameter It is reasonable to assume that subchronic or chronic effects are more related to integrated measures of exposure (AUC values) than to intermittent measures of exposure (maximally attained tissue concentrations; C_{max} values). In cases where the data do not support one or the other, AUC values should be relied upon. Because AUC is the inverse of clearance, measures of clearance or in vitro–determined intrinsic clearance may be converted to reciprocals and used for species comparisons of internal dosimetry.

Experimental Data The following points should be considered: (1) the population or types of samples used should be adequately representative of the test animal and human populations considered, (2) the route of administration in animal TK studies should be the same as the anticipated human exposure—if not, route-to-route extrapolation procedures should be evaluated carefully; (3) doses and concentrations studied should be those (in animals) that are closest to the POD and (in humans) as close to the equivalent exposure as possible; and (4) the number of subjects or samples should be adequate to identify the central tendency of the distribution, and unless data exist to the contrary, a lognormal distribution should be assumed.

Predictions of low-dose conditions are often based on use of the model to simulate exposure conditions below those used to develop and evaluate model predictions. Because multiple parameter values and sets of values in PBPK models may adequately simulate a given TK data set, consideration should be given to the parameter values that best simulate the exposures closest to those of concern for human health risk assessment. Clewell et al. (2005) demonstrate differential effects of parameter values when extrapolating below the range of observations, compared to simulating higher exposures. Optimal experimental data for extrapolation are those resulting from exposures most closely related to the exposure of concern.

Quantitation Quantification of TK differences for uncertainty factor development should demonstrate the ratio of values for the TK parameter (e.g., AUC of parent chemical in liver) at the central tendency of the distribution and a given (i.e., 95th, 99th) percentile of the distribution. When the population displays a bimodal distribution of values for the TK parameter, quantification should be determined by the ratio between the central tendency for the general population and a given percentile of the distribution of the susceptible subpopulation. The ratio is used as a replacement for the default value for the TK component of the uncertainty factor.

Both the IPCS and the EPA guidance agree that the basis for interspecies extrapolation should be at the level of the internal dose. The value for the dose metric associated with the animal POD should be the subject of human TK data and investigations. The value of the dose metric is to be demonstrated such that the external dose associated with that level of the dose metric is determined. Under the EPA's RfC methodology, this external dose (concentration) is presented directly, and UFs are subsequently applied to that value to develop the RfC. The IPCS guidance is somewhat different, although numerically equivalent (but not exclusive to the inhalation route): species differences are quantified the same, but differences between species in the *external* dose are used to determine the value for the TK component:

$$\mathrm{AK_{AF}} = \frac{D_\mathrm{A}}{D_\mathrm{H}} \tag{4}$$

where, $\mathrm{AK_{AF}}$ the animal kinetic adjustment factor is the TK component of interspecies extrapolation, D_A is the external dose or concentration in the animal at the POD, and D_H is the human dose resulting in the same level of the dose metric as at D_A. The IPCS approach differentiates *uncertainty factors*, factors with default values, from *adjustment factors*, factors with data-derived chemical-specific values.

Consistent with the IPCS's approach to data evaluation, the most recent EPA cancer guidelines (U.S. EPA, 2005) call for a critical evaluation of data before invoking default values for uncertainty factors. At present, the EPA's risk assessment forum has convened a technical panel to evaluate the feasibility and benefits of developing EPA guidance on developing data-derived uncertainty factors.

Extension of Toxicokinetic Analysis to Support Population-Based Risk Estimates

A key point for consideration is the fraction of the human population that should be covered by regulatory exposure limits. This is important, because economic consequences need to be considered in setting regulatory levels. Measures to reduce emissions or environmental levels can be very costly. Economic costs to control exposures can be compared to the likelihood of chemical-induced illness in a population; a population-based TK analysis can indicate the likelihood of a given fraction of responders in a population at a given exposure level. When the human health conditions anticipated are identified and assigned an economic value, a more meaningful comparison of costs and benefits can be developed. However, an estimation of uncertainty in both methods to estimate the costs of prevention or remediation as well as a risk analysis should be carried out. Although this is an important point, it cannot be covered by risk assessment alone, inasmuch as it (the decision establishing the regulatory limit) is a risk management decision and cannot be addressed solely through a TK analysis. Thus, the results of the risk analysis might best be presented so that multiple points within the distribution of human values can be compared.

Examples of the Use of Toxicokinetic Data in IRIS Assessments

Ethylene Glycol Monobutyl Ether: Interspecies Toxicokinetics Ethylene glycol monobutyl ether (EGBE; 2-butoxyethanol) has an established reference dose of 0.5 mg/kg·day on the EPA's integrated risk information system (IRIS) database. This value is based on the results of a subchronic drinking water study in mice and rats, where changes in mean corpuscular volume (MCV) were determined to be the critical effect. The point of departure in the rat was determined by using PBPK modeling to convert external doses to an internal dose metric, applying BMD analysis to the relationship between the dose metric observed and the critical erythrocyte effect, and identifying a $BMDL_{05}$ expressed in terms of the dose metric. PBPK modeling then demonstrated a human equivalent dose of 5.1 mg/kg·day. To this value, a combined uncertainty factor of 10 was applied, which comprised a single UF value of 10 for variation in sensitivity within the human population (UF_H).

Animal Toxicity and Dose–Response Relationship No chronic studies were available, and an NTP 1993 91-day drinking water study in male and female rats and mice was chosen as the critical study for risk assessment. In the rat study

(NTP, 1993), groups of 10 male and 10 female F344 rats were exposed to 0, 750, 1500, 3000, 4500, and 6000 ppm EGBE via the drinking water for 13 weeks. The lowest dose equated to 58.6 mg/kg·day for females and 54.9 mg/kg·day for males. Body and organ weights were measured. In addition, clinical, hematologic, gross, and histopathologic examinations were conducted. Decreases in body weight were observed in female rats exposed to the two highest dose levels. Hematologic changes were observed at all dose levels after 13 weeks and were indicative of mild to moderate anemia. These changes included reduced red blood cell (RBC) count, hemoglobin content, and hematocrit, and increased reticulocyte count (RTC) and mean corpuscular volume. Males demonstrated a lower LOAEL value then females based on drinking water intake relative to body weight for erythrocyte effects (and like females, no NOAEL). However, BMD analysis of hematological changes in males and females identified a lower $BMDL_{05}$ in females, indicating their greater sensitivity, which served as the basis for selecting the female as the most sensitive sex.

RBCs are being replaced continually. When lysed in the circulation, this leads to problems with oxygen transport and to potential kidney damage. EGBE exposure results in RBC swelling [increased mean corpuscular volume (MCV)], hemolysis, and decreased RBC count. Decreases in RBC count are sometimes offset by increases in MCV, so that *hematocrit* (a measure of erythrocyte volume relative to plasma volume) is not altered. MCV demonstrated the steepest dose–response curve among these effects, was deemed the most sensitive endpoint, and so was selected as the critical effect. Although changes in RTCs sometimes represent the largest measurable differences between exposed animals and unexposed controls, this parameter is highly variable (covariance = 30 to 60%) and does not always exhibit a clear dose–dependent trend (NTP, 1993, 1998). Changes in MCV and RBC count are the earliest measurable responses for both oral and inhalation exposures to EGBE. Because MCV changes were deemed more sensitive than RBC count, partially on the basis of the steepness of the dose–response relationship, MCV changes were selected as the critical effect.

The dose–response relationship was developed by combining PBPK modeling with BMD analysis. For reasons described later, C_{max} of the butoxyacetic acid (BAA) metabolite was chosen as the dose metric associated with the critical effect. In this analysis, observations of increases in the critical effect (increases in MCV) were paired with their respective external doses, but PBPK model–predicted BAA C_{max} levels in female rats were used as a measure of "dose" in the BMD analysis. This analysis indicated that 64 μM represented the $BMDL_{05}$ for increased mean MCV. In comparison, the LOAEL (a NOAEL was not demonstrated) was identified on the basis of external dose and was converted to BAA C_{max} using the model; the C_{max} value associated with the LOAEL was 104 μM. The C_{max} value for BAA of 64 μM in the female rat was chosen as the point of departure for species extrapolation.

Choice of Dose Metric Because the effect was observed in blood, some measure of exposure of blood, rather than systemic target, was chosen. BAA is the principal metabolite of EGBE and was thought to play a role in erythrocyte toxicity. Carpenter et al. (1956) demonstrated that concentrations of 0.1% BAA induced hemolysis in vitro, whereas hemolysis was not induced by EGBE until concentrations of approximately 2.5% were reached. RBCs of humans were resistant to this effect, whereas RBCs of rats demonstrate the effect in vitro. Species differences in sensitivity were again demonstrated when rats and humans were exposed to EGBE vapors in vivo (Carpenter et al., 1956), even though BAA has been identified as a metabolite of EGBE in human studies and used as a biomarker for EGBE exposure in humans (see Medinsky et al., 1990; Haufroid et al., 1997). For this reason, BAA levels rather than EGBE levels were deemed appropriate measures of exposure. The choice between AUC and C_{max} as the most appropriate dose metric was informed by data that compared erythrocyte effects after similar doses by gavage and drinking water (Ghanayem et al., 1987; Medinsky et al., 1990). Although similar doses via these two different methods of oral exposure would result in similar AUC values, the C_{max} attained would be much higher following a bolus gavage exposure than from an intermittent drinking water exposure. Further, regression of AUC values and C_{max} values against observed erythrocyte toxicity data revealed a better correlation with C_{max} than with AUC. This finding is in agreement with the observation of increased toxicity following gavage administration versus drinking water exposure. These lines of evidence led to the selection of the C_{max} of BAA in blood as the dose metric of interest.

Species Extrapolation: Basis PBPK models were developed for rats and humans. The model contained eight compartments: lungs and arterial blood, rapidly perfused organs, slowly perfused organs, fat, skin, muscle, liver, and GI tract. Separate models were developed for EGBE and BAA, and were linked via metabolic formation of BAA in the liver; the BAA model also contained a kidney compartment to include urinary elimination. Partition coefficient values for EGBE and BAA were determined for human blood and rat blood and solid tissues. The human EGBE blood–air PC value (7965) was taken from a study by Johanson and Dynesius (1988). Human tissue–blood PC values were developed by combining human blood–air PC values with rat solid tissue–air PC values, as is routinely done, citing species similarities in solid tissue partitioning (Thomas, 1975). The formation of BAA was assumed to occur only in the liver, using the rat liver perfusion data of Johanson (1986) scaled to the human.

All other metabolic routes for EGBE (formation of EG and glucuronide conjugate) were combined in the rat model since they were used only to account for the total disposition of EGBE in the rat metabolism studies and not for cross-species extrapolations. Because these non-BAA metabolites are not found in humans, this metabolic pathway was not activated in the human model.

Human tissue–blood partition coefficients were assumed to be equal to those of the rat. Protein binding of BAA in blood and saturable elimination of BAA by the kidneys were necessary components to describe the BAA kinetic data in rats and humans, as discussed above. Since direct measurements of protein binding were not available, these parameters were arbitrarily set to the molar equivalent values reported for phenolsulfonphthalein. Urinary elimination rates were optimized by the PBPK model to fit the rat data of Ghanayem et al. 1990. Protein binding was held constant between species, but renal elimination constants were scaled by (body weight)$^{0.74}$

Species Extrapolation: Application A PBPK model was employed to develop levels of the dose metric (blood BAA C_{max}) associated with actual animal external doses, and benchmark dose modeling was applied to reveal the relationship between the critical effect observed (increased MCV) and PBPK model–predicted levels of the dose metric (blood BAA C_{max}). A $BMDL_{05}$ concentration of 64 μM was identified in female rats. Because the $BMDL_{05}$ was determined as an internal dose, no corresponding external dose had been studied. The PBPK model was run to identify an external dose of 38.3 mg/kg·day in the rat that would produce this $BMDL_{05}$ level of the BAA metabolite (Richard Corley, personal communication, 2006). Next, a human PBPK model was employed to translate this level of the dose metric to a human equivalent oral dose of 5.1 mg/kg·day. In the present IRIS file, the rat PBPK model was run only to develop the POD; then the human PBPK model translated the POD to a human equivalent dose. In the IRIS file, the corresponding rat external dose was not identified, and species extrapolation was performed using the PBPK model as the DAF to develop the HED.

Uncertainty Factor Derivation A comparison of the external doses would yield a chemical-specific adjustment factor (or data-derived uncertainty factor) for $UF_{A\text{-}TK}$ of 38.3/5.1 = 7.5; a default value of 3.2 would be retained for species differences in TD. (*Note*: For EGBE, interspecies toxicodynamic data do exist and have been used to replace the default factor of 3.2 for TD with a value of 1. This case study focused only on TK.) Thus, the interspecies extrapolation would result in a factor of (7.5)(3.2) = 24, compared to a default value of 10. Based on default body weight scaling (U.S. EPA, 2006b) and a study-specific female rat body weight of 0.188 kg, these values would result in a scaling factor of $372^{0.75} = 84.76$, a human equivalent dose of 8.7 mg/kg·day, an apparent value of 4.4 for $UF_{A\text{-}TK}$ and a value of (4.4)(3.2) = 14.1 for UF_A.

Boron and Compounds: Intraspecies Toxicokinetics Boron combines to form many compounds; perhaps the most prevalent is boric acid. This compound has simple TK and is excreted unchanged as borate in the urine. Toxicity studies have defined testicular effects in dogs and developmental effects in rats as potential critical effects. Because of concerns with the adequacy of the dog study, and because of a more advanced data set for rat developmental toxicity, including a benchmark dose analysis, the assessment of the risks for boron and compounds identified developmental toxicity (decreased fetal body weight) as the critical effect. Circulating levels of boron were used as a measure of exposure, and differences in the renal clearance of boron between pregnant rats and pregnant humans were used for interspecies extrapolation. Differences in renal elimination parameters (glomerular filtration rates) among pregnant women served as the basis on which to develop a replacement of the default value for the TK component of UF_H.

In the assessment available on IRIS, UF_A and UF_H were each divided evenly into TK and TD components. This subdivision of UF_H into TK and TD components was the subject of much debate, and the even subdivision was the result of compromise. Other points that met with substantial discussion were the choice of dose metric (ultimately, clearance was employed), the attribution of boron's renal clearance to filtration mechanisms in the kidney, the lack of EPA guidance on the subdivision of uncertainty factors and what fraction of the human population to protect.

This case study is focused on human interindividual toxicokinetic variability, for which compound-related data on glomular filtration rate (GFR) served as the basis for the development of a nondefault value for the uncertainty factor (2 vs 3.16). The default half-order of magnitude partition of uncertainty factors (i.e., $UF_A$and UF_H) for toxicokinetics and toxicodynamics is based primarily on lack of knowledge; if there is no evidence to the contrary, an equal contribution from each source of uncertainty is assumed. The examination of species differences in boron distribution to extravascular fluids and renal elimination (renal clearance observed in pregnant rats and pregnant humans, not reviewed here) served as the basis for the replacement of the default value for $UF_{A\text{-}TK}$, while critical evaluation of the human interindividual variation of underlying renal clearance mechanism (GFR) served as the basis on which to replace the default value for the TK component of UF_H. Because no data were available to inform a mode or mechanism of action for boron, the default values for the TD component of both $UF_A$and UF_Hremain; they are $10^{0.5}$, or 3.16 for each.

Intraspecies Extrapolation: Basis First and foremost, no data existed to identify a mode of action. Because the compound (boric acid) is not metabolized further, the toxicity must relate to the parent compound. The compound is distributed with total body water, attaining tissue/blood concentration ratios largely reflective of tissue water content. Renal elimination appears to be by passive diffusion, based on hemodialysis studies in humans and the physicochemical properties of the molecule (small diameter, not protein bound). Because passive renal filtration mechanisms seem to be responsible for boron's renal clearance, measures of glomerular filtration rate (GFR; a measure of renal clearance) variance in pregnant women and supposed at-risk pregnant women were used to characterize human interindividual variability.

Intraspecies Extrapolation: Application For humans, results of oral dose studies indicated nearly complete (95%) absorption; this was not modeled as a variable parameter among humans. Variability was determined based on published studies on the variability of renal GFR among three study populations of pregnant women. Although some data on boron clearance were available on a relatively small number of pregnant human subjects (Pahl et al., 2001), the data were very limited (nonrecorded dietary exposures). It was determined that this data set was insufficient to determine human variability in boron clearance with any degree of certainty. However, data were available on the variability of GFR in pregnant humans from three different studies (Dunlop, 1981; Krutzen et al., 1992; Sturgiss et al., 1996). These data sets were available as means and standard deviations, and so were used as such. The sigma method was previously applied to data on GFR to estimate human variability on boron clearance (Dourson et al., 1998). That approach determines variability as the difference between the mean measured value and the value at a certain number of standard deviations from the mean. Whereas Dourson et al. (1998) defined variability as the difference quantified at two standard deviations below the general population mean, the EPA's boron risk assessment employed the value at three standard deviations below the mean to quantify variability in GFR among pregnant humans.

The formula modified from Dourson et al. (1998) for AF_{HK} (*adjustment factor, human kinetics*, in IPCS parlance) is

$$AF_{HK} = \frac{\text{GFR-avg}}{\text{GFR-avg} - 3\ \text{SDgfr}} \tag{5}$$

where GFR-avg and SDgfr are the mean and standard deviation of the GFR (mL/min) for the general healthy population of pregnant women. The selection of 3 SD is based on a statistical analysis of published GFR data (3 SD rather than 2 SD gave greater coverage of the published range of variability), with more consideration being given to the full range of GFR values likely to be found in a population of pregnant women. In the aggregate, the data suggest that a lower bound GFR 2 SD below the mean does not provide adequate coverage of the susceptible subpopulation. Although no conclusive information exists from controlled-dose studies in humans, it may be possible that the variability in boron clearance might be greater than GFR variability. Therefore, AF_{HK} must also account for any residual uncertainty in using GFR as a surrogate.

Further rationale for moving from two to three standard deviations was based on consideration of the sensitive subpopulation, pregnant women with decreased renal function (GFR) due to a medical condition identified as preeclampsia. There are no fully validated (dependable) studies that characterize the distribution of GFR values among women in this condition; however, limited data from 12 preeclamptic women were available.

Sensitive Subpopulations By virtue of their lower GFR, pregnant women diagnosed with preeclampsia could be considered to be a sensitive subpopulation, at least on the toxicokinetic scale. Toxicodynamic sensitivity is presumably independent of toxicokinetic sensitivity. The onset of preeclampsia generally occurs after week 20 of pregnancy and is characterized by acute hypertension, often accompanied by edema and proteinuria. Women with preeclampsia are at increased risk for premature separation of the placenta from the uterus and acute renal failure, among other adverse health effects. The fetus may become hypoxic and is at increased risk of low birth weight or perinatal death. Preeclampsia has recently been estimated to affect 3to5% of pregnant women (Skjaerven et al., 2002). With almost 4 million successful pregnancies per year in the United States (Ventura, 1999), or about 3 million at any one time, the size of the preeclamptic population at any given time could be in the range 150,000 to 200,000 women.

Because GFR is a continuous variable, because women can have a low GFR independent of preeclampsia, because pregnant women with low GFR may not be experiencing preeclampsia, and because the data set describing GFR distribution among preeclamptic women was so weak, potentially sensitive women (those with preeclampsia) were treated as a portion of the general population—a unimodal distribution of GFR values, with sensitive values occurring in the lower tail of the distribution, was undertaken.

Considering the results of Krutzén et al. (1992) in the context of the sigma method, a reduction of 2 standard deviation(SD) from the healthy population mean to establish the lower bound (which results in a GFR slightly higher than the mean of the preeclamptic GFR) would appear to be insufficient for adequate coverage of the susceptible population. The use of 3 SD below the healthy GFR mean gives coverage in the sensitive subpopulation to about 1 SD below the mean preeclamptic GFR.

Uncertainty Factor Derivation As no single study is considered to be definitive for assessment of population GFR variability, AF_{HK} is determined from the average of the individual sigma-method values for each of the three studies (Table 2). The mean GFR and standard deviation values in Table 2 are based on average GFR across the entire gestational period, except for the, estimate by Krutzén et al. (1992), which was for the third trimester only. The average sigma-method value from the three studies is 1.93. Considering a small residual uncertainty in the use of GFR as a surrogate for boron clearance, the average sigma-method value of 1.93 is rounded upward to 2.0 and established as the value for AF_{HK}. The data on preeclamptic women presented by Krutzén et al. (1992) were considered insufficient to act as a base for the interindividual AF_{HK} factor. Use of the mean (128 mL/min) and standard deviation (33 mL/min) in this sensitive subgroup of preeclamptic women probably overestimates the spread of GFR values below the mean, due to the likelihood of a lognormal distribution of GFR values and the contribution of measurement variability (beyond biological variability) to the statistical confidence limits. Given these considerations, the two-fold interindividual variability factor derived from three standard deviations below the mean of three studies for pregnancy GFR (mean = 161.5 mL/min; mean − 3 SD = 85.8) is considered preferable for providing adequate coverage to women predisposed to adverse birth outcomes due to renal complications.

The resulting sigma values were 1.54, 1.97, and 2.29 for data from Dunlop (1989), Krutzen et al. (1992), and Sturgiss et al. (1996), respectively. The mean of these values is 1.93, which was rounded to a value of 2.0 to account for some uncertainty in the method employed. This factor (the GFR value at 3 SD below the population mean) was also supported by the relationship between GFR and serum creatinine levels; excessive creatinine levels are indicators of

TABLE 2. Sigma-Method Value Calculation for $UF_{H\text{-}TK}$ (AF_{HK})[a]

Study	Mean GFR (SD) (mL/min)	Mean GFR −3 SD	Sigma-Method Value
Dunlop 1981	150.5 (17.6)[b]	97.7	1.54
Krutzén et al. 1992	195 (32)[c]	99	1.97
Sturgiss et al. 1996	138.9 (26.1)[d]	60.6	2.29
Averages:	161.5	85.8	1.93

[a]Mean GFR ÷ (mean GFR − 3 SD).
[b]Serially averaged observations across three time periods (16, 26, and 36 weeks) for 25 pregnant women.
[c]Third-trimester values for 13 pregnant women.
[d]Serially averaged observations across two time periods (early and late pregnancy) for 21 pregnant women (basal index plus basal control individuals).

renal impairment and represent a risk to human health. The default value of 3.16 for $UF_{H\text{-}TK}$ (AF_{HK}) was replaced with a data-derived adjustment factor of 2.0.

Vinyl Chloride: Interspecies Toxicokinetics, Route Extrapolation Vinyl chloride (VC) has an established oral reference dose of 3 E-3 mg/kg·day in the EPA's integrated risk information system. This value is based on the results of a chronic feeding study in Wistar rats. PBPK modeling was used to develop a human equivalent dose. Instead of the default 10-fold factor to cover animal-to-human extrapolation, UF_A was divided into TK and TD components, and a PBPK modeling approach served as the dosimetric adjustment factor in the inhalation reference concentration methodology (U.S. EPA, 1994). This case study demonstrates the application of data to inform interspecies toxicokinetic extrapolation ($UF_{A\text{-}TK}$).

Animal Toxicity and Dose–Response Relationship Til et al. (1983, 1991a, 1991b) reported the results of two-year rodent bioassays with VC in feed. Groups of 100 or 50 male and female Wistar rats were exposed to 0, 0.014, 0.13, or 1.3 mg/kg·/day in feed, which, to minimize volatilization, was available only 4 hours/day. The VC content of the feed was measured, before and following feeding, and doses were calculated to account for an approximate 20% loss to volatilization during feeding. Multiple hepatic effects were noted, including several that were deemed neoplastic or preneoplastic. Pathologists were able to delineate and determine incidences for two effects not thought to represent neoplastic or preneoplastic changes. Liver cell polymorphisms and proliferative bile duct epithelium cysts served as the basis for identifying the liver as the critical target tissue for noncancer effects in the chronic bioassay. The same changes were observed in a second study, but the doses employed in that study (Feron et al., 1981) were higher than those employed by Til and co-workers. Because of a lack of confidence in the outcome from a benchmark dose modeling approach based on external dose, a traditional (NOAEL) approach to dose–response evaluation for events not associated with carcinogenicity was performed. The point of departure for species extrapolation was the NOAEL for these events: 0.13 mg/kg·day.

Choice of Dose Metric A reactive, short-lived metabolite that achieves only low steady-state concentrations is thought to be responsible for the toxic effects of VC (Bolt, 1978). Two metabolites are derived from oxidation of VC via cytochrome P450 forms (CYP); these are each reactive and short-lived. Experiments that manipulated the longevity of these CYP-derived metabolites demonstrated an inverse relationship between metabolite longevity and protein and nucleotide binding (Guengerich et al., 1981). Thus, the metabolism of VC to reactive intermediates was demonstrated to be a critical determinant of toxicity. Because of the short-lived nature of the metabolite(s), a measure of their concentration in the target tissue (liver, the site of their formation) was deemed the appropriate dose metric. This concept and approach has also been applied to methylene chloride (Andersen et al., 1987) and chloroform (ILSI, 1997).

Species Extrapolation: Basis PBPK models were developed for rats and humans and used to extrapolate dosimetry between species (Clewell et al., 1995a,b). These models included four compartments and flow-limited kinetic behavior. Values for physiological parameters were taken from the peer-reviewed literature, although it should be noted that there are multiple sets of values, with none deemed "right" or "wrong." Partition coefficients for rat tissues were measured experimentally; metabolic rate constants were obtained by model fitting to species and strain-specific closed-chamber gas uptake data. Rat predictions also demonstrated acceptable fits to data describing total VC metabolized and hepatic glutathione content (which is diminished via VC metabolism). For the human PBPK model, metabolic rate constants were derived by allometric scaling of the rodent PBPK model and fitting predictions to observations of human closed-chamber VC concentrations. These models were subjected to external panel peer review and deemed useful for quantitative application in risk assessment.

Species Extrapolation: Application Importantly for dose extrapolation, these models demonstrated a linear relationship between applied dose and the internal liver metabolite–based dose metric [amount of metabolite in the liver (AML)] up to doses approximating 25 mg/kg·day. This allowed linear interpolation to be used to identify levels of external doses associated with specific amounts of the internal dose, rather than specific iterations via PBPK modeling. The level of the dose metric (AML) associated with the rat NOAEL (0.09 mg/kg·day) was determined to be 3.0 mg of VC metabolized per liter of liver (3.0 mg/L). This level of internal exposure in a 70-kg human was determined to result from a drinking water exposure of 0.09 mg/kg·day. Thus, doses of 0.13 and 0.09 mg/kg·day in the rat and human, respectively, are toxicokinetically equivalent.

Route-to-Route Extrapolation Vinyl chloride toxicity is manifest as liver alterations whether via inhalation or oral exposure. A well-conducted chronic dietary (oral) study and a 10-month inhalation study demonstrated similar liver lesions. Available evidence supported a mode of action involving formation of reactive metabolites in the liver; this was presumed to hold true regardless of route. Neither the inhalation nor the oral study indicated portal-of-entry effects. Both the oral and inhalation studies yielded data sufficient to develop PBPK models for both the inhalation and oral exposure routes, and use of these models for route-to-route extrapolation was based on the premise that equal amounts of the hepatic metabolite will result in the same level of injury regardless of the route of parent chemical exposure. The oral PBPK model translated the external NOAEL (POD) of 0.13 mg/kg·day to 3.0 mg of metabolite formed per liter of liver tissue. The inhalation PBPK model was exercised to simulate a continuous human exposure that would result in this same level of liver dose metric; that external concentration (the HEC_{NOAEL}) was determined to be 2.5 mg vinyl chloride per cubic meter.

Uncertainty Factor Derivation From a conceptual standpoint, these data indicate a data-derived value for interspecies toxicokinetic extrapolation of 0.13 mg/kg·day ÷ 0.09 mg/kg·day = 1.4. Combined with the default value of 3.2 for TD, a data-derived uncertainty factor for interspecies extrapolation (UF_A) would be (1.4)(3.2) = 4.48, in place of the default value of 10. The application of body weight scaling as a default approach could be performed using the EPA default (chronic) value for Wistar rat body weight of 0.462 kg (U.S. EPA, 1988). Based on a rat NOAEL of 0.13 mg/kg·day, the human equivalent dose would be 0.17 mg/kg·day and the derived uncertainty factor for $UF_{A\text{-}TK}$ would be 0.13/0.17 = 0.76; UF_A would have a value of (0.76)(3.2) = 2.4.

CONCLUSIONS

In this chapter we discussed how toxicity data can be interpreted for quantitative application in risk assessment, beyond the fundamental development of the POD in terms of an animal-applied or animal-ingested dose. A combination of the increasing power of TK tools and the increasing abundance of technically developed TK data offers the advantage of replacing default values for uncertainty factors with data-derived uncertainty factors. The establishment of guidance in this area and in the area of developing and evaluating toxicokinetic models for application in risk assessment should bolster the already increasing rate at which default uncertainty factors are replaced. Continued and increased dialogue among toxicologists, toxicokineticists, and risk assessors is required to take full advantage of the possibilities available.

Disclaimer

Acknowledgments

This work was undertaken as part of the author's duties for the U.S. government. The efforts are guided by documentation for NCEA–Cincinnati project 11660. The case studies contained in this document were developed by the author as part of his duties with the data-derived uncertainty factor technical panel convened by the U.S. EPA's risk assessment forum. Many of the concepts captured in this chapter resulted from panel deliberations and have been recapitulated from existing work. The author is grateful to Anna Lowit and members of the technical panel for their many discussions and insightful comments. The contributions of Bette Zwayer and Lana Wood in document preparation are gratefully acknowledged. This chapter has been developed and cleared under U.S. EPA peer review policy.

REFERENCES

Andersen ME, Clewell HJ III, Gargas ML, Smith FA, Reitz RH. 1987. Physiologically based pharmacokinetics and the risk assessment process for methylene chloride. *Toxicol Appl Pharmacol* 87(2): 185–205.

Andersen ME, Krishnan K, Conolly RB, McClellan RO. 1992. Mechanistic toxicology research and biologically-based modeling: partners for improving quantitative risk assessments. *CIIT Act* 12(1): 1–7.

Andersen ME, Clewell HJ III, Krishnan K. 1995. Tissue dosimetry, pharmacokinetic modeling, and interspecies scaling factors. *Risk Anal* 15(4): 533–537.

Arms AD, Travis CC. 1988. *Reference Physiological Parameters in Pharmacokinetic Modeling*. EPA/600/6–88/004. Office of Research and Development, U.S. Environmental Protection Agency, Washington, DC.

Bernillon P, Bois FY. 2000. Statistical issues in toxicokinetic modeling: a Bayesian perspective. *Environ Health Perspect* 108(suppl 5): 883–893.

Bolt HM. 1978. Pharmacokinetics of vinyl chloride. *Gen Pharmacol* 9(2): 91–95.

Brown RP, Delp MD, Lindstedt SL, Rhomberg LR, Beliles RP. 1997. Physiological parameter values for physiologically based pharmacokinetic models. *Toxicol Ind Health* 13(4): 407–484.

Carpenter CP, Keck GA, Nair JH III, Pozzani UC, Smyth HF Jr, Weil CS. 1956. The toxicity of butyl cellosolve solvent. *AMA Arch Ind Health* 14(2): 114–131.

Chiu WA. 2007. Statistical issues in PBPK modeling. *In: Toxicokinetics and Risk Assessment*. Lipscomb JC, Ohanian EV, eds. Informa Healthcare Publishers, New York.

Clark LH, Setzer RW, Barton HA. 2004. Framework for evaluation of physiologically-based pharmacokinetic models for use in safety or risk assessment. *Risk Anal* 24(6): 1697–1717.

Clewell HJ III, Gentry PR, Gearhart JM, et al. 1995a. The development and validation of a physiologically based pharmacokinetic model for vinyl chloride and its application in a carcinogenic risk assessment for vinyl chloride. ICF Kaiser report prepared for EPA/OHEA and OSHA/DHSP.

Clewell HJ III, Covington TR, Crump KS, et al. 1995b. The application of a physiologically based pharmacokinetic model for vinyl chloride in a noncancer risk assessment. ICF/Clement report prepared for EPA/NCEA under contract 68-D2–0129.

Clewell HJ III, Andersen ME, Barton HA. 2002. A consistent approach for the application of pharmacokinetic modeling in cancer and noncancer risk assessment. *Environ Health Perspect* 110(1): 85–93.

Clewell HJ III, Gentry PR, Kester JE, Andersen ME. 2005. Evaluation of physiologically based pharmacokinetic models in risk assessment: an example with perchloroethylene. *Crit Rev Toxicol* 35(5): 413–433.

Collins JM. 1987. Prospective predictions and validations in anti-cancer therapy. In. *Pharmacokinetics in Risk Assessment*. National Academies Press, Washington, DC, pp 431–440. Accessed at: http://darwin.nap.edu/books/0309037751/html/431.html.

Corley RA, Reitz RH. 1990. Dose–route extrapolations in quantitative toxicology: physiologically based pharmacokinetics and pharmacodynamics of chloroform. In: *Principles of Route-to-Route Extrapolation for Risk Assessment: Proceedings of the Workshops*; March and July; Hilton Head, SC and Durham, NC. Gerrity TR, Henry CJ, eds. Elsevier Science, New York, pp. 195–216.

Covington TR, Robinan GP, Van Landingham CB, Andersen ME, Kester JE, Clewell HJ III. 2007. The use of Markov chain Monte Carlo uncertainty analysis to support a public health goal for perchloroethylene. *Regul Toxicol Pharmacol* 47: 1–18.

Dixit R, Ward P. 2007. *Use of classical pharmacokinetic evaluations in drug development and safety assessment*. In: *Toxicokinetics and Risk Assessment*. Lipscomb JC, Ohanian EV, eds. Informa Healthcare Publishers, New York,

Dourson ML, Maier A, Meek B, Renwick A, Ohanian E, Poirier K. 1998. Boron tolerable intake: re-evaluation of toxicokinetics for data-derived uncertainty factors. *Biol Trace Elem Res* 66(1–3): 453–463.

Dunlop W. 1981. Serial changes in renal haemodynamics during normal human pregnancy. *Br J Obstet Gynaecol* 88(1): 1–9.

ECETOC (European Centre for Ecotoxicology and Toxicology of Chemicals). 2003. *Derivation of Assessment Factors for Human Health Risk Assessment*. ECETOC, Brussels, Belgium.

Evans MV, Andersen ME. 2000. Sensitivity analysis of a physiological model for 2,3,7,8-tetrachlorodibenzo-it *p*-dioxin (TCDD): assessing the impact of specific model parameters on sequestration in liver and fat in the rat. *Toxicol Sci* 54(1): 71–80.

Feron VJ, Hendriksen CF, Speek AJ, Til HP, Spit BJ. 1981. Lifespan oral toxicity study of vinyl chloride in rats. *Food Cosmet Toxicol* 19(3): 317–333.

Gargas ML, Andersen ME, Clewell HJ III. 1986. A physiologically based simulation approach for determining metabolic constants from gas uptake data. *Toxicol Appl Pharmacol* 86(3): 341–352.

Gargas ML, Burgess RJ, Voisard DE, Cason GH, Andersen ME. 1989. Partition coefficients of low-molecular-weight volatile chemicals in various liquids and tissues. *Toxicol Appl Pharmacol* 98(1): 87–99.

Gentry PR, Haber L, McDonald T, Zhao Q, Covington T, Nance P, Clewell HJ III, Lipscomb JC, Barton HA. 2005. Data for physiologically-based pharmacokinetic modeling in neonatal animals: physiological parameters in mice and Sprague–Dawley rats. *J Child Health* 2: 363–411.

Ghanayem BI, Burka LT, Matthews HB. 1987. Metabolic basis of ethylene glycol monobutyl ether (2-butoxyethanol) toxicity: role of alcohol and aldehyde dehydrogenases. *J Pharmacol Exp Ther* 242(1): 222–231.

Ghanayem BI, Sanders JM, Clark AM, Bailer J, Matthews HB. 1990. Effects of dose, age, inhibition of metabolism and elimination on the toxicokinetics of 2-butoxyethanol and its metabolites. *J Pharmacol Exp Ther* 253(1): 136–143.

Guengerich FP, Mason PS, Stott WT, Fox TR, Watanabe PG. 1981. Roles of 2-haloethylene oxides and 2-haloacetaldehydes derived from vinyl bromide and vinyl chloride in irreversible binding to protein and DNA. *Cancer Res* 41(11 pt 1): 4391–4398.

Haddad S, Gad SC, Tardif R, et al. 1995. Statistical approaches for the validation of physiologically-based pharmocokinetic (PBPK) models. Abstract. *Toxicologist* 15: 48.

Haufroid V, Thirion F, Mertens P, Buchet JP, Lison D. (1997). Biological monitoring of workers exposed to low levels of 2-butoxyethanol. *Int Arch Occup Environ Health* 70(4): 232–236.

IGHRC (Interdepartmental Group on Health Risks from Chemicals). 2005. Guidelines on route-to-route extrapolation of toxicity data when assessing health risks of chemicals. Draft v6.0, August. Medical Research Council Institute for Environment and Health, University of Leicester, Leicester, UK. Accessed at: http://www.hse.gov.uk/aboutus/hsc/iacs/acts/watch/051005/18annexe1.pdf.

ILSI (International Life Sciences Institute). 1997. *An Evaluation of EPA's Proposed Guidelines for Carcinogen Risk Assessment Using Chloroform and Dichloroacetate as Case Studies*. ILSI HESI Expert Panel, Washington, DC.

IPCS (International Programme on Chemical Safety). 2005. *Chemical-Specific Adjustment Factors for Interspecies Differences and Human Variability: Guidance Document for Use of Data in Dose/Concentration–Response/Assessment*. Harmonization Project Document 2. World Health Organization, Geneva, Switzerland.

Jepson GW, Hoover DK, Black RK, McCafferty JD, Mahle DA, Gearhart JM. 1994. A partition coefficient determination method for nonvolatile chemicals in biological tissues. *Fundam Appl Toxicol* 22(4): 519–524.

Johanson G. 1986. Physiologically based pharmacokinetic modeling of inhaled 2-butoxyethanol in man. *Toxicol Lett* 34(1): 23–31.

Johanson G, Dynesius B. 1988. Liquid/air partition coefficients of six commonly used glycol ethers. *Br J Ind Med* 45(8): 561–564.

Krishnan K, Nong A, Haddad S. 2007. In silico predictions of partition coefficients for PBPK modeling. *In: Toxicokinetics and Risk Assessment*. Lipscomb JC, Ohanian EV, Eds. Informa Healthcare Publishers, New York.

Krutzén E, Olofsson P, Back SE, Nilsson-Ehle P. 1992. Glomerular filtration rate in pregnancy: a study in normal subjects and in patients with hypertension, preeclampsia and diabetes. *Scand J Clin Lab Invest* 52(5): 387–392.

Lehman AJ, Fitzhugh OG. 1954. 100-fold margin of safety. *Assoc Food Drug Off US Q Bull* 18: 33–35.

Lipscomb JC. 2004. Evaluating the relationship between variance in enzyme expression and toxicant concentration health risk assessment. *Human Ecol Risk Assess* 10: 443–455.

Lipscomb JC, Ohanian E, eds. 2007. *Toxicokinetics and Risk Assessment*. Informa Healthcare Publishers, New York.

Lipscomb JC, Teuschler LK, Swartout J, Popken D, Cox T, Kedderis GL. 2003. The impact of cytochrome P450 2E1-dependent metabolic variance on a risk-relevant pharmacokinetic outcome in humans. *Risk Anal* 23(6): 1221–1238.

Medinsky MA, Singh G, Bechtold WE, Bond JA, Sabourin PJ, Birnbaum LS, Henderson RF. 1990. Disposition of three glycol ethers administered in drinking water to male F344/N rats. *Toxicol Appl Pharmacol* 102(3): 443–455.

Meek E, Renwick AR. 2007. Guidance for the development of chemical specific adjustment factors: integration with mode of action frameworks. In: *Toxicokinetics and Risk Assessment*. Lipscomb JC, Ohanian EV, eds. Informa Healthcare Publishers, New York.

NAS (National Academy of Sciences). 1980 *Drinking Water and Health*. National Research Council, National Academies Press, Washington, DC.

______. 1983. *Risk Assessment in the Federal Government: Managing the Process*. National Research Council, National Academies Press, Washington, DC.

NTP (National Toxicology Program). 1993. *Technical Report on Toxicity Studies of Ethylene Glycol Ethers 2-Methoxyethanol, 2-Ethoxyethanol, 2-Butoxyethanol Administered in Drinking Water to F344/N Rats and B6C3F1 Mice*. NTP TR 26. NIH Publ. 93–3349. Public Health Service, National Institutes of Health, U.S. Department of Health and Human Services, Research Triangle Park, NC.

______. 1998. *Technical Report on the Toxicology and Carcinogenesis Studies of 2-Butoxyethanol (CAS No. 111–76-2) in F344/N Rats and B6C3F1 Mice (Inhalation studies)*. NTP TR 484 NIH Draft Publ. 98–3974. Public Health Service, National Institutes of Health, U.S. Department of Health and Human Services, Research Triangle Park, NC.

Pahl MV, Culver BD, Strong PL, Murray FJ, Vaziri ND. 2001. The effect of pregnancy on renal clearance of boron in humans: a study based on normal dietary intake of boron. *Toxicol Sci* 60(2): 252–256.

Ramsey JC, Andersen ME. 1984. A physiologically based description of the inhalation pharmacokinetics of styrene in rats and humans. *Toxicol Appl Pharmacol* 73(1): 159–175.

Reddy M, Yang RS, Andersen ME, Clewell HJ III. 2005 *Physiologically Based Pharmacokinetic Modeling: Science and Applications*. Wiley-VCH, New York.

Skjaerven R, Wilcox AJ, Lie RT. 2002. The interval between pregnancies and the risk of preeclampsia. *N Engl J Med* 346(1): 33–38.

Sturgiss SN, Wilkinson R, Davison JM. 1996. Renal reserve during human pregnancy. *Am J Physiol* 271(1 pt 2): F16–F20.

Thomas, VVF-B. 1975. *Biological-Mathematical Modeling of Chronic Toxicity*. AMRL-TR-75–5. U.S. Air Force, Aerospace Medical Research Laboratory, Aerospace Medical Division, Wright-Patterson AFB, Ohio.

Til HP, Immel HR, Feron FJ. 1983. *Lifespan Oral Carcinogenicity Study of Vinyl Chloride in Rats*. TNO Report no. V 83.285/291099, Netherlands Organization of Applied Scientific Research, Division for Nutrition and Food Research TNO, Zeist, The Netherlands.

Til HP, Feron VJ, Immel HR. 1991a. Lifetime (149-week) oral carcinogenicity study of vinyl chloride in rats. *Food Chem Toxicol* 29(10): 713–718.

Til, HP, Immel HR, Feron VJ. 1991b. *Lifespan Oral Carcinogenicity Study of Vinyl Chloride in Rats*. Final Report. TNO Report V 83.285/291099, TSCATS Document FYI-AX-0184–0353, Fiche 0353. CIVO Institutes, The Netherlands.

U.S. EPA. 1988. *Recommendations for and Documentation of Biological Values for Use in Risk Assessment*. EPA/600/6-87/008. NTIS PB88-179874/AS. Prepared by the Office of Health and Environmental Assessment, Environmental Criteria and Assessment Office, Cincinnati, OH, for the Office of Solid Waste and Emergency Response, U.S. EPA, Washington, DC.

———. 1991. *Alpha2u-Globulin: Association with Chemically Induced Renal Toxicity and Neoplasia in the Male Rat*. EPA/625/3-91/019F. NTIS PB92-143668. Risk Assessment Forum, U.S. EPA, Washington, DC.

———. 1993. Reference Dose (RfD): *Description and Use in Health Risk AssessmentsM*. Office of Health and Environmental Assessment, Environmental Criteria and Assessment, U.S. EPA, Washington, DC. Accessed at: http://www.epa.gov/iris/rfd.htm.

———. 1994. *Methods for Derivation of Inhalation Reference Concentrations and Application of Inhalation Dosimetry*. EPA/600/8-90/066F. Office of Health and Environmental Assessment, U.S. EPA, Washington, DC.

———. 2002. *A Review of the Reference Dose and Reference Concentration Processes*. EPA 630 P-02/002F. Risk Assessment Forum, U.S. EPA, Washington, DC.

———. 2005. *Guidelines for Carcinogen Risk Assessment*. EPA/630/P-03/001B. Risk Assessment Forum, U.S. EPA, Washington, DC. Accessed at: http://www.thecre.com/pdf/20050404.

———. 2006a. *Integrated Risk Information System (IRIS)*. Office of Research and Development, National Center for Environmental Assessment, U.S. EPA, Washington, DC. Accessed at: http://www.epa.gov/iris/.

———. 2006b. *Harmonization in Interspecies Extrapolation: Use of* $BW^{3/4}$ *as Default Method in Derivation of the Oral RfD* (External Review Draft). EPA/630/R-06/001. Office of Research and Development, National Center for Environmental Assessment, U.S. EPA, Washington, DC. Accessed at: http://cfpub.epa.gov/ncea/cfm/recordisplay.cfm?deid = 148525.

———. 2006c. *Exposures and Internal doses of Trihalomethanes in Humans: Multi-route Contributions from Drinking Water*. EPA/600/X-06/006. Office of Research and Development, National Center for Environmental Assessment, U.S. EPA, Cincinnati, OH.

———. 2006d. *Approaches for the Application of Physiologically Based Pharmacokinetic (PBPK) Models and Supporting Data in Risk Assessment* (Final Report). EPA/600/R-05/043F. Office of Research and Development, National Center for Environmental Assessment, U.S. EPA, Washington, DC. Accessed at: http://cfpub.epa.gov/ncea/cfm/recordisplay.cfm?deid = 157668.

Ventura SJ, Mosher WD, Curtain SC, et al. 1999. Highlights of trends in pregnancies and pregnancy rates by outcome: estimates for the United States, 1976–96. *Natl Vital Stat Rep* 47(29): 1–12. Accessed at: http://www.cdc.gov/nchs/data/nvsr/nvsr47/nvs47_29.pdf.

Voisin EM, Ruthsatz M, Collins JM, Hoyle PC. 1990. Extrapolation of animal toxicity to humans: interspecies comparisons in drug development. *Regul Toxicol Pharmacol* 12(2): 107–116.

6

HEALTH RISK ASSESSMENT OF CHEMICAL MIXTURES IN DRINKING WATER

RICHARD C. HERTZBERG
Emory University, Atlanta, Georgia

GLENN E. RICE, LINDA K. TEUSCHLER, AND J. MICHAEL WRIGHT
National Center for Environmental Assessment, U.S. Environmental Protection Agency, Cincinnati, Ohio

JANE E. SIMMONS
National Health and Environmental Effects Research Laboratory, U.S. Environmental Protection Agency, Research Triangle Park, North Carolina

Accurate assessment of potential human health risk(s) from multiple-route exposures to multiple chemicals in drinking water is needed because of widespread daily exposure to this complex mixture. Hundreds of chemicals have been identified in drinking water, with the mix of chemicals varying with geographic location, source water, disinfection scenario, and surrounding land uses and industries (Richardson, 1998; U.S. EPA, 2006a). Mixtures in drinking water can consist of many chemical classes [e.g., pesticides, pharmaceuticals, metals, organic solvents, and disinfection by-products (DBPs)]. Such contaminants may be present in liquid, vapor, or aerosol forms and can enter the body via ingestion, respiration, or dermal penetration.

Risk Assessment for Chemicals in Drinking Water, Edited by Robert A. Howd and Anna M. Fan

DRINKING WATER MIXTURE CONCERNS

For certain groups of chemicals, such as the DBPs, positive data from epidemiological or toxicological studies raise concern for human health. Although inconsistent results have been reported across epidemiological studies, associations have been observed between DBP exposures and increased duration of menstrual cycles (Windham et al., 2003), as well as an increased risk of term low-birth-weight, small-for-gestational-age infancy (Gallagher et al., 1995; Wright et al., 2004) and birth defects (Klotz and Pyrch, 1999; Dodds and King, 2001). A recent prospective cohort study found little evidence of spontaneous abortions associated with exposure to most individual DBPs (Savitz et al., 2005), but previous studies support an association between DBP and spontaneous abortions (Waller et al., 1998) and stillbirths (Dodds et al., 2004; Toledano et al., 2005). In toxicological studies on DBP mixtures, a variety of effects have been observed, including mutagenicity, carcinogenicity, hepatotoxicity, nephrotoxicity, developmental toxicity, neurological effects, and changes in pharmacokinetic behavior (as summarized by Simmons et al., 2001). Thus, for DBP mixtures, both types of data may be useful in drinking water mixture risk assessments to identify human health hazards, quantify dose response and exposure, estimate risks, or characterize variability and uncertainty.

The protocols and methods typically used to evaluate chemical mixtures are best applied to simple defined mixtures consisting of only a few chemicals. Such methods, used for site assessments or toxicological studies, are often not sufficient to estimate health risk for complex drinking water mixtures. Actual drinking water exposures involve multiple chemicals, usually at very low levels, many of which may be unknown. For example, approximately 50% of the total organic halide material making up drinking water DBPs generally consists of an unknown number of unidentified chemicals (Miltner et al., 1990; Richardson, 1998; Weinberg, 1999). In addition, the toxicological properties of drinking water mixtures, such as the toxic mode of action that causes health effects, may be uncertain for many of the known chemicals. Because the potential for joint toxic action is thought to depend on dose levels and toxicological action of the mixture components, it is difficult to assess the likelihood of observing either additivity (e.g., dose addition or response addition) or some interaction effect (e.g., synergism or antagonism).

Regulatory Importance

At the U.S. Environmental Protection Agency (EPA), the Office of Pesticide Programs (OPP), Office of Water (OW), and Office of Solid Waste and Emergency Response (OSWER) address drinking water mixtures as part of their environmental assessments under the Food Quality Protection Act of 1996 (U.S. Congress, 1996a), the Safe Drinking Water Act Amendments of 1996 (U.S. Congress, 1996b), and the Comprehensive Environmental Response, Compensation, and Liability Act (U.S. Congress, 1980), respectively. OPP evaluates mixtures of pesticides that share a common mechanism of toxicity. Drinking water is one route of exposure that is included in OPP's aggregate human exposure estimates to multiple pesticides. Recent evaluations include the organophosphates, *N*-methyl carbamates, triazines, and chloroacetanilides (U.S. EPA, 2006b).

OW is concerned with mixtures of DBPs and of contaminant candidate list chemicals (e.g., organotins, pesticides, metals, pharmaceuticals). OW conducts its regulatory assessments by setting maximum contaminant levels, defined by the EPA as:

- *Maximum contaminant level goal (MCLG):* The level of a contaminant in drinking water below which there is no known or expected risk to health. MCLGs allow for a margin of safety and are nonenforceable public health goals.
- *Maximum contaminant level (MCL):* The highest level of a contaminant that is allowed in drinking water. MCLs are set as close to MCLGs as feasible using the best available treatment technology and taking cost into consideration. MCLs are enforceable standards.

MCLs and MCLGs are derived for single contaminants, with a few notable exceptions (U.S. EPA, 2006a). The four regulated trihalomethanes (THMs) and five haloacetic acids (HAAs) are drinking water DBPs that are each regulated as a group, with MCLs of 80 and 60 μg/L for each chemical mixture, respectively. Similarly, the MCL for the complex mixture of polychlorinated biphenyls (PCBs) is set at 0.5 μg/L and is used regardless of the actual levels and proportions of the mixture components found in the environmental sample. Additionally, the existing epidemiological data on DBP mixtures have been used for hazard identification for various cancers and reproductive and developmental effects, as well as a source of information for regulatory benefits analysis (U.S. EPA, 1998a, 2005a).

OSWER evaluates drinking water mixtures at Superfund site assessments using additivity approaches (U.S. EPA, 1989a). For low exposure levels when no interaction information or whole mixtures toxicity data are available, response addition and dose addition are the recommended default methods when the component chemicals in a mixture show dissimilar toxicity and similar toxicity, respectively (U.S. EPA, 2000). Because dose addition (in the form of a hazard index) and response addition (as the sum of component risks) are relatively easy to apply, using single chemical toxicity and exposure information, these two additivity assumptions are the most commonly used in site assessments and are recommended in the Superfund guidance (U.S. EPA, 1989a).

Overview of Chemical Mixtures Risk Issues

Chemical mixtures risk assessments incorporate the traditional elements of the risk assessment paradigm (NRC, 1983): hazard identification, exposure assessment, dose–response assessment, and risk characterization. An important difference from single chemical risk assessment is that the steps are no longer conducted independently, nor in a set sequence, but involve information sharing and cross-evaluation, particularly between the exposure and toxicity assessment steps. Mixture toxicity depends not only on total dose, but also on the composition (i.e., the component proportions). Because it is only feasible to test a small set of mixture compositions, the toxicity data to be used should match the composition in the exposure assessment as closely as possible. This means that dose–response information needs to be considered during the exposure assessment, and characteristics of the exposure assessment need to be incorporated into the compiling of toxicity information and then reflected in the dose–response assessment.

An overarching question critical to conducting the dose–response and exposure assessments for chemical mixtures is the level of analysis required in relation to the risk assessment goal, such as the amount of health protection needed. For example, conducting a mixtures risk assessment when chemical concentrations in the drinking water are extremely low might be considered by some to be conservative because they assert that joint toxic action is unlikely when chemical exposures are all well below their individual no-effect levels. This obviously depends on the contaminants, their interaction mechanisms, and their toxicities. The goal of such an assessment could be to screen for hazards (e.g., by considering worst-case scenarios). In this case, a mixtures approach might be chosen to estimate the maximum potential risk so that situations not exceeding the acceptable risk criterion can be excluded from further consideration. Conversely, when humans are exposed to higher levels of chemicals in drinking water, the goal might be to produce best estimates of risk by calculating central tendency risk estimates for the mixture exposure, factoring in the potential for toxicological interaction effects, or using physiologically based pharmacokinetic (PBPK) modeling to estimate internal tissue doses and more clearly define chemical interactions. In this case, a mixtures approach would be chosen to produce the most accurate risk estimate for the exposure, resulting in estimates that are health protective on average.

Consistent and clear terminology is critical to a discussion of chemical mixtures risk assessment methodology. Tables 1 and 2 articulate the differences among the many terms used to describe chemical mixtures and the types of interactions that may occur among chemicals. Table 1 presents chemical mixtures definitions in terms of specific criteria including the complexity of the mixture, similarity of biological activity, similarity of chemical structure or mixture composition, environmental source of the mixture, and toxic endpoint. Table 2 provides definitions for terms that describe various types of toxicological interactions, including forms of additivity, antagonism, synergism, and other toxicological phenomena. Tables 1 and 2 can be used in a risk assessment to classify available toxicity and exposure data in order to choose the most relevant chemical mixture method.

Regardless of the chemical mixtures approach being used, human health risk assessment is typically challenged to extrapolate toxicity information from animal to human, short-term to chronic duration, high- to low-dose effects, adult to child, average to sensitive response, and target organ to a specific mechanism of action. Where data are lacking to address these issues, the gaps are filled

TABLE 1. Definitions of Chemical Mixtures

Chemical mixture: any set of multiple chemical substances that may or may not be identifiable, regardless of their sources, that may jointly contribute to toxicity in the target population. May also be referred to as a *whole mixture* or as the *mixture of concern*.

Components: single chemicals that make up a chemical mixture that may be further classified as systemic toxicants, carcinogens, or both.

Simple mixture: a mixture containing two or more identifiable components, but few enough that the mixture toxicity can be characterized adequately by a combination of the components' toxicities and the components' interactions.

Complex mixture: a mixture containing so many components that any estimation of its toxicity based on its components' toxicities contains too much uncertainty and error to be useful. The chemical composition may vary over time or with different conditions under which the mixture is produced. Complex mixture components may be generated simultaneously as by-products from a single source or process, produced intentionally as a commercial product, or may coexist because of disposal practices. Risk assessments of complex mixtures are preferably based on toxicity and exposure data on the complete mixture. Gasoline is an example of a complex mixture of hydrocarbons.

Similar components: single chemicals that cause the same biological activity or are expected to cause a type of biological activity based on chemical structure. Evidence of similarity may include similarly shaped dose–response curves, or parallel log dose–probit response curves for quantal data on the number of animals (people) responding, and the same mechanism of action or toxic endpoint. These components are expected to have comparable characteristics for fate, transport, physiologic processes, and toxicity.

Similar mixtures: mixtures that are slightly different but are expected to have comparable characteristics for fate, transport, physiological processes, and toxicity. These mixtures may have the same components but in slightly different proportions, or have most components in nearly the same proportions with only a few different (more or fewer) components. Similar mixtures cause the same biological activity or are expected to cause the same type of biological activity due to chemical composition. Similar mixtures act by the same mechanism of action or affect the same toxic endpoint. Diesel exhausts from different engines are an example.

Chemical classes: groups of components that are similar in chemical structure and biological activity, and that frequently occur together in environmental samples, usually because they are generated by the same commercial process. The composition of these mixtures is often well controlled, so that the mixture can be treated as a single chemical. Dibenzodioxins are an example.

Source: U.S. EPA (2000).

using professional judgment, surrogate data on similar but better studied chemicals, statistical or biologically based modeling, or conservative default values based on uncertainty analyses. In animal-to-human extrapolations, target organ concordance is sought but is not necessarily required to make a judgment, particularly when the goal is health protection. To calculate cancer slope factors, or

TABLE 2. Definitions of Joint Toxic Action Between Chemicals

Additivity: when the effect of the combination is estimated by the sum of the exposure levels or the effects of the individual chemicals. The terms *effect* and *sum* must be defined explicitly. Effect may refer to the measured response or the incidence of animals adversely affected. The sum may be a weighted sum (*see* Dose additivity) or a conditional sum (*see* Response additivity).

Antagonism: when the effect of the combination is less than that suggested by the component toxic effects. Antagonism must be defined in the context of the definition of "no interaction," which is usually dose or response addition.

Chemical antagonism: when a reaction between the chemicals has occurred and a new chemical is formed. The toxic effect produced is less than that suggested by the component toxic effects.

Chemical synergism: when a reaction between the chemicals has occurred and a different chemical is formed. The toxic effect produced is greater than that suggested by the component toxic effects and may be different from effects produced by either chemical by itself.

Complex interaction: when three or more compounds combined produce an interaction that cannot be assessed according to the other interaction definitions.

Dose additivity: when each chemical behaves as a concentration or dilution of every other chemical in the mixture. The response of the combination is the response expected from the equivalent dose of an index chemical. The equivalent dose is the sum of component doses scaled by their toxic potency relative to the index chemical.

Index chemical: the chemical selected as the basis for standardization of toxicity of components in a mixture. The index chemical must have a clearly defined dose–response relationship.

Inhibition: when one substance does not have a toxic effect on a certain organ system, but when added to a toxic chemical, it makes the latter less toxic.

Masking: when the compounds produce opposite or functionally competing effects at the same site or sites so that the effects produced by the combination are less than suggested by the component toxic effects.

No apparent influence: when one substance does not have a toxic effect on a certain organ or system, and when added to a toxic chemical, it has no influence, positive or negative, on the toxicity of the latter chemical.

No observed interaction: when neither compound by itself produces an effect, and no effect is seen when they are administered together.

Potentiation: when one substance does not have a toxic effect on a certain organ or system, but when added to a toxic chemical, it makes the latter more toxic.

Response additivity: when the toxic response (rate, incidence, risk, or probability of effects) from the combination is equal to the conditional sum of component responses as defined by the formula for the sum of independent event probabilities. For two chemical mixtures, the body's response to the first chemical is the same whether or not the second chemical is present.

Synergism: when the effect of the combination is greater than that suggested by the component toxic effects. Synergism must be defined in the context of the definition of "no interaction," which is usually dose or response addition.

Unable to assess: when the effect cannot be placed in one of the foregoing classifications. Common reasons include lack of proper control groups, lack of statistical significance, and poor, inconsistent, or inconclusive data.

Source: U.S. EPA (2000).

in general, to use dose–response models to estimate either safe levels or risks, body weight scaling is one approach used to calculate toxicologically equivalent doses across species. The following equation is based on allometric scaling laws that have been shown to relate several biologic measure of physiology to body weight raised to a power (Travis and White, 1988):

$$d_h = \frac{d_a}{\left(\frac{W_a}{W_h}\right)^k} \tag{1}$$

where

d_h = human equivalent dose (mg/day)
d_a = animal dose (mg/day)
W_h = human body weight (kg)
W_a = animal body weight (kg)
k = scaling factor, typically set to 0.75

In setting safe levels, uncertainty factors are applied, one of which accounts for the animal-to-human extrapolation (used instead of allometric scaling for these calculations). For example, the EPA determines the following route-specific risk values for threshold toxicants (U.S. EPA, 2006c):

- *Reference dose (RfD):* an estimate (with uncertainty spanning perhaps an order of magnitude) of a daily oral exposure to the human population (including sensitive subgroups) that is likely to be without an appreciable risk of deleterious effects during a lifetime.
- *Reference concentration (RfC):* an estimate (with uncertainty spanning perhaps an order of magnitude) of a continuous inhalation exposure to the human population (including sensitive subgroups) that is likely to be without an appreciable risk of deleterious effects during a lifetime.

To calculate an RfD, a no-observed-adverse-effect level (NOAEL) from an animal study is divided by one or more uncertainty factors to account for lack of knowledge when making extrapolations. Uncertainty factors typically applied include a 10-fold factor for animal-to-human and a 10-fold factor for average human-to-sensitive human extrapolations.

The discussion of mixtures approaches in this chapter is oriented toward the practice of the EPA. Of the several guidelines written by the EPA on risk assessment, only a few will be discussed here, primarily the *Supplementary Guidance for Conducting Health Risk Assessment of Chemical Mixtures* (U.S. EPA, 2000).

Several other risk assessment guidelines have been published by the EPA (see Table 3) and should be consulted for additional details on exposure assessment, both modeling and monitoring, and specific health endpoints, such as cancer, developmental effects, and neurotoxicity.

In the remainder of this chapter we address current practices as well as new and emerging chemical mixtures risk assessment approaches that may be used to

TABLE 3. Selected U.S. EPA Risk Assessment Guidance Documents

Risk Assessment Guidelines of 1986, including chemical mixtures, mutagenicity, cancer, exposure assessment, and developmental effects (U.S. EPA, 1986, 1987)
Risk Assessment Guidance for Superfund (U.S. EPA, 1989a)
Guidelines for Developmental Toxicity Risk Assessment (U.S. EPA, 1991)
Guidelines for Reproductive Toxicity Risk Assessment (U.S. EPA, 1996)
Guidelines for Neurotoxicity Risk Assessment (U.S. EPA, 1998b)
Guidelines for Exposure Assessment (U.S. EPA, 1992b)
Supplementary Guidance for Conducting Health Risk Assessment of Chemical Mixtures (U.S. EPA, 2000)
General Principles for Performing Aggregate Exposure and Risk Assessments (U.S. EPA, 2001)
Guidance on Cumulative Risk Assessment of Pesticide Chemicals That Have a Common Mechanism of Toxicity (U.S. EPA, 2002b)
Guidelines for Carcinogen Risk Assessment (U.S. EPA, 2005b)
Supplemental Guidance for Assessing Susceptibility from Early-Life Exposure to Carcinogens (U.S. EPA, 2005c)

address drinking water mixtures. DBPs are utilized in many of the examples in this chapter. DBPs are a complex mixture that occurs in all chemically disinfected drinking waters. The U.S. population is exposed ubiquitously to these mixtures. These exposures have been associated with cancer and reproductive and developmental effects in epidemiologic studies. Given the exposures and the possible toxicity, DBPs are an important class of mixtures. In the following section we discuss the estimation of exposures to multiple chemicals in drinking water. In subsequent sections we present information on toxicological concepts for joint toxicity of drinking water mixtures, the various chemical mixtures risk assessment methods currently in the literature, and new and emerging risk assessment approaches for evaluating drinking water mixtures.

ESTIMATING EXPOSURES TO MULTIPLE CHEMICALS IN DRINKING WATER

In this section we summarize the conduct of exposure assessments for chemical mixtures. Following a description of the goals of such an assessment, a series of assessment issues are discussed, showing how the choice of exposure assessment method is influenced by the level of assessment detail required, the resources available to evaluate the exposure, the choice of dose–response method, and the availability of analytic chemistry methods. Through these discussions, the differences between single chemical and chemical mixtures exposure assessments are highlighted.

Exposure assessments for chemical mixtures and single chemicals share common goals. Both typically evaluate the sources, magnitude, frequency, duration,

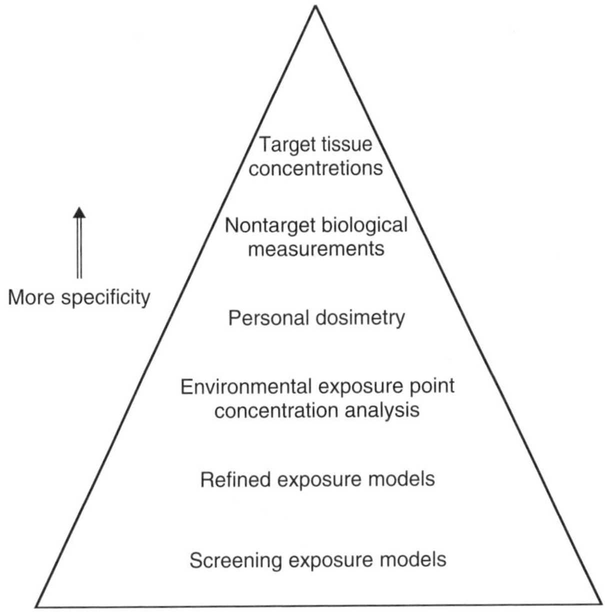

Figure 1. Hierarchy of exposure measurements.

and route of exposure to chemicals that are experienced by a population. Both make use of various approaches to examine exposures. These include:

- Direct measurement of compounds at the point of contact [e.g., measures of pollutant concentrations in tap water are integrated with contact time estimates, which can be based on questionnaires (U.S. EPA, 1997) or analyses of household water use patterns through flowmeters (Wilkes et al., 1992)]
- Predictive modeling that can be used to estimate the concentrations of chemicals in tap water and models that predict contacts with the contaminants over time (Wilkes, 1998)
- Biomonitoring that can be used to analyze past exposures through internal indicators (Wallace, 1997; Weisel et al., 1999)

The output of an exposure assessment can vary with the level of specificity required, resulting in a range of outcomes, from a screening-level analysis to measurements of internal dosage (Figure 1). Chemical mixtures assessments are complicated by the need to evaluate exposures to multiple chemicals, potentially through different exposure routes, at differing frequencies, and over varying exposure durations. Although the exposure analysis and dose–response analysis must be linked in any assessment of risk, the complexities of a mixtures risk assessment underscore this need.

Chemical Selection

The selection of chemicals to include in the mixtures analysis is a critical decision. If the dose–response assessment for the mixture is conducted by combining data on the individual chemical components, the compounds to evaluate in the exposure assessment could be defined in a relatively straightforward manner based on the information needed in the dose–response assessment. The chemical analytic methods may not be well developed for some mixtures; these methods may rely on a single chemical component of the mixture or on the total quantity present. For example, an assessment of arsenic in the drinking water may or may not include information on speciation.

Some drinking water mixtures, such as DBPs, which form as a consequence of reactions among the disinfectants used to treat the water, natural organic matter (NOM) and bromide, can be quite complex. The question then arises as to how best to analyze exposure to such a mixture (i.e., what measures best capture the chemical mixture composition and concentration, given the available resources and level of scientific understanding of the chemistry and toxicology of the mixture). At times, various surrogates are employed. Some analyses may consider only a single surrogate chemical, such as chloroform, that is used to represent exposure to (and sometimes the risk posed by) the mixture. Other evaluations have examined a small subset of compounds: for example, the four regulated trihalomethanes (THM4), the most abundant class of DBPs, are used as a surrogate measure of exposure. Other studies may use a measure such as total organic halide and attempt to evaluate some individual compounds or just measure the total halide concentration. The choice of chemicals to include in the analysis will depend on the type of analysis (screening vs. sophisticated), the resources available, the dose–response methods being considered, and the availability of analytical measurement methods, reliable mathematical models, or exposure biomarkers. The data quality associated with any measure of exposure clearly needs to be considered and, if possible, should be consistent with the quality of the data used in the dose–response assessment.

Environmental Fate

When evaluating the environmental fate of a mixture, changes occur to the environmental mixture over time as a result of (1) transport through an individual environmental compartment (e.g., drinking water, air inside a home), (2) transfer between environmental compartments, or (3) transformation mediated by biological, chemical, or physical agents. These changes affect exposure to the mixture and its toxicity. While single chemical assessments must consider the changes in the pollutant concentration and changes in the form of the pollutant during transport, analyses of mixtures are complicated by potential formation of additional components, differential rates of degradation among the mixture components, and different properties of components which lead to changes in the composition of the mixture in various media. In this section, after describing the origins of various mixtures, an example is given to illustrate the influence of environmental fate on exposures to DBPs.

The origins of many drinking water mixtures are in the source waters, such as mixtures of cyanotoxins or organic pollutants such as the polycyclic aromatic hydrocarbons (U.S. EPA, 1992a) and tri- and tetrachloroethylene (Ziglio, 1985; Vartiainen et al., 1993). (Many of these pollutants found in source waters are chemically changed by oxidants added to drinking water during chemical disinfection; i.e., occurrence of a group of compounds in source waters does not necessarily lead to exposures to the same contaminants in drinking waters.) Other mixtures arise during chemical disinfection; some DBPs result from the reactions of chemical oxidants with natural organic matter (NOM). Changes in water temperature, the levels and types of NOM in the water, seasonal changes in treatment, and time in the drinking water distribution system can lead to marked changes in the types and levels of DBPs present in drinking waters. Previous studies have shown that while THM concentrations can increase with length of time in the distribution system, HAAs may decrease due to biological degradation (Chen and Weisel, 1998; Clark, 1998). Finally, some mixtures (e.g., phthalates, lead, organotins) may originate in the pipes used to transport potable waters. As water passes through the pipes, these contaminant mixtures may be released or leached from the pipes.

While all DBPs are at least somewhat soluble, some are also volatile. For example, chloroform and bromodichloromethane have been measured in the indoor air and exhaled breath and blood samples of subjects following showering (Backer et al., 2000; Gordon et al., 2006). Other DBPs, such as the haloacetic acids, are not volatile (U.S. EPA, 2003a). Although inhalation exposure predictions are unlikely to be a main route of exposure to these DBPs (U.S. EPA, 2003a), some haloacetic acids are reportedly found bound to particulates in the air after showering (Xu and Weisel, 2003). Thus, the mixture of DBPs encountered through the consumption of tap waters (i.e., the DBP mixture in a glass of water) probably differs from the mixture of DBPs encountered through inhaled indoor air (Weisel and Jo, 1996; Backer et al., 2000; U.S. EPA, 2003a; Jo et al., 1990). Such exposure route–dependent differences in mixture composition may be important in the risk analysis.

Routes of Exposure

Individuals are exposed to various components through different routes. While single chemical assessments must consider the various routes of exposure, analyses of mixtures are complicated by the need to understand exposures through the various routes and the changes in toxicity due to absorption across different barriers. Both mixture and single chemical exposure assessments must contend with the individual variability of exposures through residential drinking water because of the differences in the ways that people use water. Ingestion exposures result from the consumption of water directly and in beverages prepared with water, dermal exposures may arise from washing, bathing, and swimming, and inhalation exposures can result, e.g., from the volatilization of water contaminants from showers, dishwashers, washing machines, and toilets.

The dose–response analysis and the exposure analysis should examine the same routes. If the dose–response analysis is used to predict risks posed by a single exposure route, but multiple routes of exposure are known to exist in the population, the risk posed by a particular mixture might be underestimated. Such is the case with drinking water DBP exposures. Predictive modeling by Wilkes et al. 1992 and others (e.g., Lin and Hoang, 2000) highlighted the significance of DBP exposures through multiple routes. Inhalation and dermal absorption have been shown to be important routes of exposure to volatile DBPs such as chloroform and the haloketones, but of less importance for nonvolatile DBPs such as the HAAs (Xu et al., 2002; Xu and Weisel, 2005a,b). The EPA and others have attempted to address multiroute exposures by relating these absorbed doses through each route to measures of internal dose and by using pharmacokinetic models to estimate the impact of exposures by various routes on internal dose measures (U.S. EPA, 2003a).

Exposure Duration

The dose–response analysis and the exposure analysis should examine the same durations of exposure, where possible. If exposures to a mixture are episodic (e.g., pesticide mixtures in contaminated well waters), but dose–response data are available only for adverse effects due to chronic exposures, the resulting risk analysis could be highly uncertain. Similarly, exposure analyses focusing on reproductive and developmental risks should examine exposures that occur prior to and during pregnancy.

Specificity of Target

Dose–response and exposure analyses should examine chemical concentrations at the same level of biological organization. If the dose–response analysis is based on a potential oral dose (e.g., a daily intake in mg/kg), the development of a sophisticated exposure model that estimates target tissue concentrations might offer little improvement to the risk assessment. Figure 2 illustrates the various dose metrics that are used in chemical risk analyses. When chemical concentrations in target tissues are available in both the exposure and dose–response assessments, there is usually higher confidence in the predicted toxicity. When there are multiple target tissues, the concentration estimates should be obtained for all of the target tissues.

Variability and Uncertainty

Characterization of variability and uncertainty is integral to all steps in risk assessment (Morgan and Henrion, 1990; NRC, 1983; Cullen and Frey, 1999). Although no single approach can be used to contend with all possible sources of variability and uncertainty in risk assessment, uncertainty analyses provide an opportunity

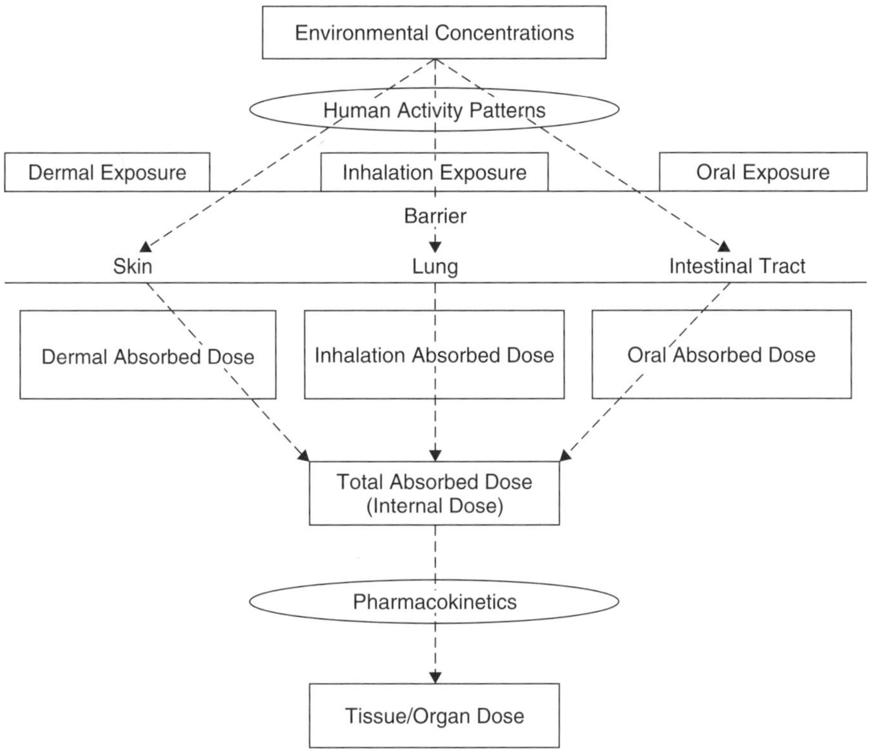

Figure 2. Dose metrics (rectangles) and modifying factors (ovals).

to evaluate the confidence that can be placed in an assessment and to identify and prioritize critical research needed to improve future risk assessments. Understanding the uncertainty inherent in a risk assessment helps risk managers to understand the range of risks they are considering. Unrealistic risk assessments could lead a risk manager to make an incorrect decision. The NRC 1983 encouraged the development of uncertainty analyses in risk assessments. The qualitative identification of sources of uncertainty and variability is currently a routine component of most human health risk analyses, regardless of their level of sophistication (e.g., screening-level analyses include an identification of sources of uncertainty and variability).

Methods to quantify these are best developed for exposure assessment. Variability refers to population heterogeneity, such as tap water consumption rates varying across individuals and over time (e.g., the same person may consume more water during hot arid weather than during cold weather). Studies have been undertaken to increase the understanding of exposure variability (e.g., drinking water intake rates) in key subpopulations (e.g., women of reproductive age, pregnant women, children) (U.S. EPA, 1997). Although tap water consumption can be measured accurately, the true values will vary and no level of precision in measurement techniques will reduce this variability across the population and over time. On the other hand, uncertainty is described as a lack of knowledge about the correct value for a specific exposure measure or estimate. This would include sources of error such as inaccuracies in the analytical methods for quantifying the level of chemicals in water. There is a true value for such a concentration in this example, but the available methods may not be sensitive enough to provide an accurate answer. Of the two, uncertainty is typically more difficult to quantify.

Most exposure assessments quantitatively simulate variability and sometimes uncertainty directly using probabilistic techniques, such as Monte Carlo methods. In this method, exposure analysts carefully examine or develop the distribution of concentrations in media to which people are exposed and distributions of exposure factors [e.g., the *Exposure Factors Handbook* (U.S. EPA, 1997) includes distributions that describe variability in a number of exposure factors across target populations, and information on the uncertainty of these estimates]. These distributions often do not distinguish between uncertainty and variability. In the method, randomly selected values from each distribution are input to the model and the results of each realization or "model run" are stored and subsequently analyzed. The results of a Monte Carlo exposure analysis typically describe both central tendency estimates of exposure to the chemicals comprising the mixture as well as the statistical dispersion around the central tendency estimate (e.g., 5th and 95th percentile values). Other procedures are also employed, depending on the goals of the analysis; these include methods such as Latin hypercube sampling, which disproportionately draws samples from the upper and lower tails of the statistical distribution (Cullen and Frey, 1999). Although beyond the focus of the discussion in this book, other techniques, such as two-dimensional Monte Carlo methods, attempt to distinguish between sources of variability and uncertainty (Morgan and Henrion, 1990; Hoffman and Hammonds, 1994; Simon, 1999).

In summary, quantifying the impacts of various sources of uncertainty in an analysis can be quite complicated. The types of issues typically evaluated in quantitative uncertainty analyses include model uncertainty, parameter uncertainty, and uncertainty in assumptions that are developed as a consequence of missing information. Identification of the sources of uncertainty in an exposure assessment can increase the level of understanding of results and can further help to determine the type of research needed to reduce it in future assessments.

Exposure Assessment in Epidemiological Studies

Exposure assessment in epidemiological studies of drinking water contaminants is very challenging given the costs and analytical limitations associated with measuring small concentrations in drinking water or in biomarkers measuring internal dose. This often results in the use of indirect measures (e.g., place of residence, proximity to source, average ambient pollution measurements) to estimate exposures in epidemiological studies. Studies that utilize environmental measurements often use surrogate measures to represent exposure to complex mixtures (e.g.,

THM4, PCBs measured as Aroclors, polycyclic aromatic hydrocarbons). Many epidemiological studies have relied on routinely collected monitoring data to estimate individual-level exposure based on town-level data assignment (or proximity to sampling location or source). The advantages of using existing environmental monitoring data are offset by the unknown relationship between the surrogate and specific chemicals of concern due to the various mixture constituents typically encountered. An additional limitation is the collinearity between co-occurring exposures (e.g., the THMs), which makes it difficult to isolate independent effects for specific chemicals in epidemiological analyses of drinking water contaminants.

Biomarkers are being used increasingly in epidemiological studies to provide more accurate estimates of internal and biologically effective dose following exposure. Biomarkers can provide data more relevant to critical periods of exposures relative to the health effects being examined, provide dose–response information, and can also help reduce exposure misclassification to specific compounds. Biomarkers allow for assessment of aggregate exposure across routes and can represent chemical mixtures, but this can complicate efforts to attribute exposures to specific media (e.g., water) or routes of exposure (e.g., ingestion). Total urinary arsenic, for example, may be representative of recent exposure to inorganic arsenic, dimethyl arsenic acid, and monomethyl arsenic acid from drinking water, but is also influenced by arsenic found in food. Along with other limitations of biomarkers (e.g., cost, invasiveness of sample collection), this lack of specificity limits the ability to determine the attributable proportion of total arsenic from certain media and results in many studies relying on more indirect measures of exposure. In addition, biomarkers generally provide limited information on which contaminants in the mixture cause the adverse effect being examined in an epidemiological study (Leaderer et al., 1993).

Quantitative Uncertainties of Exposure Assessment in Epidemiological Studies

The use of surrogate measures of chemical mixtures in epidemiological studies results in exposure measurement error in exposure estimates for individual subjects. This can lead to exposure misclassification, which may bias study findings. Nondifferential exposure misclassification (i.e., equal error between cases and controls or the two groups being compared) tends to bias results toward the null, reducing the statistical power of a study to detect associations that may exist. Confounding is another threat to validity in epidemiological studies and can result from unmeasured or poorly measured co-exposure data or from inadequate statistical adjustment. Confounding can bias an effect estimate in either direction, which increases the difficulty in delineating the role of co-exposures on a health outcome due to a drinking water mixture. Another concern for mismeasured co-exposures is the inability to examine effect measure modification adequately. *Effect measure modification*, also referred to as a *statistical interaction* or *departure from additivity*, is defined as heterogeneity of effects across a third variable (Rothman and Greenland, 1998). Examples of effect measure modification are an increased risk of disease found among the elderly or another sensitive population, or an unexpectedly high response (synergistic relationship) detected with co-occurring exposures for a particular health endpoint. The ability to examine effect measure modification is decreased when co-occurring exposures or other covariates are mismeasured. These phenomena can affect the internal and external validity of epidemiological data and add to the uncertainty in epidemiological findings.

Exposure assessment for DBP mixtures has improved since the original focus on estimating exposures for examination of carcinogenic effects. Whereas early DBP studies relied on drinking water source and disinfection type to determine exposure, the next generation of studies utilized existing monitoring data such as THM4 as a surrogate for exposure to DBP mixtures. Despite improvements in exposure assessment through the use of specific DBP concentrations and collection of individual-level water use information to estimate exposure, the potential for exposure misclassification remains in epidemiological studies (Zender et al., 2001; Whitaker et al., 2003; King et al., 2004; Wright et al., 2006). Although the magnitude and direction of exposure misclassification from the use of surrogate measures is difficult to predict, attenuation of relative risk estimates due to nondifferential misclassification could be substantial in studies examining drinking water contaminants (Bachand and Reif, 2000; Waller et al., 2001; Wright and Bateson, 2005). This research reinforces the need to integrate temporal, spatial, and interindividual variability of DBP mixtures into exposure estimates that accurately quantify aggregate exposure. More recent studies of reproductive and developmental outcomes have examined the role of individual water-use activities to better quantify aggregate exposures to DBP mixtures (Waller et al., 1998; Dodds et al., 2004; Savitz et al., 2005). Additional simulation-based research has established uptake factors for swimming, bathing, and showering to better quantify route-specific exposures (Whitaker et al., 2003). These uptake factors have been used to examine the effect of total THM exposures (integrated across the different water use activities) on spontaneous abortions and could be useful for determining total DBP exposure metrics in other epidemiological studies (Savitz et al., 2005).

Previous epidemiological studies of DBPs have used general population estimates of uptake factors in lieu of biomarker data on individual study subjects. Biomarker studies have shown that blood THM measurements correlate well with water concentrations (Miles et al., 2002). Other exposure characterization studies have demonstrated the utility of using exhaled breath measurements for THMs and urinary measurements for haloacetic acids (Kim et al., 1999; Weisel et al., 1999; Bader et al., 2004). Although biomarkers can increase the accuracy of exposure estimates and provide measures of internal dose, additional research is needed to develop more practical biomarkers for quantifying DBP exposures in large epidemiological studies. Research is also needed to better characterize the relationship between DBP surrogates and individual DBPs so that the components of the mixture or specific mixture scenarios needed to produce toxicological effects can be identified.

Drinking water mixtures are being considered increasingly in epidemiological studies, but most continue to be limited by available exposure assessment data (San Sebastian et al., 2002; Hopenhayn et al., 2003; Stewart et al., 2003).

For example, whereas most of the earlier research focuses on binary mixtures, a recent study has examined the risk of non-Hodgkin's lymphoma in relation to select pesticides, total dioxins, total PCBs, total furans, and individual congeners measured in plasma (De Roos et al., 2005). This study examined exposure biomarkers and did not determine the contributions of various exposure media to these plasma concentrations. Despite these improvements in the ability to measure concentrations in potential target tissues, exposure assessment often remains the main limitation in epidemiological studies of drinking water contaminants.

TOXICOLOGICAL CONCEPTS FOR JOINT TOXICITY

Classical Toxicological Interactions

Types of interactions among mixture components that can affect toxicological response to the whole mixture include chemical–chemical reactions and toxicokinetic and toxicodynamic interactions. Using the terminology of the EPA (2000, Table 5.2), the impact of joint exposure on toxicological response can be additive (e.g., dose-additive, where chemicals act as dilutions of each other and cause toxicity by the same mode of action), less than additive (e.g., dietary zinc that inhibits cadmium toxicity through toxicokinetic interactions that reduce the amount of dietary cadmium absorbed), or greater than additive (e.g., enhanced carcinogenicity for co-exposures to asbestos and tobacco smoke). Interaction effects may result from events taking place at many possible loci in the body, including the site of toxic action or during absorption, distribution to tissues, metabolism, excretion, or tissue repair. Any or all of these processes can vary with route of exposure, age, sex, health status, nutritional status, and so on. With the almost infinitely large number of chemical mixtures in the environment, systematic studies relevant to the toxicology of these chemical mixtures using conventional methodologies and approaches are impossible; the development of predictive and alternative toxicology methods is imperative. An evolving approach is the utilization of physiologically based pharmacokinetic/pharmacodynamic (PBPK/PD) modeling, coupled with model-oriented toxicology experiments (Tardif et al., 1997). Tissue dosimetry for key PK and PD processes is achievable with simple mixtures and the known components of complex mixtures.

Biological interaction, especially inhibition, has been known and quantified for many years. Initially, the focus was on enzyme kinetics, specifically those modeled by the Michaelis–Menten (M-M) equation. (Many mechanisms do not generate M-M kinetics, so use of these terms should be limited to those cases that have been demonstrated.) The three classical types of enzyme inhibition are competitive, uncompetitive, and noncompetitive. For each type, parts of the M-M formula for a single chemical are altered to represent the modification caused by the second chemical. For example, competitive enzyme inhibition is depicted by changing the Michaelis constant. The single-chemical M-M formula is

$$v = \frac{V_{\max}[\mathrm{A}]}{K_m + [\mathrm{A}]} \tag{2}$$

where A is the substrate, v is the rate of the reaction, $V_{\max}$ is the limiting rate at very high concentrations of substrate, and K_m is the Michaelis constant (or Michaelis concentration), that concentration where $v = V_{\max}/2$.

Inhibition involving toxic chemicals is exemplified well by altered metabolism. Using the notation more common in toxicology [see, e.g., Krishnan et al. (2002) who discuss competitive inhibition], metabolic inhibition can be considered as the alteration of the metabolic rate (v_1) of one chemical by the presence of a second chemical. One of the most common types of metabolic inhibition is competitive inhibition, where chemical 2 competes directly with chemical 1 for the same binding sites. Similar to the M-M kinetics described above, this competitively inhibited rate depends on the two constants $V_{\max_1}$ and K_{m_1} for chemical 1, the concentrations of chemicals 1 and 2 (C_1 and C_2, respectively) at the site of metabolism, and also on the inhibition constant $K_{i_{21}}$. Similar to the definition of K_m, $K_{i_{21}}$ is the concentration C_2 such that 50% inhibition of the metabolic rate takes place. The equation for competitive inhibition, using C_1 and C_2 to represent the two chemical concentrations, is the same form as formula (2) for a single chemical, but with K_m changed:

$$v_1 = \frac{V_{\max_1} C_1}{K_{m_1}\left(1 + \frac{C_2}{K_{i_{21}}}\right) + C_1} \tag{3}$$

Extending competitive inhibition to more than two chemicals is straightforward:

$$v_1 = \frac{V_{\max_1} C_1}{K_{m_1}\left(1 + \frac{C_2}{K_{i_{21}}} + \frac{C_3}{K_{i_{31}}} + \frac{C_4}{K_{i_{41}}} + \cdots\right) + C_1} \tag{4}$$

The K_m value of chemical 1 is now altered by several inhibition constants, one for each of the interacting chemicals. The advantage of such models is that the tissue concentrations are described, as is the way in which those concentrations are altered by additional chemicals. The disadvantage of relying only on these models for interactions is that joint toxicity often reflects combinations of various types of kinetic interactions with interactions in the toxicodynamics. The risk assessment approach given in the EPA guidelines, discussed below, is at a less detailed level, that of observed toxicity and external dose, so that all interactive kinetics and dynamics processes are reflected in the model.

Dose Dependence

The dose–response relationship for chemical mixtures, as for single chemicals, typically involves a segment at the lower end of the dose range, where no response above background is detected, followed by a region where increased dose results in increased response. These two dose regions are separated by an apparent dose threshold, that point or dose where one unit lower, no response is observed, and one unit higher, a significant increase over background response can be detected.

For biological responses that are "maximal" in nature (examples are mortality or percent of animals with tumors), there is an additional point higher up on the dose–response curve, beyond which any further increase in dose results in less incremental response. Eventually, the dose level reaches a plateau beyond which is no increased response. Thus, without consideration for the presence or absence of nonadditive interactions, the dose–response relationship for any given chemical mixture can be established experimentally in the laboratory and described in terms of intercept, slope, and if applicable, plateau (or maximum).

When taking such an approach to consideration of the joint toxic action of mixtures of DBPs, defined mixtures of known composition can be constructed and evaluated. Useful information may be gained on the toxicity of a particular mixture, but the extension of this knowledge to other untested mixtures that contain the same DBPs, but at differing proportions, requires consideration of the influence of mixing ratio on toxicity, not just total dose.

The influence of mixing ratio on the dose–response relationship is particularly important for DBPs, especially those regulated together as a mixture, the four regulated THMs (chloroform, bromodichloromethane, dibromochloromethane, bromoform), and the five regulated HAAs (chloroacetic acid, dichloroacetic acid, trichloroacetic acid, bromoacetic acid, dibromoacetic acid). These chemicals are present in chemically disinfected water at differing absolute concentrations and different proportions relative to one another, depending on complex interactions between source water characteristics and disinfection treatment scenario characteristics. Studies that examine the influence of mixing ratio on toxic outcome are difficult to conduct in vivo, given the large numbers of treatment groups needed. The multiple-purpose design approach for DBP mixtures (Teuschler et al., 2000; Simmons et al., 2004) included, as one of many goals, examination of the influence of mixing ratio on the hepatotoxicity of the four regulated THMs. To achieve this objective, separate in vivo experiments were conducted for each of the six possible binary combinations of the four THMs. Each experiment included a vehicle control group, three dose levels each of the individual THMs, and the binary combination at two different molar-based mixing ratios. A 1 : 1 mixing ratio was included in all experiments, so that comparisons could be drawn between the various binary combinations. In addition, each experiment contained a mixture whose mixing ratio was based on the relative proportions of the THMs representative of disinfection by chlorination (Krasner et al., 1989). Results of data analyses to date indicate little deviation from dose additivity (Simmons et al., 2001, 2004). Analyses in progress to examine the influence of mixing ratio indicate that mixing-ratio influences on additivity, where present, are seen at the highest dose level, with deviations from dose additivity, in the direction of less than additive toxicity, more commonly observed at the 1 : 1 mixing ratio than at the environmentally relevant mixing ratio.

Evaluation of the influence of mixing ratio on the dose–response relationship is more readily accomplished with in vitro assays than with in vivo assessments. An in vitro Chinese hamster ovary cytotoxicity assay (Plewa et al., 2004) was used to examine the influence of mixing ratio on the dose–response relationship of the five regulated HAAs, the nine common HAAs (the five regulated HAAs plus tribromoacetic acid, bromodichloroacetic acid, dichlorobromoacetic acid, and bromochloracetic acid) and the nine common HAAs plus iodoacetic acid. Seven different mixtures were evaluated, including three mixtures comprised of equimolar mixing ratios of the five, nine, and ten HAAs, two mixtures (a five-HAA, a nine-HAA) based on the relative proportions of HAAs in water disinfected by postchlorination and two mixtures (a five-HAA, a nine-HAA) based on the relative proportions of the HAAs in water disinfected by ozonation with postchlorination (ozone/chlorine). The dose-dependent relationship between total mixture dose and decreased cell density was influenced markedly by mixing ratio. Based on the concentration that resulted in a 50% decrease in cell density ($\%C_{1/2}$ value), the 10-HAA equimolar mixture was the most toxic and the nine-HAA chlorine mixture was the least toxic; there was an ~7.5-fold difference in toxicity between these two mixtures. Considering the environmentally relevant mixing ratios, the five-HAA ozone–chlorine mixture was more toxic than the five-HAA chlorine, the nine-HAA ozone–chlorine, or the nine-HAA chlorine mixtures (Simmons et al., 2006). These data strongly suggest that the relationship between dose and response will depend on the relative proportions (the mixing ratio) of the DBPs present in disinfected water. As differing disinfection scenarios result in both different absolute concentrations and different relative proportions of DBPs, understanding the influence of mixing ratio on dose–response relationships for key health endpoints will improve assessments of the change in risk associated with a change in disinfection scenario and will aid in understanding when two DBP mixtures may be judged sufficiently similar.

The influence of mixing ratio can be visualized by comparing the observed mixture response to that predicted under a simple joint toxicity formula that uses the dose–response data on the individual mixture components. Dose (or concentration) additivity is one of the more common such joint toxicity formulas. Among the many ways to evaluate dose addition is the *isobole*, a common graphical method used for many years to determine whether an interaction, such as supra-additivity, is present (see Figure 3). The isobole for a two-chemical mixture is the graph of the various combinations of doses ($dose_1$, $dose_2$) at which a fixed response is observed (Gessner, 1995). In other words, the x-coordinate is the dose of chemical 1 and the y-coordinate is the dose of chemical 2 such that the joint exposure ($dose_1$, $dose_2$) produces that same fixed response. For example, in Figure 3, the straight-line isobole represents the mixture doses in mg/kg that elicit a 10% response in the test animals. [The equation for that isobole line is the same form as Berenbaum's interaction index (Berenbaum, 1985) and is similar to the hazard index.] If a point, say (2000,50), is on the isobole, the dose combination of 2000 mg/kg of chemical 1 and 50 mg/kg of chemical 2 will yield a 10% response in test animals. Note that this decision tool can be applied to any fixed response measure, whether percent responding in a group, deficit of functionality, severity of a lesion, or any measure of toxicity that is constant along the isobole.

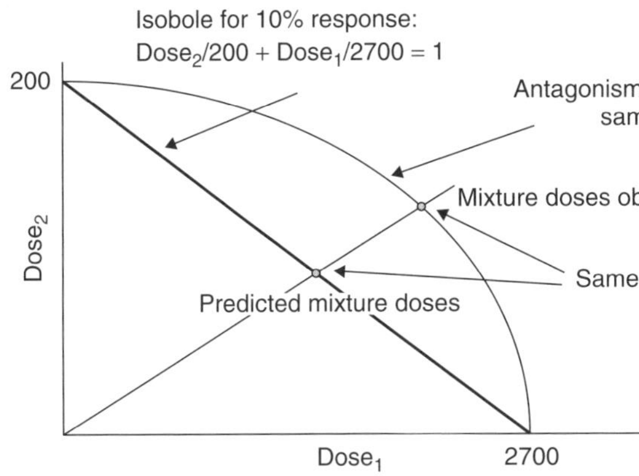

Figure 3. Isobologram.

An important question relates to the dose dependence of nonadditive interactions. A classic isobolographic experiment by Gessner and Cabana (1970) demonstrated a nonadditive interaction between chloral hydrate (a DBP) and ethanol that was present in a portion of the isobol; in the remainder of the isobol, deviation from additivity was not observed. Deviations from additivity were observed with combinations containing higher dose levels of chloral hydrate (and thus lower dose levels of ethanol). In the isobolographic experiment, it is usually difficult to separate the effect(s) of mixing ratio from the influence of dose. Ray design experiments (e.g., the "same component ratio" points in Figure 3), where for any given ray, the mixing ratio is held constant and the total mixture dose is varied, clearly separate the influence of the total dose of the mixture from the influence of the mixing ratio when more than one ray is included in the experimental design. Analyses of the multiple-purpose THM and the in vitro HAA data sets are in progress to determine the influence of dose on deviation from additivity. Results from these analyses appear to indicate that nonadditivity, when detected, is a function of higher dose regions and that dose additivity is a reasonable default at lower dose regions of the dose response curve tested.

CHEMICAL MIXTURES RISK ASSESSMENT METHODS

Procedures

EPA's risk assessment guidance for chemical mixtures (U.S. EPA, 1986, 2000) recommends a three-tiered approach to quantitative health risk assessment of a chemical mixture based on available data. First, if toxicity data on the mixture of concern are available, the risk assessment should be conducted using these preferred data. Second, if toxicity data on a "sufficiently similar" mixture are available, then, provided that the mixture of concern and the proposed surrogate mixture are judged to be similar, the risk assessment for the mixture of concern may be conducted using health effects data on the similar mixture. Third, if toxicity data are available only for the mixture's component chemicals, the risk assessment may be conducted using a simple additivity approach. Finally, if toxicological interactions data are available for the chemicals of concern, such information should be incorporated into the assessment—if not quantitatively, then at least as a qualitative evaluation of the risk.

In the past few years, theory and methods for conducting chemical mixtures health risk assessments have evolved to take advantage of new toxicological and chemical composition data (e.g., ATSDR, 2002; U.S. EPA, 2000, 2003a). Figure 4 is a flowchart that serves as a general guide for selection of the available chemical mixtures risk assessment methods by evaluating data on health effects and chemical composition for the mixture of concern. Once the type of data to be used is determined (i.e., component versus whole mixture data), key questions must be answered to select a risk assessment method. Information on the actual mixture of concern is often weak or nonexistent. The issues for a whole mixture assessment then center around the accuracy of information on chemical composition and the quality of toxicological or epidemiological data for similar mixtures. Important issues for a component-based assessment of a mixture include (1) whether or not the components share a similar toxic mode of action or act independently, (2) whether their dose–response curves are similar, and (3) how to interpret available toxicological interactions data on the components. The final step is the combining of exposure data with the dose–response data and toxicity information, resulting in a risk characterization using one of the methods shown in the bottom row of Figure 4.

Additivity Methods

For exposures to a mixture of several components with doses in a low region where toxicological interactions are not expected to occur, methods based on additivity concepts are often used as default procedures to estimate risk or hazard. Common types of additivity include: *simple similar action* for chemicals that share a common toxic mode of action, and *simple dissimilar action* for chemicals that cause a common health effect by different toxic mode of action (Feron and Groten, 2002). When simple similar action is operational, the total dose of the chemicals in the mixture is of concern. In this case, a risk assessment method based on dose addition may be used where the doses of the chemical components are scaled for relative toxicity and summed for use in estimating risk [e.g., relative potency factors (RPFs), the hazard index (HI)]. When simple dissimilar action is operational, the toxic actions of multiple chemicals in the body may be thought of as independent events that cause the same toxic outcome; thus, the sum of toxic responses, rather than doses, is of concern. The most commonly used method, response addition, involves combining the probabilistic risks of an adverse effect for all the mixture components. These simple models based on additivity are useful for hazard screening and interim estimates of risk, and have been shown to describe satisfactorily the response for certain simple mixtures. These approaches, however, might not be appropriate for all toxic effects or all

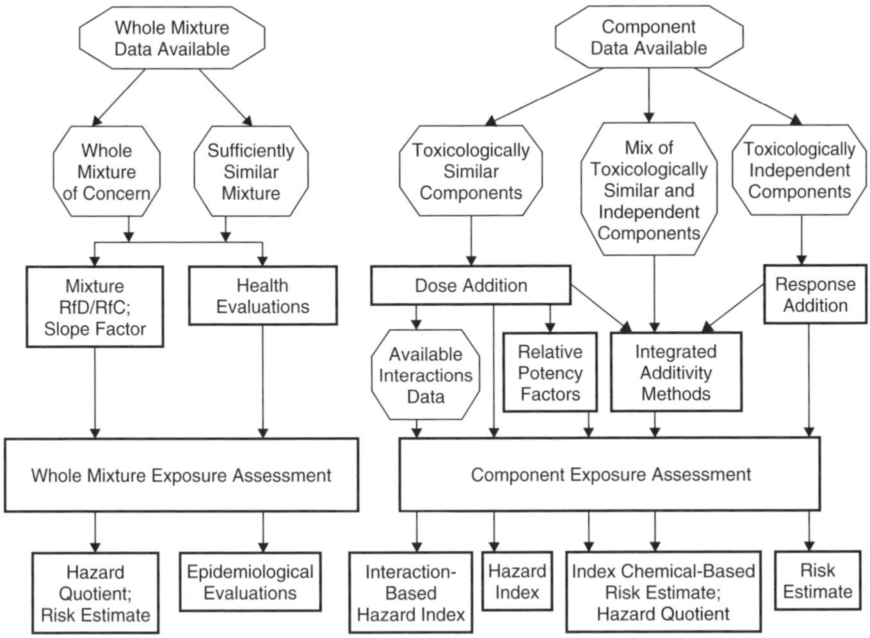

Figure 4. Flowchart of chemical mixtures risk assessment methods.

conditions. For example, the relative toxic potency between two chemicals might change for different types of toxicity or toxicity by various exposure routes (U.S. EPA, 2000).

Several studies have demonstrated that dose (or concentration) addition often predicts reasonably well the toxicities of mixtures composed of a substantial variety of both similar and dissimilar compounds (Pozzani et al., 1959; Smyth et al., 1969, 1970; Murphy, 1980; Ikeda, 1988; Feron et al., 1995), although exceptions have been noted. For example, Feron et al. (1995) discuss studies where even at the same target organ (the nose), differences in mode of action led to other than dose-additive response. Dose-additive models may be an adequate default procedure for chemicals affecting the same target organ but may not be the most biologically plausible approach if the compounds do not have the same toxic mode of action. Consequently, depending on the nature of the risk assessment and the available information on modes of action and patterns of joint action, the most reasonable additivity model should be used.

Choosing among these additivity methods may be difficult in practice because toxicological data on the toxic mode of action for various chemicals often lack depth or are unavailable, making judgments regarding toxic mode of action difficult. As noted in the *Supplementary Guidance for Health Risk Assessment of Chemical Mixtures* (U.S. EPA, 2000), "the evidence for either dose addition or response addition as a good approximation for a mixture risk assessment is not strong, and clearly is not comprehensive in representing the varying types of chemicals considered in environmental risk assessment. Whenever evidence exists that clearly disagrees with both dose and response addition, then alternative approaches should be considered." Despite these reservations, default additivity methods have utility in addressing potential health risks until improved data and more advanced methods are produced. The application of default procedures implicitly acknowledges that exposures and risks typically involve multiple chemicals and allows for the screening of potential mixtures risks when a single-chemical assessment may underestimate health risk.

The similarity of toxicological action can be considered as a continuum of information ranging from a high level of detail regarding the molecular basis of the toxic effect (mechanism of action) to knowledge of key cellular and biochemical events (mode of action), to a low level of knowledge regarding a general toxicological effect at the target organ level. Several publications have identified the mode of action for single chemicals and determined its relevance to human toxicity (e.g., Sonich-Mullin et al., 2001; Meek et al., 2003; IPCS, 2005; Seed et al., 2005). As the level of specific knowledge accumulates, the certainty of mode of action for each mixture component will increase, and the identification of toxicological similarity across a mixture's components will improve, resulting in confidence that a mixtures risk assessment method is being applied appropriately.

A second issue for methods based on dose addition concerns the assumption of similarly shaped dose–response curves for mixture components, although consideration of slope shape is secondary to mode of action analyses. The strictness of this requirement varies for different risk assessment approaches. One classical support for dose addition has been when the dose–response curves for the component chemicals are parallel, where the response data are binary (presence or absence of an effect). Using geometric principles, strictly parallel means "being everywhere equidistant and not intersecting." Binary response data described by parallel logdose probit models is one example where the dose–response curves are, in fact, parallel lines. True parallelism, however, does not hold for most dose–response curves, such as those that are bounded by the probability limits of zero and one, causing the dose–response curves to converge as these boundaries are approached. More generally, dose addition is supported if the dose–response curves are congruent (i.e., similar in shape, such that if one were lifted off the page and placed on top of the other, it would fit perfectly on top). An example of this would be several logistic functions with the same slope parameter but different intercept terms. Congruent curves are required when the risk assessment method uses an index chemical approach, such as the RPF methods. For this approach, the components' doses are scaled for relative potency to equivalent doses of an index chemical and added together to represent the entire mixture dose in index chemical equivalents. Because this total mixture dose is evaluated using the index chemical dose–response curve, the individual component curves need to share a similar shape with the index chemical. In practice, statistical work is needed to show that similarly shaped curves can be supported with empirical evidence and that these curves can be categorized as congruent, at least within the range of doses of interest for the risk assessment. Dose–response curves can be investigated for congruence using measures such as the same functional form, similar slope parameters, or a common maximum response. In contrast, for response addition, there is no need for a restriction on the dose–response curves.

Hazard Index Noncancer assessments of drinking water at Superfund sites have applied dose addition in the form of a hazard index (HI) as a screening-level check for potential health risks (U.S. EPA, 1989a). The HI is calculated as the sum of hazard quotients (HQs) for the chemical components of the mixture. A HQ is typically calculated as the ratio of a chemical's exposure level to its reference level, such that values larger than 1 are of potential concern. Example reference levels used previously in HI formulas include the virtually safe dose, ED_{10}, and official risk values such as the EPA's reference dose (RfD) or reference concentration (RfC). For a group of n chemicals in a mixture and using the RfD as a safe or allowable level, the HI for oral exposure is calculated:

$$\text{HI} = \sum_{i=1}^{n} \frac{E_i}{\text{RfD}i} \qquad (5)$$

where

E_i = is the exposure level of the ith chemical and
RfD_i = is the reference dose of the ith chemical.

A similar index for inhalation exposure uses the RfC for the reference level.

The HI is usually calculated for groups of chemicals whose effects are observed in a common target organ. The HI is interpreted similarly to the HQ (i.e., the more HI exceeds 1, the greater is the concern for toxicity). Note that the HI provides an indication of risk but is not an explicit probabilistic risk estimate. It is used by the EPA as a decision point to assist in determining remediation methods. Other groups and agencies use the HI method but replace the RfD term in the denominator with other toxicological benchmarks.

As an expression of dose addition, the formula for the HI is subject to several important uncertainties (U.S. EPA, 2000). First, the assumption of common toxic mode of action might not apply because only commonality of the target organ is considered. Second, the use of a safe level, such as a lower bound on the toxicity threshold, might not be an accurate measure of toxic potency. Toxicity data, usually from poorly conducted or poorly reported studies, result in a lower safe level because of larger uncertainty factors or use of lower confidence bounds on dose. Finally, the use of RfDs as safe levels may result in an overestimate of the degree of concern because the RfD is based on the chemical's critical effect or the most sensitive effect from available toxicological data. For example, if chemical A affects the liver (its critical effect) and also the kidney, while chemical B just affects the kidney, then a HI for renal toxicity would use the liver RfD for chemical A, a dose lower than what would be a threshold for renal toxicity. The resulting HI would then be larger than it should be if only renal toxicity data were used. A solution is to generate RfD surrogates that are organ specific. These have been termed target organ toxicity doses (TTDs) (derived for each target organ of concern using RfD methodology) and recommended as an improvement over the RfD in target organ–specific HI calculations (Mumtaz et al., 1997; U.S. EPA, 2000).

Relative Potency Factors To evaluate actual risk or estimate safe levels for a group of components, an approach based on dose addition uses RPFs for the dose scaling. The RPF approach relies on both the existence of good toxicological dose–response data for at least one component of the mixture (referred to as the *index chemical*) and scientific judgment as to the toxicity of the other individual compounds in the mixture and of the mixture as a whole. The applicability of RPFs may be limited to certain types of effects or to a specific effect because of data limitations; RPF application may also be limited to a specific route of exposure or exposure duration. [Note that the toxicity equivalence factors (TEFs), used for EPA's dioxin assessment, are a special case of the RPF approach (U.S. EPA, 1989b).] In the RPF approach, the doses of the related compounds are converted into index chemical equivalent doses by scaling the exposure level of each compound by its toxicity relative to the index chemical. This scaling factor or proportionality constant is based on an evaluation of the results of a typically small set of toxicological assays or analyses of the chemical structures. This constant, called the RPF, represents the relative toxicity with respect to the index chemical.

In the RPF approach, the index chemical equivalent doses are computed as the product of the concentration measured for the mixture component and its RPF. These index chemical equivalent doses are summed to express the mixture exposure in terms of an equivalent exposure to the index chemical, as shown for n chemicals:

$$R_m = f_1\left[\sum_{i=1}^{n}(\mathrm{RPF}_i D_i)\right] \tag{6}$$

where

R_m = risk posed by chemical mixture
$f_1(\cdot)$ = dose–response function of index chemical 1
D_i = dose of the ith mixture component ($i = 1, \ldots, n$)
RPF_i = toxicity proportionality constant relative to index chemical for the ith mixture component ($i = 1, \ldots, n$)

Risk can be quantified by comparing the mixture's index chemical equivalent dose to the index chemical's dose–response curve. This estimate of index chemical equivalent exposure should be considered an interim and approximate decision-making tool. The RPFs must be defined as to the scope of toxicological effects that are covered and the degree of similarity in chemical structure and mode of action that can be inferred from a summation of the adjusted exposure levels.

Response Addition *Response addition* operates by first estimating the probabilistic risk of observing a toxic response for each chemical component in the mixture. Then the component risks are summed to estimate total risk from exposure to the mixture, assuming independence of toxic action (i.e., the toxicity of one chemical in the body does not affect the toxicity of another chemical). This can be thought of as an organism receiving two (or more) independent insults to the body, so the risks are added under the statistical law of independent events. For a chemical mixture with $i = 1, \ldots, n$ components, each with a risk (r_i), the mixture risk (R_{m}) is equal to 1 minus the product of the risks of not seeing an effect for the mixture components:

$$R_m = 1 - \prod_{i=1}^{n}(1 - r_i) \tag{7}$$

When the individual risk estimates (the r_i) are small, this equation can be simplified to equal the sum of the r_i: thus the name *response addition*. Toxic effects that are described by the proportion of exposed animals showing toxicity are often determined for mixtures using response addition. For example, the probabilistic risk of cancer in a given dose group is typically estimated by the proportion of responders in that group. At Superfund site assessments, total cancer risk is calculated by summing the individual cancer risks for the carcinogens in the drinking water mixture (U.S. EPA, 1989a). Risks are appropriately aggregated for cancers across various target organs because the result is interpreted as the risk of any cancer, and the cancers from each chemical component are considered to be independent events in the body.

Interaction Methods

A common concern for toxicologists and risk assessors evaluating chemical mixtures is the potential for toxicological interactions to occur from co-exposures. The EPA (U.S. EPA, 2000) defines toxicological interactions as responses that deviate from those expected under some definition of additivity. Although, as shown in Table 2, many terms are used to describe such action (e.g., inhibition, potentiation, masking), the most commonly used terms are synergism (i.e., effects are greater than additive) and antagonism (i.e., effects are less than additive). It must be emphasized that in risk characterization antagonism is not used the same way as inhibition. Antagonism implies only a lesser joint response than predicted from dose addition. The presence of antagonism does not justify lowering of risk estimates of an affected chemical, say by increasing its reference dose. An antagonist is also toxic. In contrast, the inhibitor chemical is not toxic by itself, but does reduce the toxicity of the second chemical. Only for inhibition could risk levels (e.g., an RfD or HQ) for the second chemical be adjusted because of reduced toxicity.

The analysis of toxicological interactions is complicated by the dose dependence of interactions, many chemical and dose-level combinations, and lack of toxicity data for higher-order combinations beyond binary mixtures. Thus, recent approaches for evaluating interactions have been proposed that use available binary toxicity data to adjust the value of the additive HI.

An excellent data source for several high-profile environmental contaminants are the six interaction profiles published by the Agency for Toxic Substances and Disease Registry (ATSDR) that evaluate health effects, dose–response, and toxicological interactions data for simple mixtures (Pohl et al., 2003). A qualitative binary weight of evidence is developed for each chemical pair, considering mechanistic evidence, strength of interactions data, influence of exposure duration, and route and sequence of exposure. When HI ≥ 0.1 can be calculated for these groups of chemicals during a site assessment, ATSDR recommends that a qualitative weight of evidence be applied to assess potential consequences of toxic joint action (Pohl et al., 2003). Based on similar evaluations of interactions data, the EPA recommends that a quantitative method, the interaction-based HI, be applied to numerically modify the HI using binary interactions data (U.S. EPA, 2000).

Interaction-Based Hazard Index In the method described in this section, the key assumption is that interactions in a mixture can be adequately represented as departures from dose addition (Hertzberg et al., 1999). The method follows an obvious approach: to begin with the dose-additive HI and modify its calculation to reflect the interaction results, using plausible assumptions to fill in the data gaps. A secondary assumption is that the influence of all the toxicological interactions in the mixture can be adequately approximated by some function of the pairwise interactions (e.g., the effect of chemical 2 on the toxicity of chemical 1). The interaction-based HI ($\mathrm{HI_{INT}}$), based on early work by Mumtaz and Durkin (1992), was developed under the following constraints:

- The method was to use readily available data, or at least information that can be feasibly obtained.
- The method would include several steps, each of which could be modified or replaced when more data or biological models became available.
- The method should be plausible, either supported by some empirical cases or supported by consensus among practicing mixtures toxicologists and risk assessors.

For n chemicals, the formula for the $\mathrm{HI_{INT}}$ is

$$\mathrm{HI_{INT}} = \sum_{j=1}^{n} \mathrm{HQ}_j \left(\sum_{k \neq j}^{n} f_{jk} M_{jk}^{B_{jk} g_{jk}} \right) \tag{8}$$

where

HI_{INT} = HI modified by binary interactions data
HQ_j = hazard quotient for chemical j (e.g., daily intake/acceptable level)
f_{jk} = fraction of all interacting chemicals: toxic hazard of the kth chemical relative to the total hazard from all chemicals potentially interacting with chemical j (thus, k cannot equal j); to calculate, the formula is

$$f_{jk} = \frac{HQ_k}{\left(\sum_{i=1}^{n} HQ_i\right) - HQ_j} \tag{9}$$

M_{jk} = interaction magnitude: the influence of chemical k on the toxicity of chemical j; to calculate, estimate from binary data (see example calculation below) or use default value = 5
B_{jk} = weight of evidence (WOE) score: assessment of the evidence that chemical k will influence the toxicity of chemical j (see Table 4)
g_{jk} = equitoxicity measures: degree to which chemicals k and j are present in equally toxic amounts; to calculate, the formula is

$$g_{jk} = \frac{\sqrt{HQ_j HQ_k}}{(HQ_j + HQ_k)/2} \tag{10}$$

In equation (4), the first summation shown is the additive HI, and the second summation shown is the modification of each term for interactions. Calculations for the first summation and variables f_{jk} and g_{jk} are fairly simple because these terms rely on developing a HQ for each chemical component and then combining them as shown. Two evaluations are required for the WOE score (B_{jk}), which is based on the available literature on toxicological interactions for chemicals j and k. For each pair of component chemicals in the mixture, one WOE is determined for the influence of chemical j on the toxicity of chemical k, and one for the reverse. This qualitative judgment is then changed into a numerical score as shown in Table 4 [see U.S. EPA (2000) for additional details].

TABLE 4. EPA Weight of Evidence for Pairwise Interactions

		Weight of Evidence Score (B_{jk})	
Category	Description of Studies	Greater Than Additive	Less Than Additive
I	Directly relevant to humans	1.0	−1.0
II	Animal studies, but relevant	0.75	−0.5
III	Plausible evidence; relevant?	0.5	0.0
IV	Additivity demonstrated or accepted because of poor data	0.0	0.0

The term M_{jk} represents the maximum interaction effect that chemical j can have on the toxicity of chemical k. As with the WOE score, the interaction magnitude need not be symmetric (i.e., the magnitude of interactive influence of chemical k on the toxicity of chemical j may be different than the corresponding magnitude of chemical j on the toxicity of chemical k). When binary mixture toxicological data are available, a simple calculation can be used to determine a value for M_{jk} from effective doses (e.g., from ED_{10}s that cause a 10% effect in the test animals). The EPA uses the proportional change in effective dose as the interaction magnitude. For example, if the ED_{10} is predicted to be 20 mg/kg·day for chemicals j and k together under an assumption of dose addition but the observed experimental value for j and k together is equal to 5 mg/kg·day, the interaction magnitude would equal $20/5 = 4$. In the absence of data, the EPA (U.S. EPA, 2000) recommends a default value of 5, based on observations of interactions data from the literature.

This method for modifying the HI is based on commonly discussed principles of toxicological interactions. The algorithm, however, does not attempt to model toxicological interaction mechanisms directly. While the interaction magnitude is based on directly observed toxicity measures, the adjustment for different component doses is simple and heuristic. Instead of estimating joint toxicity or risk, the interaction HI method should be regarded as a method for modeling "concern" for toxicological interactions, which reflects issues of magnitude as well as likelihood. In this respect, the interaction HI is interpreted in the same way as is the common additive HI, and the approach corresponds more closely with the current use of uncertainty factors in the risk assessment of single chemicals than with an attempt to model interactions biologically. As more interaction studies are completed and more interaction mechanisms and modes of interaction are understood, this method will be refined further.

Complex Mixture Methods

The chemical composition of a complex mixture (e.g., PCBs, drinking water disinfection by-products) and the variability in chemical composition represent a tremendous challenge for estimating dose–response and characterizing risk. Methods for evaluating a complex mixture include direct toxicological testing or epidemiological evaluation of the environmental mixture, testing of its concentrate or of a laboratory concoction that mimics its formation, using surrogate information on a sufficiently similar mixture, or evaluating single components or simple subsets of the complex mixture. In all cases, chemical characterization and toxicological assessment play an important role in conducting the risk assessment.

Whole Mixture Reference Values If epidemiological or toxicological data are available on the whole mixture of concern, a toxicity value, such as an RfD, RfC, or cancer slope factor (upper limit on lifetime cancer risk per unit exposure), can be determined for the complex mixture by treating it as if it were a single substance. When such a toxicity value is derived, a description of the relevance of that value to environmental mixtures is needed. For example, the EPA's integrated risk information system (IRIS) contains RfDs for the whole mixtures Aroclor 1016 and Aroclor 1254 and an inhalation cancer slope factor for coke oven emissions (U.S. EPA, 2006c). After a review of the spectrum of effects found in available studies on Aroclor 1016, a critical effect of reduced birth weights in a monkey reproductive bioassay (Barsotti and van Miller, 1984) was selected to establish an RfD of 7×10^{-5} mg/kg·day. This assessment was supported by a series of reports that evaluated perinatal toxicity and long-term neurobehavioral effects of Aroclor 1016 in the same groups of infant monkeys (Levin et al., 1988; Schantz et al., 1989, 1991). An uncertainty factor (UF) of 100 was used: a threefold factor is applied to account for sensitive persons; a threefold factor for extrapolation from rhesus monkeys to humans; a threefold factor for limitations in the database, particularly relative to the issue of male reproductive effects; and a threefold factor for extrapolation from a subchronic exposure to a chronic RfD (where "three" indicates one-half of a log10 unit, so that $3 \times 3 = 10$). The NOAEL was selected and UFs applied as if Aroclor 1016 were a single chemical. However, uncertainty was expressed regarding use of the RfD given that the substance is a mixture.

> The database for PCBs in general is extensive. Studies examining Aroclor 1016 have been performed in rhesus monkeys, mice, rats, and mink. However, despite the extensive amount of data available, only medium confidence can be placed in the database at this time. It is acknowledged that mixtures of PCBs found in the environment do not match the pattern of congeners found in Aroclor 1016, therefore the RfD is only given medium confidence. For those particular environmental applications where it is known that Aroclor 1016 is the only form of PCB contamination, use of this RfD may rate high confidence. For all other applications only medium confidence can be given.

The procedure for deriving such values (U.S. EPA, 2000) is as follows:

1. Collect and evaluate data on the whole mixture or similar mixture. Epidemiological data are preferred with supporting toxicology data, but a good toxicology study is also acceptable in the absence of human data.
2. Evaluate stability within a mixture. Consider the variability in the chemical components and their relative proportions.
3. Assess sufficient similarity between mixtures (if applicable to the risk assessment). Assess the similarity across the mixtures' components and their relative proportions. Consider the toxicological similarity of the mixtures or of their common components. Note whether the mixtures are emitted from common sources or produced by similar industrial processes.
4. Conduct dose–response assessment. Use the same procedures as for single chemical toxicity values on IRIS (U.S. EPA, 2006c) (e.g., reference doses, reference concentrations, cancer slope factors).

5. Characterize uncertainties. Discuss the relevance of health effects data to the environmental exposures. Consider the stability of the mixture and environmental fate.

Sufficiently Similar Mixtures In the development of chemical mixtures risk assessment methods, *sufficient similarity* is a key concept for evaluation of a complex mixture. It is applied when inadequate toxicity data are available directly on a mixture of concern, but toxicity data can be acquired on a mixture composed of similar chemical components in similar proportions. If the two mixtures are judged to be sufficiently similar, the toxicity data for the latter can be used as surrogate data in conducting a quantitative risk assessment for the mixture of concern. The EPA (U.S. EPA, 2000) proposed this general concept for the evaluation of complex mixtures, defining *sufficient similarity* as two mixtures close in chemical composition where there are small differences in their components and in the proportions of their components. Specific guidance, however, is needed regarding statistical and toxicological criteria for determining with confidence whether a pair of chemical mixtures are truly sufficiently similar.

Sufficient similarity assumes that the toxicological consequences of exposure to two mixtures (i.e., a mixture of concern and a mixture for which toxicological data are available) are identical or indistinguishable from one another. In practice, ad hoc procedures require some degree of chemical similarity or at least an understanding of how chemical differences between the mixtures affect their toxicology. Professional judgments of sufficient chemical similarity are based on an evaluation of the composition of two mixtures by noting if significant and systematic differences exist in the quantities of components and their proportions in the mixture. In addition, if information exists on differences in environmental fate, uptake and pharmacokinetics, bioavailability, or toxicological effects for either of these mixtures or their components, it should be considered in the determination of sufficient similarity.

The sufficient similarity of drinking water DBPs from various source waters and treatment types is an important consideration for health risk assessment of this environmental mixture. As discussed above, the types and quantities of DBPs present in drinking water vary depending on factors such as the natural organic matter present in the raw water, the oxidizing agent used, the temperature of the water, the pH of the water, and the time since the water was treated. In addition, mixture components are subject to physical, chemical, and biological processes in the distribution system, which change the original mixture's composition; thus, communities are exposed to different DBP mixtures, due to the receipt of drinking water from various source waters or different chemical disinfectants. Significant differences in the composition of DBP mixtures may change the human health risk associated with those mixtures.

If criteria for the sufficient similarity of DBP mixtures were established, they could be used to evaluate health risks across community systems where at least one system has been the subject of an epidemiological study; the question to be answered is whether the results of the study would be generalizable to the other community. Further, the criteria could be applied to evaluate whether toxicological results from a complex DBP mixture study would be applicable to a given type of source water and disinfectant. If the water in a community were very different from those that had already been studied, this would provide an impetus for further testing of the unique water to characterize the constituents for toxicity. Otherwise, the toxicity of the DBPs in the treated water could be judged to be sufficiently similar to one or more treated waters that had already been evaluated.

NEW APPROACHES FOR ASSESSING RISK FROM EXPOSURE TO DRINKING WATER MIXTURES

New approaches continue to be developed to improve drinking water mixture health risk assessments. Technical issues addressed most recently include linking external exposure models that incorporate human activity patterns and residential characteristics with PBPK models to produce internal dose estimates of multiple route exposures (U.S. EPA, 2003a; Teuschler et al., 2004), improvements in dose–response modeling for chemical mixtures with mixed modes of action using the RPF approach (U.S. EPA, 2003b), toxicological and chemical characterization of complex DBP mixtures (Simmons et al., 2002), and evaluation of potential exposure misclassification in epidemiological studies (Wright and Bateson, 2005; Wright et al., 2006). Emphasis has been on evaluating low-dose, environmentally relevant exposures and mixing ratios for the chemical mixtures of interest and investigating the potential contributions to risk posed by unidentified material in the mixture. This section highlights an example application of a new approach using DBP mixtures.

Integrating Dose and Response for Mixed Modes of Action and Multiple-Route Exposures to Mixtures

The EPA has conducted research to examine real-world aggregate drinking water exposures and to develop a method for deriving cumulative risk estimates for multiple-route exposures to DBP mixtures (U.S. EPA, 2003a; Teuschler et al., 2004). Application of this approach advances the science of mixtures risk assessment, offering a method for combining information when a mixture contains mixed toxic modes of action and for linking external exposure modeling results with estimates of internal dose. Three specific results from this research include:

1. External exposure modeling was conducted and linked with physiologically based pharmacokinetic (PBPK) modeling to produce internal dose measures from dermal, oral, and inhalation exposures for use in risk assessment. This new approach provided human exposure distributions for integrating health risks for multiroute DBP exposures.

2. These internal doses were estimated for an adult female and an adult male, each of reproductive age, and for a child (age 6), accounting for human activity patterns that affect contact time with drinking water. Exposure distributions were estimated for 13 major DBPs, including most of the DBPs currently regulated by the EPA.
3. A new risk assessment method, the cumulative relative potency factors (CRPF) approach, integrates the principles of dose addition and response addition to produce multiple-route chemical mixture risk estimates using internal doses. The CRPF approach evaluates human health risks using total internal doses and single-chemical oral toxicological dose–response data based on knowledge or assumptions regarding toxic modes of action. The CRPF approach is a new method that can be generalized for the evaluation of other drinking water mixtures, whose components act by more than one toxic mode of action.

Various dose metrics can be used for DBP mixtures depending on physicochemical characteristics and available data. Internal dose estimates were generated beginning with environmental concentrations and linking external exposure modeling outputs with pharmacokinetic models. Models selected for this effort were the Total Exposure Model (TEM) (Wilkes, 1998) and the Exposure-Related Dose Estimating Model (ERDEM) (U.S. EPA, 2002a) software. Combining these two models into one analysis provided the ability to estimate organ and target tissue doses (estimated using ERDEM) as a function of external measures (estimated using TEM), such as human behaviors, environmental factors, and other exposure-related parameters. Figure 5 illustrates the flow of information in and out of the two models. In this research, TEM provided 24-hour time history simulations using data on chemical properties, human activity patterns, human drinking water intake, and building characteristics. These time histories were joined with pharmacokinetic data and became input to the PBPK model, ERDEM. Multiroute exposure internal doses were then estimated.

As shown in Table 5, results from TEM show the importance for risk assessors to know the percentage of exposure route contribution for these 13 dominant DBPs. From an exposure perspective, the inhalation route is most important for the trihalomethanes, and the oral route of exposure is most important for the haloacetic acids and acetonitrile compounds. Using results from ERDEM, Table 6 illustrates the types of results the models produce, in this case the distribution of 48-hour absorbed doses of the DBP, bromodichloromethane, for an adult female for the kidney, ovaries, liver, blood, and total absorbed dose. These estimates can then be used in the CRPF approach to conduct a multiple-route mixtures risk assessment based on internal doses.

Strategy of the CRPF Approach

The CRPF approach is an emerging method that combines the principles of dose addition and response addition to assess mixtures risk for multiple-route

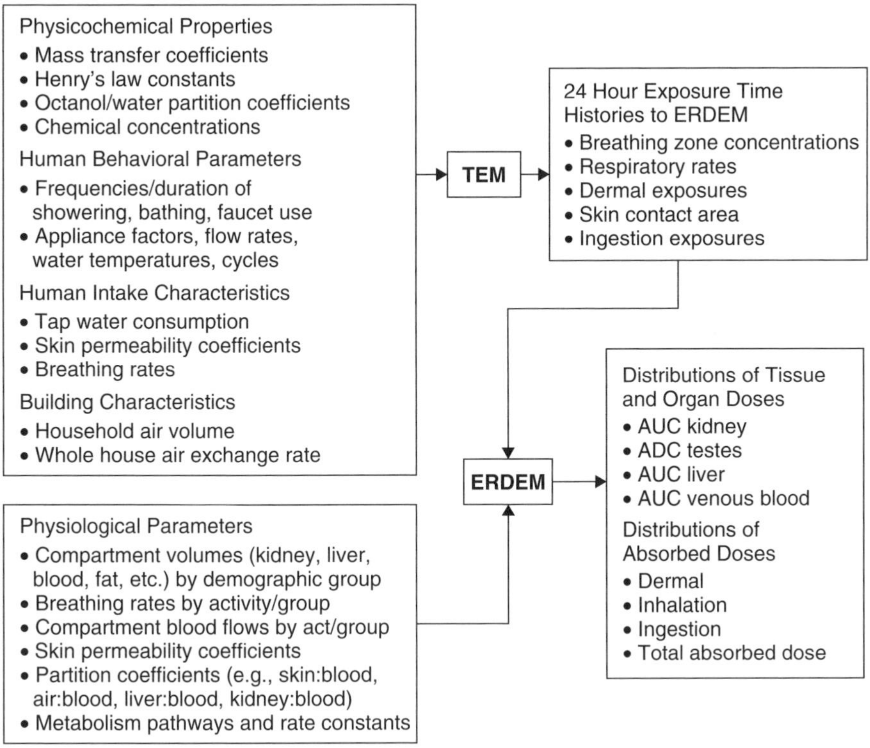

Figure 5. Flow of information between TEM and ERDEM.

exposures. The CRPF approach can be used for most health effects. The basic strategy for the CRPF approach is as follows:

1. Group chemicals into common toxic mode of action subclasses.
2. Conduct dose–response modeling of toxicology data for each individual chemical; develop relative potency factors (RPFs) within each subclass (e.g., using ratios of effective doses or slope parameters); estimate index chemical equivalent doses (ICEDs) for DBPs within each subclass.
3. Estimate each subclass risk using the RPF method (assumes the same toxic mode of action within each subclass.
4. Estimate the total mixture risk using response addition (assumes statistical and toxicological independence between subclasses).

Figure 6 illustrates how animal toxicology data are used to build a human dose–response model and generate RPFs, and then integrated with the exposure modeling results to estimate individual chemical risks and mixture risks based on the index chemical. Because animal dose–response data are typically available for only a single exposure route (usually oral), practical implementation of

TABLE 5. TEM Exposure Route Contributions for 13 of the Major DBPs for an Adult Female

	Contribution % to Total by Route			
Chemical	Dermal	Ingestion	Inhalation	Dose (mg/day)
Trihalomethanes				
Chloroform ($CHCl_3$)	9	9	81	1.52×10
Bromodichloromethane (BDCM)	4	13	83	4.13×10^{-1}
Dibromochloromethane (DBCM)	5	15	80	2.56×10^{-1}
Bromoform ($CHBr_3$)	7	27	67	1.12×10^{-1}
Haloacetic acids				
Chloroacetic acid	3	97	0	1.13×10^{-2}
Dichloroacetic acid (DCA)	0	100	0	7.03×10^{-2}
Trichloroacetic acid (TCA)	0	100	0	7.48×10^{-2}
Bromoacetic acid	3	97	0	2.22×10^{-2}
Dibromoacetic acid	3	97	0	9.54×10^{-3}
Bromochloroacetic acid	3	97	0	2.02×10^{-2}
Acetonitrile compounds				
Dichloroacetonitrile	2	95	2	4.53×10^{-3}
Trichloroacetonitrile	3	96	1	3.13×10^{-4}
Dibromoacetonitrile	3	97	0	1.8×10^{-3}

TABLE 6. ERDEM 48-Hour Absorbed Doses (mg/kg) of Bromodichloromethane for an Adult Female

	Percentile				
Demographic Group[a]	5th	10th	50th	90th	95th
AUC kidney (mg/L·hr)	5.36×10^{-5}	0.00013	0.00103	0.00424	0.00723
AUC ovaries (mg/L·hr)	7.39×10^{-5}	0.00018	0.00142	0.00584	0.00994
Total absorbed dose (mg)	0.0206	0.0328	0.177	0.703	1.22
AUC liver (mg/L·hr)	3.33×10^{-5}	4.41×10^{-5}	0.000217	0.000794	0.00135
AUC venous blood (mg/L·hr)	4.85×10^{-5}	0.000107	0.000778	0.00319	0.00539

[a]AUC, area under the curve generated by the PBPK model.

the CRPF approach for multiple exposure routes requires route extrapolations. Limited inhalation or dermal toxicity data are available for the DBPs. Thus, the CRPF approach has been implemented using dose–response information on the oral route only. For each chemical, two adjustments are made to the animal toxicology data in order to extrapolate to human internal dose: (1) administered animal doses are adjusted to internal animal doses using bioavailability factors, and (2) internal animal doses are then adjusted to internal human equivalent doses

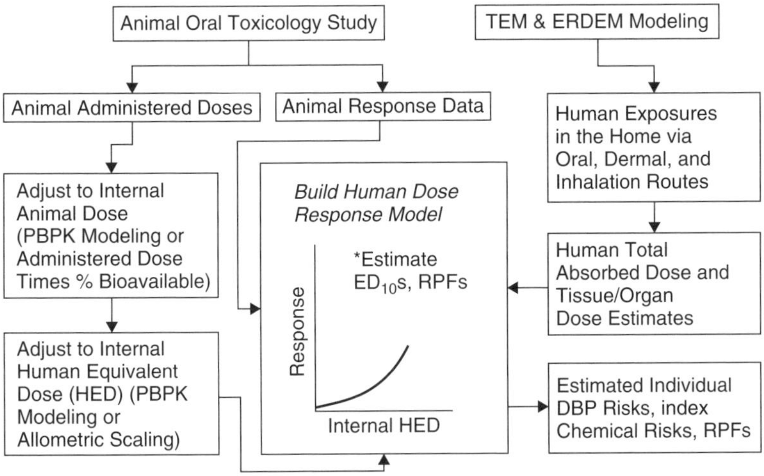

Figure 6. Dose–response modeling using single-chemical oral toxicology data for each DBP; assumes no portal-of-entry effects. EDx is the effective dose causing a response in $x\%$ of the population of interest.

using allometric scaling. An alternative approach to making these adjustments is to estimate the internal human tissue doses using PBPK modeling. Using these data and assuming that the animal and human responses are the same, based on human internal doses, dose–response curves are developed for each chemical.

Figure 7 and Table 7 illustrate how the CRPF approach may be used to estimate risk from exposure to a mixture of two subclasses, genotoxic and nongenotoxic

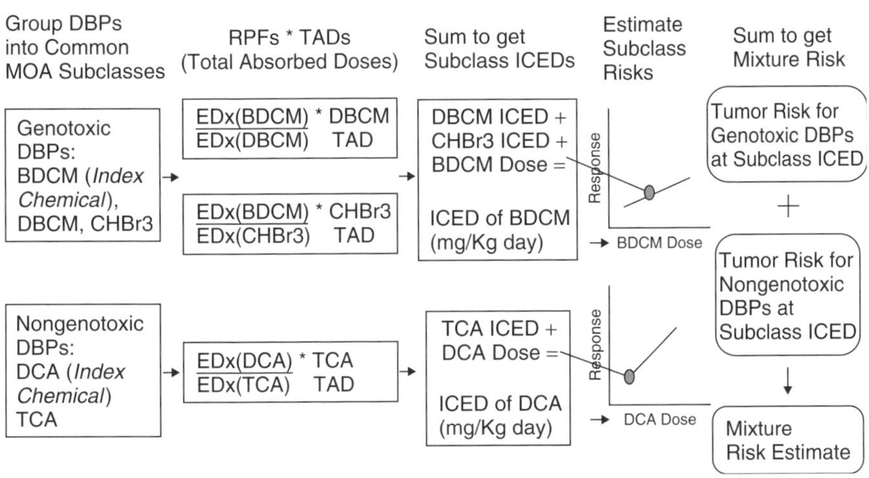

Figure 7. Schematic of CRPF to illustrate DBP mixture cancer risk.

TABLE 7. Illustration of CRPF Approach for Central Tendency Cancer Risk Estimates (Assumes 100% Bioavailability)

DBP	95% Upper Bound Slope Factor (SF)[a] (MLE Slope Factor) (mg/kg·day)$^{-1}$	RPF (SF_i/SF_1)[b]	Total Absorbed Dose for 70-kg Male		ICED (mg/kg·day)		Subclass Risk: MLE Slope Factor × Subclass ICED
			50% mg/day	50% mg/kg·day	Component	Subclass	
Genotoxic subclass							
BDCM	6.20×10^{-2} (5.70×10^{-3})	1.00	8.43×10^{-2}	1.20×10^{-3}	1.20×10^{-3}	2.32×10^{-3}	1.32×10^{-5}
DBCM	8.40×10^{-2}	1.35	5.49×10^{-2}	7.84×10^{-4}	1.06×10^{-3}		
$CHBr_3$	7.90×10^{-3}	0.13	3.00×10^{-2}	4.29×10^{-4}	5.46×10^{-5}		
Nongenotoxic subclass							
DCA	4.80×10^{-2} (1.50×10^{-2})	1.00	3.14×10^{-2}	4.49×10^{-4}	4.49×10^{-4}	12.84×10^{-4}	1.93×10^{-5}
TCA	8.40×10^{-2}	1.75	3.34×10^{-2}	4.77×10^{-4}	8.35×10^{-4}		
$CHCl_3$	RfD = 0.01	—	3.02×10^{-1}	4.31×10^{-3}	—		
					Total mixture average cancer risk:		3.25×10^{-5}

[a]Slope factors for BDCM, DBCM, DCA, and $CHBr_3$ are from IRIS (U.S. EPA, 2006). MLE slope factor for BDCM is from the same dose–response model as the 95% upper bound slope factor. The slope factor for TCA, derived from data in Bull and Kopfler (1991), is included here to illustrate the CRPF approach only and does not represent an EPA peer-reviewed, endorsed value. This illustration assumes that exposures below the $CHCl_3$ RfD of 0.01 mg/kg·day do not contribute to carcinogenicity.

[b]SF_1 slope factor for index chemical; SF_i slope factor for ith chemical in the subclass.

DBPs. The CRPF approach uses information on mode of action to assign chemicals to common mode of action subclasses. These subclasses differ with respect to mode of action, but the toxicological endpoint (or outcome) is the same. Risk for each subclass is then estimated by applying an RPF analysis. For each subclass, an index chemical is selected that has high-quality dose–response data that acts (or is judged to act) through the same mode of action as the other members of the subclass for the effect and route of concern. Index chemical equivalent doses (ICEDs) are calculated using the RPF approach (U.S. EPA, 2000).

The ICED has the same mathematical interpretation as the toxicity equivalents (TEQs), such as those used currently with dioxinlike chemicals. The 2,3,7,8-TCDD TEQ refers to the quantification of dioxin concentrations based on the congeners' equivalent 2,3,7,8-TCDD toxicity (U.S. EPA, 1989b), calculated using the dioxin TEFs. ICED is applied when the mixture components use RPFs. The TEQ is then a special case of the ICED.

The risk posed by the subclass can be estimated using the dose–response information for the index chemical (e.g., find the value of ICED on the dose axis of the index chemical's dose–response curve and read off the corresponding risk). For each subclass, the RPF approach uses dose addition to estimate risk for the toxicologic outcome common across the subclasses. However, because each subclass differs in mode of action, their risks are independent of each other (i.e., the toxicity caused by one subclass does not influence the toxicity caused by the other subclass). This condition meets the criteria required to apply response addition; the subclass risk estimates are added to yield a risk estimate for the total DBP mixture. In Table 7, central tendency estimates of cancer risk are calculated for a 70-kg adult male by combining dose–response information with the total absorbed dose estimates from the exposure model, TEM (U.S. EPA, 2003a). The 50th percentile doses (mg/day) were converted to mg/kg·day doses (dividing by 70 kg) and then multiplied by the RPF for each DBP to obtain component ICEDs. The sum of the component ICEDs forms each subclass ICED. The product of the subclass ICEDs and the maximum likelihood estimate (MLE) of the slope factor for the subclass index chemical provides a central tendency estimate of cancer risk for that subclass (see footnote *a* of Table 7). The subclass risks are then added to obtain the final total central tendency estimate of cancer risk for the whole mixture. The final step of such an effort is to fully characterize the uncertainties that exist as a product of the analysis. This risk characterization should include uncertainties in the CRPF process, including discussions regarding subclass development, choice of index chemical, and the strength of the exposure assessment.

In summary, in this research, realistic multiple-route exposure estimates are developed for specific populations that account for human activity patterns and water-use patterns, and physicochemical properties of DBPs. The scientific basis for evaluating DBP mixture health risks is improved beyond methods based on individual DBP concentrations and single-route risk analyses. The CRPF method provides a way to assess mixtures whose components cause a common health effect through different toxic modes of action. This research established the feasibility of analyzing multiroute exposures to DBP mixtures using a CRPF approach. Important data gaps include:

- Careful treatment of mode of action and common mode of action subclasses for the major DBPs for developmental and reproductive effects and cancer.
- Development of physicochemical properties and external exposure factors for several major DBPs.
- Corresponding methods for combining independent chemical groups when response is not probability (risk) but some other metric, such as a measured biological effect.
- PBPK modeling of the pregnant uterus and DBP doses to the fetus.
- Improved uncertainty and sensitivity analyses.

CONCLUSIONS

Drinking water contains hundreds of chemicals in combinations that vary with geographic locations, source waters, disinfection processes, and surrounding land uses and industries. Most of these chemicals are present in small concentrations, yet toxicological and epidemiological data exist that raise concerns for potential human health impacts from drinking water exposures. Thus, careful analyses of drinking water mixture exposures, dose response, and toxicity are important to evaluate. Generalizations of drinking water safety or toxic risk are then difficult, with the best conclusions drawn from risk assessments of the joint toxicity and aggregate exposure of the specific finished drinking water of concern.

In this chapter the importance of developing credible information on drinking water exposures, health effects information, risk characterization, and uncertainty analysis has been discussed. Examination of drinking water exposures needs to include factors such as fate and transport, chemical degradation, route and pathway of exposures, and human activity patterns and exposure classification. The importance of the exposure route (i.e., dermal absorption, ingestion, and inhalation) has also been demonstrated, showing that different groups of chemicals may dominate certain routes of exposure. It has also been shown that mixture toxicity and risk depends not only on the total mixture dose, but also on the component proportions, or mixing ratios, and that the potential for toxic joint action needs to be considered.

Several chemical mixtures risk assessment methods are in use by risk assessors, as shown in Figure 4, and newer, improved approaches continue to be developed. For example, it is desirable to develop sufficient similarity concepts into a practical tool for assessing the complex DBP mixture that could then be applied for comparing exposure scenarios in locations where epidemiological studies have been conducted with scenarios in other communities. The cumulative relative potency factor approach is one useful way to address the health risk from multichemical, multiroute exposures that can highlight the chemicals and routes of most concern. The approach for combining independent groups of similar chemicals (dose additive within each group) is one more step toward an evaluation of the actual mixture of concern using information on all of the mixture component chemicals. Many of these approaches are new and need evaluation with a variety of environmental mixtures as well as further exploration of their statistical properties to improve experimental design and data evaluation. However, great progress has occurred in data collection and risk assessment approaches for evaluating drinking water chemical mixtures to ensure confidence in the safety of drinking water for the population.

Disclaimer

The views expressed in this chapter are those of the authors and do not necessarily reflect the views and policies of the U.S. Environmental Protection Agency. Mention of trade names or commercial products does not constitute endorsement or recommendation for use.

REFERENCES

ATSDR (Agency for Toxic Substances and Disease Registry). 2002. *Guidance Manual for the Assessment of Joint Toxic Action of Chemical Mixtures*. U.S. Department of Health and Human Services, Atlanta, GA.

Bachand AM, Reif JS. 2000. Effects of exposure misclassification on studies of reproductive outcomes. *Proceedings of the International Workshop for Disinfection By-products in Epidemiologic Studies*, Ottawa, Ontaris, Canada, May 7–10.

Backer LC, Ashley DL, Bonin MA, Cardinali FL, Kieszak SM, Wooten JV. 2000. Household exposures to drinking water disinfection by-products: whole blood trihalomethane levels. *J Exp Anal Environ Epidemiol* 10(4): 321–326.

Bader EL, Hrudey SE, Froese KL. 2004. Urinary excretion half life of trichloroacetic acid as a biomarker of exposure to chlorinated drinking water disinfection by-products. *Occup Environ Med* 61(8): 715–716.

Barsotti DA, van Miller JP. 1984. Accumulation of a commercial polychlorinated biphenyl mixture (Aroclor 1016) in adult rhesus monkeys and their nursing infants. *Toxicology* 30: 31–44.

Berenbaum MC. 1985. The expected effect of a combination of agents: the general solution. *J Theor Biol* 114(3): 413–431.

Chen WJ, Weisel CP. 1998. Concentration changes of halogenated disinfection by-products in a drinking water distribution system. *Am Water Works Assoc J* 90: 151–163.

Clark RM. 1998. Chlorine demand and TTHM formation: a second-order model. *J Environ Eng* 124(1): 16.

Cullen AC, Frey HC. 1999. *Probabilistic Techniques in Exposure Assessment: A Handbook for Dealing with Variability and Uncertainty in Models and Inputs*. Plenum Press, New York.

De Roos AJ, Hartge P, Lubin JH, Colt JS, Davis S, Cerhan JR, Severson RK, Cozen W, Patterson DG. Jr, Needham LL, Rothman N. 2005. Persistent organochlorine chemicals in plasma and risk of non-Hodgkin's lymphoma. *Cancer Res* 65(23): 11214–11226.

Dodds L, King WD. 2001. Relation between trihalomethane compounds and birth defects. *Occup Environ Med* 58(7): 443–446.

Dodds L, King W, Allen AC, Armson BA, Fell DB, Nimrod C. 2004. Trihalomethanes in public water supplies and risk of stillbirth. *Epidemiology* 15(2): 179–186.

Feron VJ, Groten JP. 2002. Toxicological evaluation of chemical mixtures. *Food Chem Toxicol* 40(6): 825–839.

Feron VJ, Groten JP, Jonker D, Cassee FR, van Bladeren PJ. 1995. Toxicology of chemical mixtures: challenges for today and the future. *Toxicology* 105(2–3): 415–427.

Gallagher MD, Nuckols JR, Stallones L, Savitz DA. 1998. Exposure to trihalomethanes and adverse pregnancy outcomes. *Epidemiology* 9(5): 484–489.

Gessner PK. 1995. Isobolographic analysis of interactions: an update on applications and utility. *Toxicology* 105(2–3): 161–179.

Gessner PK, Cabana BE. 1970. A study of the interaction of the hypnotic effects and of the toxic effects of the chloral hydrate and ethanol. *J Pharmacol Exp Ther* 174(2): 247–259.

Gordon SM, Brinkman MC, Ashley DL, Blount BC, Lyu C, Masters J, Singer PC. 2006. Changes in breath trihalomethane levels resulting from household water-use activities. *Environ Health Perspect* 114(4): 514–521.

Hertzberg RC, Rice GE, Teuschler LK. 1999. Methods for health risk assessment of combustion mixtures. In: *Hazardous Waste Incineration: Evaluating the Human Health and Environmental Risks*. Roberts S, Teaf C, Bean J, eds. CRC Press, Boca Raton, FL, pp 105–148.

Hoffman FO, Hammonds JS. 1994. Propagation of uncertainty in risk assessments: the need to distinguish between uncertainty due to lack of knowledge and uncertainty due to variability. *Risk Anal* 14(5): 707–712.

Hopenhayn C, Ferreccio C, Browning SR, Huang B, Peralta C, Gibb H, Hertz-Picciotto I. 2003. Arsenic exposure from drinking water and birth weight. *Epidemiology* 14(5): 593–602.

Ikeda M. 1988. Multiple exposure to chemicals. *Regul Toxicol Pharmacol* 8(4): 414–421.

IPCS (International Programme on Chemical Safety). 2005. IPCS framework for analyzing the relevance of a cancer mode of action for humans. Public review draft. April. Accessed at: http://www.who.int/ipcs/methods/harmonization/areas/frameworktext-review.pdf.

Jo WK, Weisel CP, Lioy PJ. 1990. Routes of chloroform exposure and body burden from showering with chlorinated tapwater. *Risk Anal* 10: 575–580.

Kim H, Haltmeier P, Klotz JB, Weisel CP. 1999. Evaluation of biomarkers of environmental exposures: urinary haloacetic acids associated with ingestion of chlorinated drinking water. *Environ Res* 80: 187–195.

King WD, Dodds L, Armson BA, Allen AC, Fell DB, Nimrod C. 2004. Exposure assessment in epidemiologic studies of adverse pregnancy outcomes and disinfection byproducts. *J Exp Anal Environ Epidemiol* 14:466–472.

Klotz JB, Pyrch LA. 1999. Neural tube defects and drinking water disinfection byproducts. *Epidemiology* 10(4): 383–390.

Krasner SW, McGuire MJ, Jacangelo JG, Patania NL, Reagan KM, Aieta EM. 1989. The occurrence of disinfection byproducts in U.S. drinking water. *J Am Water Works Assoc* 81: 41–53.

Krishnan K, Haddad S, Beliveau M, Tardif R. 2002. Physiological modeling and extrapolation of pharmacokinetic interactions from binary to more complex chemical mixtures. *Environ Health Perspect* 110(suppl 6): 989–994.

Leaderer BP, Lioy PJ, Spengler JD. 1993. Assessing exposures to inhaled complex mixtures. *Environ Health Perspect* 101(suppl 4):167–177.

Levin ED, Schantz SL, Bowman RE. 1988. Delayed spatial alternation deficits resulting from perinatal PCB exposure in monkeys. *Arch Toxicol* 62: 267–273.

Lin TF, Hoang SW. 2000. Inhalation exposure to THMs from drinking water in south Taiwan. *Sci Total Environ*. 246(1): 41–49.

Meek ME, Bucher JR, Cohen SM, Dellarco V, Hill RN Lehman- McKeeman LD, Longfellow DG, Pastoor T, Seed J, Patton DE. 2003. A framework for human relevance analysis of information on carcinogenic modes of action. *Crit Rev Toxicol* 33(6): 591–653.

Miles AM, Singer PC, Ashley DL, Lynberg MC, Mendola P, Langlois PH, Nuckols JR. 2002. Comparison of trihalomethanes measured in tap water and blood. *Environ Sci Technol* 36(8): 1692–1698.

Miltner RJ, Rice EW, Stevens AA. 1990. Pilot-scale investigation of the formation and control of disinfection by-products. Presented at the AWWA Annual Conference, Cincinnati, OH. *AWWA Annu Conf Proc* 2: 1787–1802.

Morgan MG, Henrion M. 1990. *Uncertainty: A Guide for Dealing with Uncertainty in Quantitative Risk and Policy Analysis*. Cambridge University Press, Cambridge, UK.

Mumtaz MM, Durkin PR. 1992. A weight-of-evidence scheme for assessing interactions in chemical mixtures. *Toxicol Indus Health* 8(6): 377–406.

Mumtaz MM, Poirier KA, Coleman JT. 1997. Risk assessment for chemical mixtures: fine-tuning the hazard index approach. *J Clean Technol Environ Toxicol Occup Med* 6(2): 189–204.

Murphy SD. 1980. Assessment of the potential for toxic interactions among environmental pollutants. In: *The Principles and Methods in Modern Toxicology*. Galli CL, Murphy SD, Paoletti R, eds. Elsevier/North-Holland Biomedical Press, Amsterdam, The Netherlands.

NRC (National Research Council). 1983. *Risk Assessment in the Federal Government: Managing the Process*. Committee on the Institutional Means for Assessment of Risks to Public Health, Commission on Life Sciences. National Academies Press, Washington, DC.

______. 1994. *Science and Judgment in Risk Assessment*. National Academies Press, Washington, DC.

Plewa MJ, Wagner ED, Jazwierska P, Richardson SD, Chen PH, McKague B. 2004. Halonitromethane drinking water disinfection by-products: chemical characterization and mammalian cell cytotoxicity and genotoxicity. *Environ Sci Technol* 38(1): 62–68.

Pohl HR, Roney N, Wilbur S, Hansen H, De Rosa CT. 2003. Six interaction profiles for simple mixtures. *Chemosphere* 53(2): 183–197.

Pozzani UC, Weil CS, Carpenter CP. 1959. The toxicological basis of threshold values, 5: The experimental inhalation of vapor mixtures by rats, with notes upon the relationship between single dose inhalation and single dose oral data. *Am Ind Hyg Assoc J* 20: 364–369.

Richardson SD. 1998. Identification of drinking water disinfection by-products. In: *Encyclopedia of Environmental Analysis and Remediation*, Vol 3. Meyers RA, ed. Wiley, New York, pp. 1398–1421.

Rothman KJ, Greenland S. 1998. *Modern Epidemiology*, 2nd ed. Lippincott-Raven, Philadelphia, PA.

San Sebastian M, Armstrong B, Stephens C. 2002. Outcomes of pregnancy among women living in the proximity of oil fields in the Amazon basin of Ecuador. *Int J Occup Environ Health* 8(4): 312–319.

Savitz DA, Singer PC, Hartmann KE, Herring AH, Weinberg HS, Makarushka C, Hoffman C, Chan R, Maclehose R. 2005. *Drinking Water Disinfection By-products and Pregnancy Outcome*. American Water Works Association Research Foundation, Denver, CO. Accessed at: http://www.awwarf.org/research/TopicsAndProjects/execSum/2579.aspx.

Schantz SL, Levin ED, Bowman RE, et al. 1989. Effects of perinatal PCB exposure on discrimination-reversal learning in monkeys. *Neurotoxicol Teratol* 11: 243–250.

Schantz SL, Levin ED, Bowman RE. 1991. Long-term neurobehavioral effects of perinatal polychlorinated biphenyl (PCB) exposure in monkeys. *Environ Toxicol Chem* 10: 747–756.

Seed J, Carnev EW, Corlev RA, et al. 2005. Overview: using mode of action and life stage information to evaluate the human relevance of animal toxicity data. *Crit Rev Toxicol* 35(8–9): 663–672.

Simmons JE, Teuschler LK, Gennings C. 2001. The toxicology of disinfection by-product mixtures: methods for multi-chemical assessment, present research efforts, and future research needs. In: *Microbial Pathogens and Disinfection By-products in Drinking Water: Health Effects and Management of Risks*. Craun GF, Hauchman FS, Robinson DE, eds. ILSI Press, Washington, DC, pp 325–340.

Simmons JE, Richardson SD, Speth TF, Miltner RJ, Rice G, Schenck KM, Hunter ES III, Teuschler LK. 2002. Development of a research strategy for integrated technology-based toxicological and chemical evaluation of complex mixtures of drinking water disinfection byproducts. *Environ Health Perspect* 110(suppl 6): 1013–1024.

Simmons JE, Teuschler LK, Gennings C, Speth TF, Richardson SO, Miltner RS, Narotsky MG, Schenck KD, Hunter ES III, Hertzberg RC, Rice GE. 2004. Component-based and whole-mixture techniques for addressing the toxicity of drinking-water disinfection by-product mixtures. *J Toxicol Environ Health A* 67(8–10): 741–754.

Simmons J, Wagner ED, Plewa MJ. 2006. Mixing ratio influences haloacetic acid (HAA) mixture toxicity. Abstract. Society of Toxicology 45th Annual Meeting, March 5–9, San Diego, CA.

Simon TW. 1999. Two dimensional Monte Carlo simulation and beyond: a comparison of several probabilistic risk assessment methods applied to a Superfund site. *Hum Ecol Risk Assess* 5(4): 823–843.

Smyth HF Jr, Weil CS, West JS, Carpenter CP. 1969. An exploration of joint toxic action, I: Twenty-seven industrial chemicals intubated in rats in all possible pairs. *Toxicol Appl Pharmacol* 14: 340–347.

———. 1970. An exploration of joint toxic action, II: Equitoxic versus equivolume mixtures. *Toxicol Appl Pharmacol* 17(2): 498–503.

Sonich-Mullin C, Fielder R, Wiltse J, Baetcke K, Dempsey J, Fenner-Crisp P, Grant D, Hartley M, Knaap A, Kroese D, Mangelsdorf I, Meek E, Rice JM, Younes M. 2001.

IPCS conceptual framework for evaluating a mode of action for chemical carcinogenesis. International Programme on Chemical Safety. *Regul Toxicol Pharmacol* 34: 146–152.

Stewart PW, Reihman J, Lonky EI, Darvill TJ, Pagano J. 2003. Cognitive development in preschool children prenatally exposed to PCBs and MeHg. *Neurotoxicol Teratol* 25(1): 11–22.

Tardif R, Charest-Tardif G, Brodeur J, Krishnan K. 1997. Physiologically based pharmacokinetic modeling of a ternary mixture of alkyl benzenes in rats and humans. *Toxicol Appl Pharmacol* 144: 120–134.

Teuschler LK, Gennings C, Stiteler WM, Hertzberg RC, Colman JT, Thiyagarajah A, Lipscomb JC, Hartley WR, Simmons JE. 2000. A multiple-purpose design approach to the evaluation of risks from complex mixtures of disinfection by-products (DBPs). *Drug Chem Toxicol* 23(1): 307–321. Erratum in: *Drug Chem Toxicol* 23(3):511.

Teuschler LK, Rice GE, Wilkes CR, Lipscomb JC, Power FW. 2004. A feasibility study of cumulative risk assessment methods for drinking water disinfection by-product mixtures. *J Toxicol Environ Health A* 67(8–10): 755–777.

Toledano MB, Nieuwenhuijsen MJ, Best N, et al. 2005. Relation of trihalomethane concentrations in public water supplies to stillbirth and birth weight in three water regions in England. *Environ Health Perspect* 113(2): 225–232.

Travis C, White R. 1988. Interspecific scaling of toxicity data. *Risk Anal* 8(1): 119–125.

U.S. Congress. 1980. Title 42: The Public Health and Welfare, Chapter 103. The comprehensive environmental response, compensation, and liability act (CERCLA). Accessed at: http://www.epa.gov/epahome/laws.htm.

_____. 1996a. 1996 amendments to the safe drinking water act : Public Law 104–182. 104th Congress, Washington, DC. Accesed at : http://www.epa.gov/safewater/sdwa/text.html.

_____. 1996b. The food quality protection act: Public Law 104-170 August 3. 1996. 104th Congress, Washington, DC. Accessed at : http://www.epa.gov/pesticides/regulating/laws/fqpa/.

U.S. EPA (Environmental Protection Agency). 1986. Guidelines for the health risk assessment of chemical mixtures. *Fed Reg*. 51(185): 34014–34025.

_____. 1987. *Risk Assessment Guidelines of 1986*. EPA/600/8-87/045. U.S. EPA, Washington, DC.

_____. 1989a. *Risk Assessment Guidance for Superfund*. Vol 1, *Human Health Evaluation Manual* (Part A). EPA/540/1-89/002. Accessed at: U.S. EPA, Washington, DC. http://nepis.epa.gov/pubtitleOSWER.htm.

_____. 1989b *Interim Procedures for Estimating Risks Associated with Exposures to Mixtures of Chlorinated Dibenzo-P- Dioxins and-Dibenzofurans (CDDs and CDFs) and 1989 Update*. EPA/625/3-89/016. Risk Assessment Forum, Washington, DC. March. Accessed at http://nepis.epa.gov/pubtitleORD.htm.

_____. 1991. *Guidelines for Developmental Toxicity Risk Assessment*, The workshop report on toxicity equivalency factors for polychlorinated biphenyl congeners. EPA/625/3-91/020 Risk Assessment Forum, U.S. EPA Washington, DC, June. Accessed at: http://nepis.epa.gov/pubtitleORD.htm.

_____. 1992a. *Drinking Water Criteria Document for Polycyclic Aromatic Hydrocarbons*. EPA/600/X-92/015 Office of Research and Development, U.S. EPA, Cincinnati, OH.

_____. 1992b. Guidelines for exposure assessment. *Fed Reg* 57(104): 22888–22938.

______. 1996. *Guidelines for Reproductive Toxicity Risk Assessment*. EPA/630/R-96/009 *Fed Reg* 61(212): 56274–56322.

______. 1997. *Exposure Factors Handbook*. EPA/600/P-95/002Fa-c Office of Research and Development, U.S. EPA, Washington, DC. Accessed at: http://nepis.epa.gov/pubtitleORD.htm.

______. 1998a. *Stage 1 Disinfectants and Disinfection Byproducts Rule*. EPA/815/F-98/010 U.S. EPA, Washington, DC. Accessed at http://www.epa.gov/safewater/mdbp/dbp1.html.

______. 1998b. Guidelines for neurotoxicity risk assessment. *Fed Reg* 63(93): 26926–26954.

______. 2000. *Supplementary Guidance for Conducting Health Risk Assessment of Chemical Mixtures*. EPA/630/R-00/002. U.S. EPA, Cincinnati, OH. Accessed at: http://nepis.epa.gov/iris/bachkgr-d.htm.

______. 2001. *General Principles for Performing Aggregate Exposure and Risk Assessments*. Office of Pesticide Programs, U.S. EPA, Washington, DC. Fax-on-Demand. Fax (202) 401-0527. Item 6043.

______. 2002a. *Exposure Related Dose Estimating Model (ERDEM) for Assessing Human Exposure and Dose*. EPA/600/R-00/011. Office of Research and Development, U.S. EPA, Washington, DC.

______. 2002b. *Guidance on Cumulative Risk Assessment of Pesticide Chemicals That Have a Common Mechanism of Toxicity*. Office of Pesticide Programs, U.S. EPA, Washington, DC. Accessed at: http://www.epa.gov/oppfead1/trac/science/cumulative_guidance.pdf.

______. 2003a. *The Feasibility of Performing Cumulative Risk Assessments for Mixtures of Disinfection By-products in Drinking Water*. EPA/600/R-03/051. Office of Research and Development, National Center for Environmental Assessment, U.S. EPA, Cincinnati, OH.

______. 2003b. *Developing Relative Potency Factors for Pesticide Mixtures: Biostatistical Analyses of Joint Dose–Response*. EPA/600/R-03/052. Office of Research and Development, National Center for Environmental Assessment, U.S. EPA, Cincinnati, OH. Accessed at: http://cfpub.epa.gov/ncea/cfm/recordisplay.cfm?deid = 66273.

______. 2005a. *Stage 2 Disinfectants and Disinfection Byproducts Rule*. EPA/815/F-05/003. Office of Water, U.S. EPA, Washington, DC. Accessed at: http://www.epa.gov/safewater/disinfection/stage2/.

______. 2005b. *Guidelines for Carcinogen Risk Assessment*. EPA/630/P-03/001B. Risk Assessment Forum, U.S. EPA, Washington, DC.

______. 2005c. *Supplemental Guidance for Assessing Susceptibility from Early-Life Exposure to Carcinogens*. EPA/630/R-03/003F. Risk Assessment Forum, U.S. EPA, Washington, DC.

______. 2006a. List of drinking water contaminants and MCLs. Office of Ground Water and Drinking Water, U.S. EPA, Washington, DC. Accessed at: http://www.epa.gov/safewater/mcl.html.

______. 2006b. Assessing pesticide cumulative risk. U.S. EPA, Washington, DC. Accessed at: http://www.epa.gov/oppsrrd1/cumulative/.

______. 2006c. Integrated risk information system (IRIS). U.S. EPA, Washington, DC. Accessed at: http://epa.gov/iriswebp/iris/index.html. Accessed March 9 2006.

Vartiainen T, Pukkala E, Rienoja T, Strandman T, Kaksonen K. 1993. Population exposure to tri- and tetrachloroethylene and cancer risk: two cases of drinking water pollution. *Chemosphere* 27 (7): 1171–1181.

Wallace LA. 1997. Human exposure and body burden for chloroform and other trihalomethanes. *Crit Rev Environ Sci Technol* 27(2): 113–194.

Waller K, Swan SH, DeLorenze G, Hopkins B. 1998. Trihalomethanes in drinking water and spontaneous abortion. *Epidemiology* 9(2): 134–140.

Waller K, Swan SH, Windham GC, Fenster L. 2001. Influence of exposure assessment methods on risk estimates in an epidemiologic study of total trihalomethane exposure and spontaneous abortion. *J Expo Anal Environ Epidemiol* 11(6): 522–531.

Weinberg H. 1999. Disinfection byproducts in drinking water: the analytic challenge. *Anal Chem News Features* 4(12):801A–808A.

Weisel CP, Jo WK. 1996. Ingestion, inhalation, and dermal exposures to chloroform and trichloroethene from tap water. *Environ Health Perspect* 104:48–51.

Weisel CP, Kim H, Haltmeier P, Klotz JB. 1999. Exposure estimates to disinfection by-products of chlorinated drinking water. *Environ Health Perspect* 107: 103–110.

Whitaker HJ, Nieuwenhuijsen MJ, Best NG. 2003. The relationship between water concentrations and individual uptake of chloroform: a simulation study. *Environ Health Perspect* 111(5): 688–694.

Wilkes CR, Small, M, Andelman JB, Giardino N, Marshall J. 1992. Inhalation exposure model for volatile chemicals from indoor uses of water. *Atmos Environ* 12(26A): 2227–2236.

Wilkes C. 1998. Case study. In: Exposure to Contaminants in Drinking Water: Estimating Uptake Through the Skin and by Inhalation. Olin S, ed. CRC Press, Boca Raton, FL pp. 183–224.

Windham GC, Waller K, Anderson M, Fenster L, Mendola P, Swan S. 2003. Chlorination by-products in drinking water and menstrual cycle function. *Environ Health Perspect* 111(7): 935–941.

Wright JM, Bateson TF. 2005. A sensitivity analysis of bias in relative risk estimates due to disinfection by-product exposure misclassification. *J Expo Anal Environ Epidemiol* 15(3): 212–216.

Wright JM, Schwartz J, Dockery DW. 2004. The effect of disinfection by-products and mutagenic activity on birth weight and gestational duration. *Environ Health Perspect* 112(8): 920–925.

Wright JM, Murphy PA, Nieuwenhuijsen MJ, Savitz DA. 2006. The impact of water consumption, point-of-use filtration and exposure categorization on exposure misclassification of ingested drinking water contaminants. *Sci Total Environ* 366(1): 65–73.

Xu X, Weisel CP. 2003. Inhalation exposure to haloacetic acids and haloketones during showering. *Environ Sci Technol* 37(3): 569–576.

——. 2005a. Dermal uptake of chloroform and haloketones during bathing. *J Exp Anal Environ Epidemiol* 15(4): 289–296.

——. 2005b. Human respiratory uptake of chloroform and haloketones during showering. *J Exp Anal Environ Epidemiol* 15:6–16.

Xu X, Mariano TM, Laskin JD, Weisel CP. 2002. Percutaneous absorption of trihalomethanes, haloacetic acids, and haloketones. *Toxicol Appl Pharmacol* 184:19–26.

Zender R, Bachand A, Reif J. 2001. Exposure to tap water during pregnancy. *J Exp Anal Environ Epidemiol* 11(3): 224–230.

Ziglio G. 1985. Human exposure to environmental trichloroethylene and tetrachloroethylene: preliminary data on population groups of Milan, Italy. *Bull Environ Contam Toxicol* 26:131–136.

7

PROTECTION OF INFANTS, CHILDREN, AND OTHER SENSITIVE SUBPOPULATIONS

George V. Alexeeff and Melanie A. Marty
California Environmental Protection Agency, Oakland, California

The Safe Drinking Water Act of 1974 (PL93-523) directed the U.S. Environmental Protection Agency (EPA) to promulgate drinking water regulations. To address the scientific basis of the standards, the EPA arranged with the National Academy of Sciences' Safe Drinking Water Committee (SDWC) to study the adverse effects attributable to water contaminants (NRC, 1977). As part of its evaluation, the SDWC laid out a framework for developing drinking water standards. For their approach to noncarcinogens, (i.e., where there was no indication of carcinogenicity), the SDWC adopted the *acceptable daily intake* (ADI) approach applied to food contaminants. This involved identifying the maximum dose producing a no-observed-adverse-effect-level (NOAEL) in animal studies, and dividing by an uncertainty factor (UF), that is, "a number that reflected the uncertainty that must be considered when experimental data in animals are extrapolated to man" (NRC, 1977, pp. 803–804). A factor of 10 was used where good chronic human exposure data were available and supported by chronic oral toxicity data in other species. A factor of 100 was used where good chronic oral toxicity data were available in some animal species. Finally a factor of 1000 was used when there was only limited chronic toxicity data. Further, 1 to 20% of the ADI was assigned to water. The standard was set for a 70-kg adult with an average daily intake of 2L of water. For carcinogens, the NRC recommended application of a mathematical model for dose–response extrapolation, without a threshold, to estimate risk from exposure.

Risk Assessment for Chemicals in Drinking Water, Edited by Robert A. Howd and Anna M. Fan

While acknowledging the heterogeneity of the human population, the SDWA guidance developed in 1977 (pp. 57–58) did not explicitly take into account susceptible populations, such as children, the elderly, or pregnant women. Over the course of the committee's work (1977 to 1989), the SDWC acknowledged that since children drink proportionately more water than adults, contaminants pose greater risks to children than to adults, however, they did not explicitly calculate drinking water levels to address this concern (NRC, 1986, p. 8).

In 1993, the NRC's Committee on Pesticides in the Diets of Infants and Children (PDIC) expanded our understanding of risk assessment of sensitive subpopulations in their landmark publication *Pesticides in the Diets of Infants and Children*. While building on the work of the SDWC, they found both quantitative and occasionally qualitative differences in toxicity of pesticides between children and adults. For noncarcinogens, the PDIC acknowledged that the common UF applied for ADI development of 100 could be divided into two 10-fold factors, one to address uncertainty in extrapolating data from animals to human and another to accommodate variation within the human population. Although the latter 10-fold factor was considered generally protective of infants and children, they thought there may be stages of development that are uniquely sensitive to a particular toxicant, thereby requiring an additional uncertainty factor. The committee further suggested that a third UF, up to10-fold, be considered to address not only fetal developmental effects, but also evidence of "postnatal developmental toxicity and when data from toxicity testing relative to children are incomplete" (NRC, 1993, p. 9). They went further to state: "In the absence of data to the contrary, there should be a presumption of greater toxicity to infants and children." For carcinogens, the PDIC stated that for cases considered by the committee, underestimation of risk assuming constant exposure was limited to a factor of about three-to five fold in all cases. The key recommendations of the NRC report were incorporated into the provisions of the Food Quality Protection Act of 1996 (PL 104-170).

Soon after the publication of the NRC report, in 1995, the EPA established an agency-wide policy to ensure that environmental health risks of children are explicitly and consistently evaluated in their risk assessments, risk characterizations, and environmental and public health standards. Part of the EPA's process to ensure that all standards are protective of any heightened risks faced by children is being undertaken through its implementation of the Safe Drinking Water Act Amendments of 1996 (PL 104-182). These amendments include a requirement ([1458(a)] Sec. 137) periodically to "conduct studies to identify subpopulations at greater risk (e.g., infants, children, pregnant women) than the general public of adverse health effects from exposure to contaminants in drinking water, and report to Congress on the results of studies."

Subsequently, in 1997, President Clinton signed Executive Order 13045 requiring each federal agency to "make it a high priority to identify and assess environmental health risks and safety risks that may disproportionately affect children; and ensure that its policies, programs, activities, and standards address disproportionate risks to children that result from environmental health risks or safety risks" (http://yosemite.epa.gov/ochp/ochpweb.nsf/content/whatwe_executiv.htm). Soon after, the EPA established the Office of Children's Health Protection (OCHP), which supports and facilitates agency efforts to protect children's health from environmental threats.

Some individual states have undertaken similar initiatives to address children's health risks. In 2004, California legislation (Health and Safety Code Section 116365.2) was passed to require specific consideration of infants, children, and other sensitive subpopulations in setting public health goals (PHGs) for drinking water contaminants. The new legislation allows special consideration to be given to those contaminants that "cause or contribute to adverse health effects in members of subgroups that comprise a meaningful portion of the general population, including, but not limited to, infants, children, pregnant women, the elderly, individuals with a history of serious illness, or other subgroups that are identifiable as being at greater risk of adverse health effects than the general population when exposed to the contaminant in drinking water." Further, the statute requires that health effects assessments specifically consider (1) exposure patterns among bottle-fed infants and children that result in disproportionately high exposure, (2) special susceptibility of infants and children in comparison to the general population, (3) effects of simultaneous exposures to compounds with the same mechanism of action; and (4) any interactions of multiple contaminants. These sensitive subpopulation issues are being taken into consideration in further development of the PHGs. In this chapter we discuss factors that contribute to our understanding of susceptible populations, with an emphasis on water contaminants.

As a result of national and local efforts, additional research and guidance has led to the identification of compounds of concern to infants and children and to the development of improved risk assessment procedures to ensure protection of sensitive subpopulations. Similar comprehensive efforts have not been undertaken for other susceptible populations, but they are considered on a case-by-case basis.

FACTORS INFLUENCING DIFFERENCES IN SUSCEPTIBILITY BETWEEN INFANTS AND CHILDREN AND ADULTS

There are a number of reasons to suspect that risk assessments based on adults may in some cases underpredict the risks of exposure to infants and children [reviewed by the California OEHHA (2001) and in Miller et al. (2002)]. The potential impact of environmental chemicals on children's health may be affected by developmental factors, including behavioral, physiological, and sociological factors that differ from those of adults. These factors can change both exposure patterns and responses to toxicants. Developmental differences between infants and children and adults that affect the disposition of chemicals or the toxicodynamics may result in significant illness or disabilities in children but not in a similarly exposed adult. Minamata disease, a result of methyl mercury poisoning from consuming contaminated seafood, is one example, where prenatal exposure produced profound neurodevelopmental toxicity in the absence of clinical findings in mothers who were similarly exposed (Rogan, 1995). Assessments that adequately describe risks from exposures to children present challenges due to the myriad changes that accompany rapid growth and development from conception through adolescence.

There is limited scientific information that addresses age-specific differences in sensitivity to toxic effects. The vast array of differences that may affect toxicologic response to xenobiotic chemicals among the fetus, newborn and infants, children, adolescents, and adults has not been cataloged. In the following sections we give examples of such differences that influence exposure and disposition of toxicants and a few words on pharmacodynamic differences. The latter concept is tied more to the concept that different targets of toxicity are present in the developing fetus, infants, and children than in adults. Although these examples are in no way a complete listing, they provide a foundation for understanding the variety of information that must be considered in a risk assessment in order to anticipate the possible toxicological risks to the growing and developing human. The EPA recently released a useful document describing a framework for child-specific risk assessment that offers information and questions to guide the risk assessor (U.S. EPA, 2006a).

Differences in Exposures: Infants, Children, and Adults

There is a growing body of evidence that children receive greater doses of environmental toxicants on a body weight basis than adults through common exposure pathways such as inhalation and ingestion. These greater doses stem not only from their higher metabolic rate (leading to higher consumption or exposures on a body weight basis), but also from unique exposures. For example, infants and children have more limited diets than adults and exhibit differences in behavioral patterns that influence exposure. We address some of these issues in the following sections.

Water Intake Exposure Exposures to drinking water are covered elsewhere in this book. It should be noted that drinking water intake rates per kilogram of body weight are highest in the youngest children and decrease steadily to adult intake rates (Ershow and Cantor, 1989; California OEHHA, 2000). Because reconstituted formula may be the infant's sole source of fluids and nutrients for at least the first few months of life, formulafed infants may receive greater doses (on a mg/kg body weight basis) of waterborne contaminants than older children and adults. For these reasons, to protect infants and children, drinking water exposure concentrations should be calculated on an intake rate per body weight basis even when there is no perceived increase in infant or childhood susceptibility. This is particularly important when the endpoint is applicable to children and/or infants, or when evaluating a carcinogen.

Breast Milk Exposure Breast milk intake is an exposure route unique to infants, and thus represents an additional source of toxicant exposure absent for older children and adults. As an infant's sole source of nutrients, a breast-fed infant would receive repeated doses of any waterborne contaminants that transfer into breast milk.

Factors that influence the transfer of chemicals from the mother's plasma into breast milk include the lipophilicity of the compound, protein binding in plasma and milk, and the pK_a value of the compound in relation to the pH of milk and plasma (Findlay, 1983; Fleishaker et al., 1987). Basic compounds can attain higher concentrations in breast milk (pH 6.6 to 7.0) than plasma (pH 7.4) by ionizing at the more acidic pH of the milk (Findlay, 1983; Morriss et al., 1986). Further, breast milk pH changes over time, rising to 7.4 at 10 months postpartum (Morriss et al., 1986); thus, drug and chemical delivery through breast milk may also change over the nursing period. Studies have observed that the concentrations in breast milk of weakly basic drugs such as codeine and morphine were over twice those in the mother's plasma (Findlay, 1983), where as caffeine and acetaminophen concentrations in the milk and plasma were equal. Salicylate, a weak acid, showed higher peak concentrations in plasma than in milk. The extent of protein binding in the plasma also influences the availability for transport into milk. Similarly, a higher affinity for milk protein than plasma protein might increase the concentration of the chemical in the milk relative to the plasma (Findlay, 1983).

Chronic exposure to highly lipophilic, poorly metabolized environmental contaminants results in the concentration of these contaminants in adipose tissue and breast milk lipid. For a toxicant such as tetrachlorodibenzo-*p*-dioxin (dioxin), an infant's intake rate from breast milk may be substantially greater than the mother's environmental intake rate. Hoover et al. (1991) estimated an average maternal daily intake of 0.93 pg/kg·day and calculated that a breast-fed infant would receive a daily dose of between 20 and 70 times this amount during the first year of life. Based on measurements of dioxins and PCBs in the diet and in breast milk, a study of families in the Netherlands demonstrated that the WHO toxic equivalent (TEQ) intake of dioxin and PCBs on a picogram/kg body weight basis is about 50 times greater for the breast-feeding infant than for the adult (Patandin et al., 1999). Furst (2006) estimated that infants in 2003 in Germany received about 64 pg/kg·day WHO-TEQ, about two orders of magnitude higher than adult dietary intake. Data from Taiwan indicate nursing infants receive about 13 pg/kg·day WHO-TEQ (Hsu et al., 2007). Regulatory efforts to reduce PCDD/PCDF and PCBs in the environment have significantly reduced exposures through breast milk over the last three decades; however, concentrations of polybrominated diphenyl ethers in breast milk have risen over that same time span (Furst, 2006).

Breast-feeding infants may also be exposed to nonlipophilic contaminants in breast milk. For example, 36 to 80% of lead blood levels in a breast-feeding infant during the first 60 to 90 days postpartum may be attributed to lead in breast milk (Gulson et al., 1998). Hallen et al. (1996) demonstrated in rats that lactational transfer of lead is considerably higher than placental transfer in recently or currently exposed animals. Mercury concentrations in breast milk have been shown to correlate with the presence of amalgam fillings and fish consumption (Oskarsson et al., 1996; Drexler and Schaller, 1998).

Behavioral Factors Influencing Exposure Because children are growing and are generally more physically active than adults (Wiley et al., 1991a, 1991b), their higher energy expenditure requires greaterww consumption of food, air, and water on a body weight basis than that for adults. This increases the dose of toxicants the children receive from inhalation, food, and fluids. The increased physical activity can also result in increased soil contact (e.g., playing soccer) relative to adults. Toddlers and young children typically crawl, lie, sit, and play on the grass or soil. Also, contaminants present in tap water can become aerosolized in showers and deposit on household surfaces. Infants and toddlers frequently put hands, toys, and other objects in their mouths as a means of exploration and when teething. Although children may only mouth body parts and objects for a few seconds, infants 6 to 12 months of age were observed to spend an average of 44 minutes per day in this activity (Groot et al., 1998). This mouthing behavior may lead to higher doses of surface contaminants in infants and toddlers (Zartarian et al., 2000) than in adults. Thus, due to their higher intake rates and behavior, children's multipathway exposures to waterborne contaminants can be higher than those of adults.

Dermal Exposure Infants and children have a greater body surface area body weight ratio than adults. This ratio is more than two times greater in a newborn infant than in an adult (Snodgrass, 1992; U.S. EPA, 1997), which may result in increased dermal exposures, on a body weight basis. This potential for increased exposure is reflected in the observation that dermally applied pharmaceuticals can be more toxic to infants (U.S. EPA, 1992). For waterborne chemicals that cross the skin, potential for increased dermal dosage is an important consideration in assessing risk from waterborne contaminants while swimming or bathing.

Differences in Disposition of Toxicants

Recent studies evaluating pharmacokinetic differences by age using information from the pharmaceutical literature have demonstrated significant differences in disposition of chemicals (Ginsberg et al., 2002; Hattis et al., 2003; Kearns et al., 2003). The largest differences appear to be between infants and adults. These differences must be considered when conducting risk assessments, as they affect the dose of the toxicant at the target organ and clearance of the substance. In general, many drugs have longer half-lives (slower clearance) in infants, due partially to immature xenobiotic metabolizing enzymes and immature renal and hepatic function. In the following sections we discuss factors that affect the disposition of toxicants and which should be considered in conducting a risk assessment.

Absorption A number of factors that influence the absorption of environmental pollutants differ by age (Miller et al., 2002; Kearns et al., 2003). Gastric pH and emptying time (Milsap and Jusko, 1994), both of which vary with age from birth through infancy, influence passive diffusion through the lining of the stomach and intestines, the mechanism by which most toxicants are absorbed into the systemic circulation. At birth the gastric acidity is weak or neutral (pH 4 to 8), fluctuates over the first few weeks, and then increases, approaching adult levels by three months of age (Miller, 1941; Avery et al., 1966; Agunod et al., 1969). Premature infants may have prolonged periods of lower gastric acidity, due to immature acid secretion (Agunod et al., 1969). The pH of the stomach influences the absorbed dose of ionizable chemicals, thus altering the potential dose to the infant.

Absorption rates for several chemicals, such as phenobarbital, digoxin, arabinose, and xylose, increase throughout the first year of life. Gastric emptying time influences the fraction of an oral dose that is absorbed into the systemic circulation. The gastric emptying rate in neonatal infants is both variable and prolonged (Siegner and Fridrich, 1975; Siegel et al., 1984). Delayed absorption observed in neonates is due partially to slower gastric emptying and gastrointestinal motility, and other factors such as lower bile acid secretion (Murphy and Signer, 1974; Heimann, 1980), which may result in decreased absorption of lipid-soluble chemicals (Besunder et al., 1988).

Skin permeability and hence dermal absorption changes as the skin develops. Dermal absorption can be higher in neonates because of increased skin hydration and an immature epidermis (Bearer, 1995). When the skin moisture content is increased, skin permeability increases (Klaassen, 1996). Because young children often mouth their hands, the skin may be more permeable to contacted toxicants. In terms of assessing risk from waterborne exposures, contamination of surfaces that children contact may result in increased exposures relative to adults. The observation of severe toxicity in infants following dermal application of hexachlorophene (Tyrala et al., 1977) and isopropanol (McFadden and Haddow, 1969) are in part reflections of the increased dermal absorption of these compounds. The affects of age-related physiological differences on dermal absorption has not been fully characterized. Thus, childhood differences affecting the total amount of drinking water contaminants absorbed include lower gastric acidity in neonates, slower gastric emptying in neonates, lower intestinal absorption in neonates compared to children, a higher surface area/body weight ratio (0.067 in newborn vs. 0.025 in adult), more permeable skin surface (in premature infants especially), and behavioral factors resulting in greater exposure.

Distribution The distribution of absorbed chemicals is affected by the concentration and types of plasma proteins, the relative water volume (both extracellular and intracellular), fat and tissue compartments of the body, and total body water (Milsap and Jusko, 1994). Total body water content is around 80% in full-term neonates versus 50 to 60% in adults (Friis-Hansen, 1961, 1971). Extracellular fluid decreases from gestation (65% of total) to puberty (20% of total) (Reed, 1996). The apparent volume of distribution (V_d) affected by the percentage of body water relates the amount of chemical in the body to its plasma concentration. The more water-soluble toxicants have higher V_d and lower clearance rates (CL) in infants than in adults. In infants, lipophilic compounds would have lower

V_d higher plasma concentrations, and potentially more rapid clearance, depending on the maturity of elimination processes, than that of water-soluble compounds. The drugs gentamycin, theophylline, and phenytoin have two-to three fold higher V_d values (L/kg body weight) in neonates than in adults. Conversely, in neonates, diazepam, which is more lipophilic, has a much lower V_d value than in adults (Milsap and Jusko, 1994).

Since bound toxicant is not available for action at the receptor, plasma protein binding is an important determinant of toxic response. The extent of binding depends primarily on the quantity of binding proteins available, the binding or affinity constant of the chemical, and the number of binding sites available (Besunder et al., 1988). Total plasma protein concentration and composition change from birth into infancy, and the greatly reduced plasma protein binding observed in neonates (e.g., for theophylline, diazepam, and phenytoin) is probably due to reduced total plasma protein concentration, the persistence of fetal albumin with lower affinity for drugs, and increased blood concentrations of endogenous substances, such as free fatty acids and bilirubin, which can inhibit protein binding of drugs or other exogenous chemicals (Morselli et al., 1980; Brodersen et al., 1983; Herngren et al., 1983). Lower levels of albumin and elevated free fatty acids and bilirubin in the neonate may result in a higher V_d value and higher unbound concentrations of trichloroacetic acid (TCA), a metabolite of perchloroethylene (PCE) and trichloroethylene (TCE), in the blood (Muller et al., 1972; Ginsberg et al., 2002). Other factors that affect tissue distribution of toxicants include higher organ/body weight ratios in infants and children than in adults, and the lack of a mature blood–brain and other tissue–blood barriers. Morphine is three to 10 times more toxic to newborn than to adult rats, due primarily to the higher permeability of the newborn brain to morphine (Rozman and Klaassen, 1996), and the distribution of methyl mercury to the brain is greater if exposure occurs in utero or neonatally than in mature animals (Takeuchi and Morikawa, 1960).

Other childhood factors affecting distribution of xenobiotics include, lower body fat/body weight, lower mass of skeletal muscle/body weight, higher relative brain and liver weights versus adult (6-fold and 2.5-fold, respectively), fetal hemoglobin present in neonates (nitrate and nitrite sensitivity), and lower blood pH in neonates.

Metabolism The development of xenobiotic metabolizing enzymes in utero and after birth affects the rates of activation of chemicals to toxic intermediates, and the detoxification and ultimately the clearance of xenobiotic compounds. Recent reviews have highlighted available information on the ontogeny of the phase I enzymes (Cresteil, 1998; Hines and McCarver, 2002). The total cytochrome P450 content of human liver microsomes is unchanged from fetal life through the first year of postnatal life and is approximately one-third the total adult content (Treluyer et al., 1991). There are fetal isoforms of the CYP enzymes, and the exogenous chemical substrate kinetics of these forms is not well characterized. In general, the level of inducibility of fetal CYP forms is unknown

(Hakkola et al., 1998). Cresteil (1998) describes three groups of neonatal cytochrome P450: CYP3A7 and CYP4A1 present in fetal liver and are active on endogenous substrates; an early neonatal group, including CYP2D6 and CYP 2E1 which surge within hours of birth; and a later-developing group, CYP3A4, CYP2Cs, and CYP1A2. Total CYP3A protein, a major cytochrome P450 enzyme responsible for biotransformation of many exogenous compounds, is relatively constant in neonates and adults, but the isoforms change (Table 1) CYP3A7 expression peaks in the neonatal liver and declines over time to undetectable levels in adult liver; CYP3A7 appears to be responsible for AHH activity in the fetus (Hakkola et al., 1998). CYP3A4 is the primary adult hepatic form of the 3A series and is very low in neonatal liver, increasing slowly after birth; at one month there is about one-third of the CYP3A4 activity as in an adult liver (Lacroix et al., 1997; Hakkola et al., 1998). CYP2E1 is an important enzyme catalyzing metabolism of a number of environmental contaminants, including benzene, trichloroethylene, and toluene. Following birth hepatic CYP2E1 increases gradually, reaching about one-third of adult levels by 1 year of age and is essentially 100% of adult levels by 10 years of age (Vieira et al., 1996; Cresteil, 1998). CYP2D6 levels are low in fetal hepatic tissue and rise after birth, reaching about two-thirds of the adult levels between infancy and 5 years of age (Treluyer et al., 1991). CYP2C9 and 2C19, the most abundant CYP2 enzymes in adult human liver, appear in the week after birth, and are about 30% of adult levels up to 1 year of age (Treluyer et al., 1997). CYP1A2 appears between 1 and 3 months of age, and reaches about half of adult levels at 1 year (Hines and McCarver, 2002). Evidence from both substrate activation and immunological detection indicates CYP1A1 is expressed in fetal liver, where it can activate such xenobiotics as benzo[*a*]pyrene and aflatoxin B1 (Shimada et al., 1996). CYP1A1 appears to be less important in adult liver but present in inducible form in extrahepatic tissues (Hakkola et al., 1998), where as CYP1B1 is present in both adult and fetal extrahepatic tissues and is important in the activation of steroids and polycyclic aromatic hydrocarbons (Hines and McCarver, 2002). Studies of pharmaceutical clearance demonstrate the ontogeny of cytochrome P450 in infants and children, including gender-based differences (e.g., caffeine demethylation) (Kearns et al., 2003).

Although there has been some study of the development of human P450 enzymes and of the phase II conjugation enzymes in the liver (reviewed in Cresteil, 1998; Hines and McCarver, 2002; McCarver and Hines, 2002; Kearns et al., 2003), there is less information about the timing of development of activity in other tissues. Watzka et al. (1999) observed sex- and age-related differences in CYP11A1 activity in the human brain, where enzyme activity increased dramatically from birth and reached adult levels by puberty; levels are significantly higher in women than in men. In the lung, animal studies have shown that exposure to environmental toxicants (side stream tobacco smoke) can induce cytochrome P450 enzymes, resulting in earlier activity (Gebremichael et al., 1995).

The impact of activation of compounds by cytochrome P450 enzymes on toxicity in early life depends on the rate of detoxification. Development of both phase I and phase II enzymes and the physiological elimination processes of the

TABLE 1. Developmental Expression of CYP P450 Forms in the Human Liver

Form	Fetus	Neonate < 4 Weeks	Infant < 12 Months	Child < 15 Years	Adult	Substrate Metabolized or Metabolic Reactions
CYP1A1	+	?	?	?	+	Aryl hydrocarbon hydroxylations, PAHs, halogenated aromatics, aromatic amines (e.g., 2-acetylaminofluorene), estradiolphenoxazone ethers
CYP1A2	−	−/+	+	+++	+++ (18)	Aromatic amines, caffeine *N*-3 demethylation (Kearns et al., 2003)
CYP1B1	−/+	−	−	−	−	PAHs, steroids in extra-hepatic tissues in fetus and adult
CYP2A6/2A13	−	?	+	++	++ (6)	Nicotine, coumarin 7-hydroxylation; more important in extrahepatic tissues; found in fetal nasal mucosa
CYP2A7	−	?	?	?	?	Nicotine?
CYP2B6/7	−	−/+	++	?	+ (<1)	Hexane, cyclophosphamide
CYP2 C (primarily 9 and 19)	+/?	+	+	+++	+++ (25)	Tolbutamide, diazepam, phenytoin, barbiturates—hydroxylation, demethylation (Treluyer et al., 1997)
CYP2D6	+/±	++/+	+	+	+ (2)	NNK, debrisoquine, nortryptiline, sparteine, dextromethorphan (Treluyer et al., 1991)

CYP2E1	−/+	±/+	++	++	++ (9)	Benzene, toluene, ethanol, trichloroethylene, nitrosamines, styrene, acetaminophen, chlorzoxazone (Vieira et al., 1996); greater amounts in developing brain than in liver
CYP2J2	+	?	?	?	++	Retinoic acid, arachidonic acid; also found in fetal olfactory mucosa (Hines and McCarver, 2002)
CYP3A4	+/−	+	++	+++	+++ (40)	Benzo[*a*]pyrene-7,8 diol, aflatoxin, 1-nitropyrene; hydroxylation of dehydroepiandrosterone (Lacroix et al., 1997)
CYP3A5	+	+	+	++	++ (10–30)	Benzo[*a*]pyrene-7,8 diol, aflatoxin, 1-nitropyrene; hydroxylation of dehydroepiandrosterone (Lacroix et al., 1997)
CYP3A7	+++(30)	++	−	−	−	Substrate specificity similar to 3A4/5; hydroxylation of testosterone and dehydroepiandrosterone (Lacroix et al., 1997)

Source: Adapted from Hakkola et al. (1998); additional information from Hines and McCarver (2002) and Kearns (2003).

[a]Symbols for detection of CYP mRNA/protein (when different): ?, unknown; −, not detected; ±, possibly present in small quantities; +, present in low concentration; ++, present in moderate concentration; +++, present in high concentration. Values in parentheses are a percentage of the total P450.

liver and kidney all affect the clearance of a xenobiotic chemical. For several drugs, Ginsberg et al. (2004) note that clearance in the neonate and young infant is lower than that in adults, a little higher than that in adults from 6 months to 2 years of age, and about the same as an adult in later years. This is partly explained by the larger liver/body weight ratio in infants, coupled with development of the metabolizing enzymes postnatally.

Capacity for repair can be greater or lesser in infants than adults. In neonatal rabbits, repair of injury to pulmonary Clara cells by toxicants activated by cytochrome P450 enzymes is decreased, resulting in permanent structural changes in the adult animal (Smiley-Jewell et al., 1998, 2000).

The activity of phase II conjugation reactions, which are usually detoxifying, is generally lower in the neonate than in the adult (Milsap and Jusko, 1994). Hence, there is concern that detoxification and elimination of chemicals is slower in infants. Conjugation with glucuronic acid by a number of UDP-glucuronyltransferases is significantly lower at birth, with overall activities 2.5-fold below adult levels (Levy et al., 1975). Expression of some of the UGT enzymes matures to adult levels in two months after birth, although glucuronidation of some drugs by the UGT1A subfamily does not reach adult levels until puberty (Snodgrass, 1992; McCarver and Hines, 2002). Removal of aniline, *N*-hydroxyarylamines, phenol, and benzene metabolites is slowed by reduced glucuronidation in neonates. Acetylation by the n-acetyltransferases and sulfation by sulfotransferases are generally high in newborn infants and somewhat comparable to adult levels, although it may vary by tissue and by specific sulfotransferase (McCarver and Hines, 2002). Neonates may conjugate drugs or environmental chemicals with sulfate rather than glucuronic acid (e.g., acetaminophen) because the latter enzymatic pathway is less mature. There are several forms of glutathione (GSH) sulfotransferases (GSTs), with GST-P_i prevalent in the fetus and decreasing postnatally. GST-alpha and GST-mu are detected in fetal liver and increase over the first few years of life to adult levels (McCarver and Hines, 2002). Blood esterase activity, which is less than half that of adults at birth, is more depressed in premature infants than in full-term infants and does not reach the latter's activity for 10 to 12 months. Low esterase activity may partly account for the prolonged effect of local anesthetics observed during delivery (Ecobichon and Stephens, 1973) and may be important in the detoxification of organophosphate pesticides (Furlong et al., 2006). Epoxide hydrolase, important in detoxifying reactive epoxide metabolites, is present in fetal liver, although at much reduced activity relative to adults (McCarver and Hines, 2002).

As a result of differing enzyme activity, some chemicals are metabolized by wholly different metabolic pathways, depending on age. In infants, theophylline is N-methylated to caffeine, a minor pathway for adults in whom the majority of theophylline dose is N-demethylated or C-oxidated to monomethylxanthines or methyluric acid. Several studies have evaluated age-related pharmacokinetic differences in humans using information about drug disposition (Renwick, 1998; Renwick et al., 2000; Ginsberg et al., 2002; Hattis et al., 2003). Calculation of internal doses as the area under the blood concentration–time curve (AUC) for the same doses (mg/kg) indicated that the major difference from adults occurs in preterm and full-term neonates and young infants (Renwick, 1998). Higher AUC internal doses in neonates and young infants versus adults were noted for seven drugs which are substrates for glucuronidation, one with substrate specificity for CYP1A2 and four with substrate specificity for CYP3A4 metabolism, and interindividual variation in elimination by these three routes did not differ by age group (Renwick et al., 2000). Ginsberg et al. (2002) evaluated pharmacokinetic information on 45 drugs in children and adults metabolized by different cytochrome P450 pathways, phase II conjugations, or eliminated unchanged by the kidney. These authors noted half-lives in infants three- to nine fold longer than those of adults. In evaluating the interindividual variability by age, Hattis et al. (2003) note that the largest interindividual variability occurs in the youngest children, apparently due to variability in development of critical metabolism and elimination pathways. Notably, these authors observe that for risk assessment modeling, unimodal distributions may be inadequate for young children and for overweight older children.

While variation in pharmacokinetics with age is an important consideration in risk assessment, additional complexity is added by genetic polymorphisms, rendering some persons more susceptible than others. For example, Perera et al. (1999) have shown that there is significant transplacental transfer of polycyclic aromatic hydrocarbons and environmental tobacco smoke constituents from mother to fetus, that PAH DNA adducts in maternal and newborn white blood cells are increased from environmental exposure, that the fetus is more sensitive to genetic damage than the mother, and that newborns with a specific restriction fragment length polymorphism (RFLP), CYP1A1 Msp1, had elevated numbers of adducts compared to those without the RFLP.

Excretion Many drugs are more slowly eliminated in neonates and infants than in adults (Kearns et al., 2003). This difference would also apply to some environmental contaminants in drinking water. Differences in volume of distribution, metabolism, and the maturity of renal and hepatic elimination processes all play a role. Renal elimination depends on the maturity of processes related to tubular reabsorption and secretion, and glomerular filtration rates. At birth, the glomerular filtration rate (GFR) is low (2 to 4 ml/min), increases in the first few days (8 to 20 ml/min), and increases slowly to adult values in 8 to 12 month-old infants (Plunkett et al., 1992; Kearns et al., 2003). Premature infants may have very low GFR (0.6 to 0.8 mL/min), which correlates with their gestational age (Milsap and Jusko, 1994; Robillard et al., 1999).

Newborn and young animals have less capacity to excrete chemicals into the bile than do adult animals. A number of drugs and prototype compounds are excreted more slowly in the bile of neonates than in adult rats, including ouabain, sulfobromophthalein and the glucuronide metabolites (Klaassen, 1973). In neonatal rats, methyl mercury is poorly eliminated in the bile, the main elimination route in adults, resulting in a longer half-life (Ballatori and Clarkson, 1982).

Thus, childhood factors that affect excretion include: lower renal function in neonates and young children with tubular secretion and GFR increasing during infancy; and slowly developing biliary function despite a larger liver/body weight ratio.

Pharmacokinetic Modeling

Disposition of toxicants may differ among infants, children, and adults due to differences in absorption, distribution, metabolism, and excretion. The disposition of environmental toxicants can be modeled using physiologically based pharmacokinetic (PBPK) models, which give predictions of internal body or tissue dosimetry. A few published studies used PBPK models with physiological inputs appropriate for different age groups, but little systematic exploration of PBPK modeling for varied age groups has been conducted. Published PBPK modeling exercises of environmental toxicants and drugs indicate that infants in the first year of life are likely to show increased internal dosages of a number of chemicals and their metabolites relative to adults (Ginsberg et al., 2002; Hattis et al., 2003). In assessing risks of environmental contaminants to infants and children, the following areas should be addressed as much as is practicable by PBPK modeling:

- PBPK modeling for human dosimetry should include infant and child anatomical and physiological profiles to address potentially critical developmental stages (e.g., newborn, 3, 6, 9 months, 1 to 2 years, 2 to 8 years, 8 to 16 years) for both male and female.
- Values based on age-related metabolism of drugs using similar enzymes should be incorporated if available; scaling from adult values may be the only alternative where data are unavailable.
- When appropriate, subpopulations that may be at greater risk from exposure to a given environmental toxicant due to genetic polymorphisms, body composition, and illness should also be included in the analysis if possible.

Pharmacodynamics Pharmacodynamics (PD) is the interaction of a chemical with a receptor and the subsequent biological response. Interactions with receptors will vary by age and can manifest as either qualitative or quantitative differences in toxicity. The typical risk assessment is generally based on toxicological or epidemiological information from mature animals or humans, with the exception of risk assessments based on developmental toxicity studies. Toxic endpoints of special concern with respect to both prenatal and postnatal exposures include cancer, asthma, endocrine effects, neurotoxicity and neurodevelopmental effects, immunotoxicity, and other developmental effects. Whenever possible, risk assessments should evaluate experimental evidence obtained in young animals (or children) to account for potential age-related changes.

Conclusions

There are a number of physiological and behavioral factors that differ between children and adults, resulting in differences in exposure and response to toxicants present in water, air, food, and soil. Both behavioral and physiological factors influence exposure at each portal of entry, absorption, distribution, metabolism, and excretion of toxicants. In addition, infants and children may have qualitatively different responses due to different target tissue sensitivities during windows of susceptibility in the developmental process. All of these considerations should be addressed as much as is practicable in risk assessments to ensure adequate protection of infants and children from exposure to environmental contaminants.

CRITICAL SYSTEMS AND PERIODS IN DEVELOPMENT

Development is a complex phenomenon involving coordination of development of organ structure and function. During development, an organism may be more sensitive to toxicants if they disrupt the coordinated developmental processes. Many developmental toxicants have been shown to alter structure and/or function in animals and humans when exposure occurs pre- or postnatally, sometimes in the absence of significant maternal toxicity. Preconceptional exposures to environmental chemicals may also be important for the health of the offspring.

Effects that occur in early development can dramatically affect children's well-being, quality of life, and survival (Wilson, 1977; Schardein, 1993; Kimmel, 2001). Birth defects are a major cause of infant and child morbidity and mortality in the United States. Thus, consistent with good public health practice because of the varying effects and timing of sensitivity for developmental toxicity, and a general lack of information on developmental windows of susceptibility that may be affected by a chemical, it is necessary for the risk assessor to consider both prenatal and postnatal exposures in order to protect infants and children. Reference doses for chemicals for which developmental toxicity is the most sensitive endpoint could be considered protective of infants and children.

In considering compounds that are developmental toxicants, the risk assessor should determine whether developmental toxicity was the most sensitive endpoint for that compound. If other toxicological endpoints are observed at exposures lower than those necessary to induce developmental toxicity, the developmental toxicity of the compound is not necessarily a driving force in developing a reference dose.

Central Nervous System

The development of the central nervous system occurs both in utero and postnatally through adolescence. Key processes in brain development include: (1) increase in brain mass (Dobbing and Sands, 1979) through proliferation of neurons and support cells, (2) migration and differentiation of neuronal cells, (3) establishment of connections and the reinforcement of selected neuronal pathways through synaptogenesis and synaptic pruning, and (4) formation of myelin (Rodier, 2004). The developing brain is uniquely sensitive to disruption, and these processes are unidirectional and occur at very specific times for different structures. During in utero development, closure of the neural tube occurs, and proliferation of neurons and migration of cortical neurons are important. Proliferation and migration continue in infancy and early childhood, along with synaptogenesis, myelination, and development of the blood–brain barrier. Structural maturation of neural pathways continues through childhood and adolescence, and involves synaptic pruning as well as an increase in the diameter and myelination of axons. Certain toxicants can have profound effects on all of these neurologic developmental processes (Rodier, 1994, 1995, 2004; Paus et al., 1999; Golub, 2000; Colborn, 2006). Chemicals that may affect multiple developmental processes resulting in permanent brain dysfunction are summarized in Table 2.

New neurons are produced from stem cells and progenitor cells that appear in the central nervous system very early in development, and are present even in adult brains to a limited extent (Jacobs et al., 2000). Since there is much more cell proliferation in younger brains, they are more sensitive to xenobiotics that interfere with cellular proliferation. Indeed, methyl mercury, which arrests mitosis, is particularly injurious to the developing brain, as demonstrated by the profound effects of prenatal exposures in Minamata disease (Rodier, 2004).

Another critical developmental process involved in restructuring is the selection of certain synaptic connections and the elimination of others. Neurotransmitters such as GABA and acetylcholine are essential to this process, and xenobiotics that interfere with the function of these neurotransmitters can also interfere with the development of appropriate synaptic connections. Chemicals can also trigger or inhibit apoptosis, which results in the elimination of some neurons that are

TABLE 2. Chemicals Associated with Disruption of Neurodevelopmental Processes

Developmental Process	Chemicals That Disrupt This Process[a]
Proliferation	Methylazoxymethanol (MAM), ethanol, methyl mercury, chlorpyrifos
Migration	MAM, methyl mercury, ethanol
Differentiation	Ethanol, nicotine, methyl mercury, **lead**
Synaptogenesis	Ethanol, **lead**, triethyltin, parathion, **polychlorinated biphenyls (PCBs)**
Gliogenesis and myelination	Thyroid hormone disruptors, ethanol, **lead**
Apoptosis	Ethanol, **lead**, methyl mercury, barbiturates, glutamine, halothane, ketamine
Neurotrophic signaling	**Aluminum**, ethanol, cholinesterase inhibitors, methyl mercury

Source: Based on Rice and Barone (2000) and Olney et al. (2000).

[a]Chemicals in bold type are common water contaminants.

needed and the retention of others that are meant to die during synaptic pruning (Olney et al., 2000). Fetal alcohol syndrome is a recognized consequence of maternal–fetal exposure during critical periods of brain development (Clarren and Smith, 1978), and is thought to partially involve enhanced apoptosis in the third trimester.

Autism, an impairment of social interaction along with abnormal speech development and unusual behaviors, is associated with exposure to thalidomide during days 20 to 24 of gestation, but not before or after this time. This window of susceptibility corresponds with the production of the first neurons forming the motor nuclei of the cranial nerves, and thalidomide exposure injures these forming nerves (Rodier et al., 1997). This type of damage during specific critical windows of susceptibility can result in irreversible injury. The effect of a chemical is dependent on the cellular process that it affects and the structures that may be undergoing that process at the time of exposure.

Arsenic and lead in drinking water may inhibit cognitive and behavioral development in human children (Calderon et al., 2001; Tsai et al., 2003; Wright et al., 2006). Animal studies support this observation. Arsenic administered orally to Wistar rats decreases acetylcholinesterase activity in the hypothalamus, cerebellum, and brainstem, and slows the ability to learn and unlearn tasks, an effect that is more pronounced in younger rats (Nagaraja and Desiraju, 1994).

Endocrine System

The endocrine system consists of glands and other structures that secrete hormones directly into the circulatory system, which in turn influence other bodily functions, including growth and metabolism. Like the central nervous system, the endocrine system is extremely complex and imperfectly understood. Due to endocrine involvement in growth and development, early life exposure to endocrine disruptors is a serious concern. Thyroid hormones are critical to brain development in utero and early childhood. Certain polychlorinated biphenyls and dibenzo-p-dioxins produce neurological impairment in animals by altering the thyroid function during critical periods of thyroid-hormone dependent brain development (Porterfield, 2000). Prenatal exposure of children to PCBs from contaminated fish and cooking oil has been associated with impaired cognitive development, decreased short-term and long-term memory, and less ability to focus (Chen et al., 1992; Jacobson and Jacobson, 1996). These effects may be due to thyroid hormone disruption. Perchlorate, a common drinking water contaminant used in the manufacture of solid propellants for rockets and missiles, inhibits iodine uptake by the thyroid. This has prompted concern about exposures of pregnant women and young infants to perchlorate in drinking water due to the association between maternal thyroid hormone levels and brain development in utero as measured by IQ, as well as the importance of adequate thyroid hormone levels for postnatal brain development (Haddow et al., 1999; Klein et al., 2001; Colborn, 2004). At least one study suggests that risk assessors need to consider a cumulative risk approach when evaluating endocrine disruptors (Crofton et al., 2005).

In that study, thyroid hormone disruption from several different polyhalogenated aromatics (dioxins and PCBs) demonstrated dose additivity at the lowest dose but synergism at higher doses.

Sex hormone disruption can also be caused by environmental toxicants, with concomitant adverse impacts on children. Estrogens are involved in growth, including bone maturation in both boys and girls prior to puberty, drive growth and development of reproductive organs and mammary tissue in girls (Aksglaede et al., 2006), and modulate intracellular signaling such as occurs during brain development (Zsarnovszky et al., 2005). Thus, xenobiotic estrogens have the potential to influence growth and development in humans. In addition, xenobiotic estrogens can predispose certain tissues to cancer later in life [i.e, diethylstilbestrol (DES)]. Studies have linked elevated DDE in breast milk with decreased duration of lactation in women possibly due to estrogenic properties of DDE (Rogan et al., 1987; Gladen and Rogan, 1995). DDE also appeared to be correlated with higher body mass in both boys and girls (Gladen et al., 2000). Chemicals that are androgenic or antiandrogenic may influence development in boys. Phthalate ester plasticizers have been associated with decreased masculinization, particularly following in utero exposure, in animal models (Higuchi et al., 2003) and in human infants (Swan et al., 2005).

Immune System

The development of the immune system is complex and is characterized by several unique events that are not part of adult immune function. Early in development, immune system cells must differentiate and migrate to virtually all organs, where they provide regulatory as well as host defense roles. The Kupffer cells in the liver, alveolar macrophages, testicular macrophages, and microglia in the brain are all examples of cells that migrate during development and are critical regulatory cells in their respective organs. Disrupting immune system development can lead to lifelong immune dysfunction. The review by Dieter and Piepinbrink (2006) integrates available information on the disruption of key immune system developmental processes by xenobiotics and the implications for human health. For example, polycyclic aromatic hydrocarbons, TCDD, and tributyltin interfere with the seeding of the thymus with pro-T cells or with thymopoiesis, resulting in thymic atrophy and decreased T-cell function postnatally (Holladay and Smialowicz, 2000). Lead, mercury, and other toxicants can selectively impair TH1 cells, increasing the risk of TH2-mediated atopy and asthma. The effects on the developing immune system are not possible to predict from toxicological studies in adults (Dieter and Piepinbrink, 2006).

Environmental contaminants that have been found to cause developmental immunotoxicity in rodents include (Holladay and Smialowicz, 2000):

- Polycyclic halogenated hydrocarbons (TCDD, PCB, and PBB)
- Polycyclic aromatic hydrocarbons

- Pesticides (hexachlorocyclohexane, chlordane, diazinon, DDT, carbofuran hexachlorobenzene, and tributyltin)
- Metals (mercury, lead, and cadmium)
- Hormonal substances (estrogens and diethylstilbestrol, testosterone, and cortisone)

AGE AT EXPOSURE AND SUSCEPTIBILITY TO CARCINOGENS

A number of animal and human studies indicate that age at which exposure to a carcinogenic occurs is an important determinant of tumorigenesis for some carcinogens including vinyl chloride, diethylstilbestrol, tamoxifen, polycyclic aromatic hydrocarbons (e.g., dimethylbenzanthracene), nitrosourea compounds (ethyl and methyl nitrosourea), nitrosamines, and alkenylbenzenes (Anderson et al., 2000; Miller et al., 2002). Indeed, a neonatal mouse model, in which animals are dosed with chemicals twice before weaning, was developed to test the carcinogenic potential of mutagenic chemicals because of the sensitivity of neonatal animals to such carcinogens (Flammang et al., 1997; McClain et al., 2001).

Human studies have shown that radiation exposure of the chest wall during puberty in girls, for example for treatment of Hodgkin's disease, greatly elevates the risk for early-onset breast cancer (Bhatia et al., 1996; Aisenberg et al., 2000). Studies of Japanese atomic bomb survivors also demonstrate increased risks for some tumors dependent on age at exposure, with generally increasing relative risk for younger age at exposure; leukemia risk increased the most for those who were younger than 10 years at the time of exposure (NRC, 1990; Thompson et al., 1994). DES is a synthetic estrogen that induces vaginal and cervical tumors in adolescent girls whose mothers took the drug during pregnancy (Preston-Martin, 1989; Hatch et al., 1998). It is also well known that chemotherapy in childhood increases the risk for cancer later in life (Kushner et al., 1998; Anderson et al., 2000). In a study of humans exposed to arsenic in drinking water in Chile, Smith et al. (2006) demonstrated that earlier exposure (in utero and childhood) results in significantly greater risks of both bronchiectasis and early-onset lung cancer relative to exposure as adults.

A number of animal studies of chemical carcinogenesis have demonstrated that short-term or single doses of prototype carcinogens in the fetal or neonatal period result in tumors later in life, and that in some cases the tumors developed with a shorter latency or in a different organ or from different cell types within the same organ than when exposure occurred in the mature animal. Age at exposure modulated the histology and origin (mesenchymal.epithelial) of renal tumors in response to dimethylnitrosamine (Hard, 1979), and the incidence and latency of liver tumors in mice in response to diethylnitrosamine (Vesselinovitch et al., 1984). When rodents are exposed to vinyl chloride early in life, a higher incidence of neoplasms is observed relative to animals exposed later in life (Drew et al., 1983; Maltoni and Cotti, 1988). The observation of increased potency of vinyl chloride following exposure at young ages was incorporated into EPA's risk assessment for vinyl chloride (Cogliano et al., 1996; U.S. EPA, 2000). The induction of malignant tumors of the brain and nervous system after a single injection of azoxymethane in rats was highly dependent on age, with the youngest animals developing the highest incidence, which decreased to zero in the adult animals (Druckrey and Lange, 1972). Transplacental carcinogenesis has also been demonstrated for a number of toxicants including urethane and arsenic (Anderson et al., 2000; Waalkes et al., 2003). In utero but not adult exposure to tamoxifen, a widely used pharmaceutical, and genistein, a phytoestrogen found in soy, in animal models results in elevated incidence of mammary and reproductive organ tumors in the offspring (Newbold et al., 1997; Carthew et al., 2000).

Traditional quantitative estimates of cancer potency which are used in cancer risk assessment are based either on carcinogenicity bioassay data in rodents, during which exposure starts when the animals are already 6 to 8 weeks of age and sexually mature, or on human occupational epidemiology data. In either case, the influence of early life exposures to carcinogens on cancers occurring later in life is missed entirely. It should be noted that the impacts of carcinogen exposure on childhood cancer are poorly understood. Cancer risk assessment as currently practiced attempts to predict risk of cancers that typically manifest in late adulthood, and does not address cancers that occur in childhood.

Based on evidence in the toxicological and epidemiological literature, the EPA recently developed guidance for risk assessment of carcinogens which includes weighting for age at exposure, particularly for genotoxic carcinogens (U.S. EPA, 2005). The EPA analyzed data from studies where it was possible to compare the potency of a carcinogen when exposure occurred in early postnatal life to the potency when given to mature animals. In the EPA's assessment of potency by age at exposure, the geometric mean of the ratio of juvenile to adult cancer potencies for 12 mutagenic carcinogens was 10 (Barton et al., 2005; U.S. EPA, 2005). For six chemicals acting through other modes of action, the geometric mean of the juvenile to adult potency ratio was about 3 (Barton et al., 2005). Based on their analysis, the EPA has instituted a policy of weighting the cancer risk from exposures to genotoxic carcinogens between 0 and <2 years of age by a factor of 10, and between 2 and <15 years of age by a factor of 3 (U.S. EPA, 2005). The guidance states that age-appropriate exposure factors should be utilized in assessing cancer risk.

DRINKING WATER STANDARDS DEVELOPED TO PROTECT SENSITIVE SUBPOPULATIONS

While drinking water goals and standards are generally developed to protect sensitive subpopulations, few are specifically based on sensitive subpopulation endpoints. Although sensitive subpopulations are mentioned when justifying standard uncertainty factors, the point of departure for drinking water standards are for the most part based on studies in healthy mature animals or adults. Among drinking water standards, those based on specific sensitive subpopulations include copper, fluoride, lead, nitrate, nitrite, and methoxychlor. Although there has been much discussion of sensitive subpopulations with regard to development of drinking water goals for perchlorate (see Chapter 11), the actual basis of the drinking water risk assessments is iodide inhibition in healthy adults (California OEHHA, 2004; NAS, 2005).

Drinking water standards for copper are designed to prevent adverse gastrointestinal effects. The EPA maximum contaminant level goal (MCLG) of 1.3 mg/L for copper in drinking water is based on preventing adverse effects in healthy adults (U.S. EPA, 1991a). The California Public Health Goal (PHG) of 170 ppb for copper in drinking water is based on gastrointestinal effects in children, the sensitive group for this chemical (California OEHHA, 1997a). The EPA's maximum contaminant level (MCL) of 4 mg/L for fluoride in drinking water was set to protect against crippling skeletal fluorosis, with a secondary MCL of 2 mg/L to protect against dental fluorosis (in mild cases, fluorosis is a slight discoloration of teeth, in more severe cases it can lead to pitting and breaking of the teeth) (U.S. EPA, 1986). Similarly, the California PHG of 1 ppm (1000 ppb) for fluoride in drinking water is based on a no-observed-adverse-effect level (NOAEL) of 1 mg/L for dental fluorosis in children (California OEHHA, 1997b). The health effects that can be caused by lead exposure in infants and children are well documented and include delays in physical or mental development and deficits in attention span and learning abilities. Similarly, the effects of lead on high blood pressure in adults are also well studied. Although the EPA has adopted an MCLG of zero for lead in drinking water, based on "occurrence of low level effects" and because EPA classifies lead as a class B2 carcinogen, the agency has not adopted an MCL for lead in drinking water because they regard the development of such a level as "not feasible" (U.S. EPA, 1991a, 2006b). A California PHG of 2 ppb for lead in drinking water is based on the neurobehavioral effects of lead in children and the hypertensive effects of lead in adults (California OEHHA, 1997c). The U.S. EPA and California MCLs of 10 ppm for nitrate and 1 ppm for nitrite in drinking water (calculated as N) are based on the protection of infants from the occurrence of methemoglobinemia, the principal toxic effect observed in humans exposed to nitrate or nitrite (U.S. EPA, 1990, 1991b; California OEHHA 1997d). The U.S. EPA and California standards for methoxychlor of 40 ppb in drinking water are based on animal data for protection of the reproductive health of pregnant women and the safety of their offspring (California DHS, 1977; California OEHHA, 1999; U.S. EPA 2006c).

Disclaimer

The opinions expressed in this chapter are those of the authors and not neccessarily those of the Office of of Enviromental Health Hazard Assessment or the California Enviromental Protection Agency.

REFERENCES

Agunod M, Yamaguchi N, Lopez R, Luhby AL, Glass GBJ. 1969. Correlative study of hydrochloric acid, pepsin, and intrinsic factor secretion in newborns and infants. *Am J Dig Dis* 14: 400–414.

Aisenberg AC, Finkelstein DM, van Leeuwen FE, Klokman WJ, Aleman BMP. 2000. Second malignancies in Hodgkin's disease. *J Clin Oncol* 18: 2186–2187.

Aksglaede L, Juul A, Leffers H, Skakkebaek NE, Andersson A-M. 2006. The sensitivity of the child to sex steroids: possible impact of exogenous estrogens. *Human Reprod Update* 12: 341–349.

Anderson LM, Diwan BA, Fear NT, Roman E. 2000. Critical windows of exposure for children's health: cancer in human epidemiological studies and neoplasms in experimental animal models. *Environ Health Perspect* 108 (suppl 3): 673–694.

Avery GB, Randolph JG, Weaver T. 1966. Gastric acidity in the first day of life. *Pediatrics* 37: 1005–1007.

Ballatori N, Clarkson TW. 1982. Developmental changes in the biliary excretion of methylmercury and glutathione. *Science* 216: 61–63.

Barton HA, Cogliano VJ, Flowers L, Valcovic L, Setzer W, Woodruff TJ. 2005. Assessing susceptibility from early-life exposure to carcinogens. *Environ Health Perspect* 113 (9): 1125–33.

Bearer C. 1995. How are children different from adults? *Environ Health Perspect* 103 (suppl 6): 7–12.

Besunder JB, Reed MD, Blumer JL. 1988. Principles of drug biodisposition in the neonate; a critical evaluation of the pharmacokinetic–pharmacodynamic interface (Part 1). *Clin-Pharmacokinet* 14: 189–216.

Bhatia S, Robison LL, Oberlin O, Greenberg M, Bunin G, Fossati-Bellani F, Meadows AT. 1996. Breast cancer and other second neoplasms after childhood Hodgkin's disease. *N Engl J Med* 334 (12): 745–751.

Brodersen R, Friis-Hansen B, Stern L. 1983. Drug-induced displacement of bilirubin from albumin in the newborn. *Dev Pharmacol Ther* 6: 217–229.

Calderon J, Navarro ME, Jimenez-Capdevill ME, Santos-Diaz MA, Golden A, Rodriguez-Leyva I, Borja-Aburto V, Diaz-Barriga F. 2001. Exposure to arsenic and lead and neuropsychological development in Mexican children. *Environ Res* A 85: 69–76.

Carthew P, Edwards RE, Nolan BM, Martin EA, Heydon RT, White IN, Tucker MJ. 2000. Tamoxifen induces endometrial and vaginal cancer in rats in the absence of endometrial hyperplasia. *Carcinogenesis* 21 (4): 793–7.

California DHS (Department of Health Services). 1977. *California Code of Regulations* Title 22, Article 5.5, Primary Standards:–Organic Chemicals, Section 65555. CDHS, Sacramento, CA.

California OEHHA (Office of Environmental Health Hazard Assessment). 1997a. *Public Health Goal for Copper in Drinking Water*. California Environmental Protection Agency, Oakland, CA. Accessed at: http://www.oehha.ca.gov/water/phg/pdf/copper_c.pdf.

——. 1997b. *Public Health Goal for Fluoride in Drinking Water*. California Environmental Protection Agency, Oakland, CA. Accessed at: http://www.oehha.ca.gov/water/phg/pdf/fluor_c.pdf.

______. 1997c. *Public Health Goal for Lead in Drinking Water*. California Environmental Protection Agency, Oakland, CA. Accessed at: http://www.oehha.ca.gov/water/phg/pdf/lead_c.pdf.

______. 1997d. *Public Health Goals for Nitrate and Nitrite in Drinking Water*. California Environmental Protection Agency, Oakland, CA. Accessed at: http://www.oehha.ca.gov/water/phg/pdf/nit2_c.pdf.

______. 1999. *Public Health Goal for Methoxychlor in Drinking Water*. California Environmental Protection Agency, Oakland, CA. Accessed at: http://www.oehha.ca.gov/water/phg/pdf/metox_f.pdf.

______. 2000. *Air Toxics Hot Spots Program Risk Assessment Guidelines Part IV*, Technical Support Document for Exposure Assessment and Stochastic Analysis. California Enviromental Protection Agency, Oakland, CA, October. Accessed at: www.oehha.ca.gov.

______. 2001. *Prioritization of Toxic Air Contaminants Under the Children's Environmental Health Protection Act*. California Environmental Protection Agency, Oakland, CA, October. Accessed at: http://www.oehha.ca.gov/air/toxic_contaminants/SB25finalreport.html.

______. 2004. *Public Health Goal for Perchlorate in Drinking Water*. California Environmental Protection Agency, Oakland, CA. Accessed at: http://www.oehha.ca.gov/water/phg/pdf/finalperchlorate31204.pdf.

Chen YC, Guo YL, Hsu CC, Rogan WJ. 1992. Cognitive development of Yu-Cheng ("oil disease") children prenatally exposed to heat-degraded PCBs. *JAMA* 268: 3213–3218.

Clarren SK, Smith DW. 1978. The fetal alcohol syndrome. *N Engl J Med* 298: 1063–1067.

Cogliano VJ, Hiatt GFS, Den A. 1996. Quantitative cancer assessment for vinyl chloride: indications of early life sensitivity. *Toxicology* 111: 21–28.

Colborn T. 2004. Neurodevelopment and endocrine disruption. *Environ Health Perspect* 112: 944–949.

______. 2006. A case for revisiting the safety of pesticides: a closer look at neurodevelopment. *Environ Health Perspect* 114: 10–17.

Cresteil T. 1998. Onset of xenobiotic metabolism in children: toxicological implications. *Food Addit Contam* 15(suppl): 45–51.

Crofton KM, Craft ES, Hedge JM, Gennings C, Simmons JE, Carchman RA, Carter WH Jr, DeVito MJ. 2005. Thyroid-hormone disrupting chemicals: evidence for dose-dependent additivity or synergism. *Environ Health Perspect* 113: 1549–1554.

Dieter RR, Piepinbrink MS. 2006. Perinatal immunotoxicity: why adult exposure assessment fails to predict risk. *Environ Health Perspect* 114: 477–483.

Dobbing J, Sands J. 1979. Comparative aspects of the brain growth spurt. *Early Hum Dev* 3: 79–83.

Drew RT, Boorman GA, Haseman JK, McConnell EE, Busey WM, Moore JA. 1983. The effect of age and exposure duration on cancer induction by a known carcinogen in rats, mice, and hamsters. *Toxicol Appl Pharmacol* 68: 120–130.

Drexler H, Schaller K-H. 1998. The mercury concentration in breast milk resulting from amalgam fillings and dietary habits. *Environ Res A* 77: 124–129.

Druckrey H, Lange A. 1972. Carcinogenicity of azoxymethane dependent on age in BD rats. *Fed Proc* 31 (5): 1482–1484.

Ecobichon DJ, Stephens DS. 1973. Perinatal development of human blood esterases. *Clin Pharmacol Ther* 14: 11–14.

Ershow AG, Cantor KP. 1989. Total water and tapwater intake in the United States: population-based estimates of quantities and sources. NCF order 263-MD-810264. National Cancer Institute, Washington, DC.

Findlay JWA. 1983. The distribution of some commonly used drugs in human breast milk. *Drug Metab Rev* 14: 653–684.

Fleishaker JC, Desai N, McNamara PJ. 1987. Factors affecting the milk-to-plasma drug concentration ratio in lactating women: physical interactions with protein and fat. *J Pharm Sci* 76: 189–193.

Flammang TJ, Von Tungeln LS, Kadlubar FF, Fu PP. 1997. Neonatal mouse assay for tumorigenicity: alternative to the chronic rodent bioassay. *Regul Toxicol Pharmacol* 26: 230–240.

Friis-Hansen B. 1961. Body water compartments in children: changes during growth and related changes in body composition. *Pediatrics* 28: 169–181.

______. 1971. Body composition during growth. in vivo measurements and biochemical data correlated to differential anatomical growth. *Pediatrics* 47: 264–274.

Furlong CE, Holland N, Richter RJ, Bradman A, Ho A, Eskenazi B. 2006. PON1 status of farmworker mothers and children as a predictor of organophosphate sensitivity. *Pharmacogenet Genom* 16 (3): 183–190.

Furst P. 2006. Dioxins, polychlorinated biphenyls and other organohalogen compounds in human milk: levels, correlations, trends and exposure through breastfeeding. *Mol Nutr Food Res* 50: 922–933.

Gebremichael A, Chang AM, Buckpitt AR, Plopper CG, Pinkerton KE. 1995. Postnatal development of cytochrome P4501A1 and 2B1 in rat lung and liver: effect of aged and diluted sidestream cigarette smoke. *Toxicol Appl Pharmacol* 135: 246–53.

Ginsberg G, Hattis D, Sonawane B, Russ A, Banati P, Kozlak M, Smolenski S, Goble R. 2002. Evaluation of child/adult pharmacokinetic differences from a database derived from the therapeutic drug literature. *Toxicol Sci* 66: 185–200.

Ginsberg G, Hattis D, Miller R, Sonawane B. 2004. Pediatric pharmacokinetic data: Implications for environmental risk assessment for children. *Pediatrics* 113: 973–983.

Gladen BC, Rogan WJ. 1995. DDE and shortened duration of lactation in a northern Mexican town. *Am J Publie Health* 85 (4): 504–508.

Gladen BC, Ragan NB, Rogan WJ. 2000. Pubertal growth and development and prenatal and lactational exposure to polychlorinated biphenyl and dichlorodiphenyl dichloroethene. *J Pediatr* 136 (4): 490–496.

Golub M. 2000. Adolescent health and the environment. *Environ Health Perspect* 108: 355–362.

Groot ME, Lekkerkerk MC, Steenbekkers MLA. 1998. *Mouthing Behavior of Young Children: An Observational Study*. Agricultural University-Wageningen, Wageningen, The Netherlands.

Gulson BL, Jameson CW, Mahaffey KR, Mizon KJ, Patison N, Law AJ, Korsch MJ, Salter MA. 1998. Relationships of lead in breast milk to lead in blood, urine, and diet of the infant and mother. *Environ Health Perspect* 106 (10): 667–674.

Haddow JE, Palomaki GE, Allan WC, Williams JR, Knight GJ, Gagnon J, O'Heier CE, Mitchell ML, Hermos RJ, Waisbren SE, Faix JD, Klein RZ. 1999. Maternal thyroid deficiency during pregnancy and subsequent neuropsychological development of the child. *N Eng J Med* 341: 549–555.

Hakkola J, Tanaka E, Pelkonen O. 1998. Developmental expression of cytochrome P450 enzymes in human liver. *Pharmacol Toxicol* 82: 209–217.

Hallen IP, Jorhem L, Oskarsson A. 1996. Placental and lactational transfer of lead in rats: a study on the lactational process and effects on offspring. *Arch Toxicol* 69: 596–602.

Hard G. 1979. Effect of age at treatment on incidence and type of renal neoplasm induced in the rat by a single dose of dimethylnitrosamine. *Cancer Res* 39: 4965–4970.

Hatch EE, Palmer JR, Titus-Ernstoff L, Noller KL, Kaufman RH, Mittendorf R, Robboy SJ, Hyer M, Cowan CM, Adam E, Colton T, Hartge P, Hoover RN. 1998. Cancer risk in women exposed to diethylstilbestrol in utero. *JAMA* 280 (7): 630–634.

Hattis D, Ginsberg G, Sonawane B, Smolenski S, Russ A, Kozlak M, Goble R. 2003. Differences in pharmacokinetics between children and adults, II. Children's variability in drug elimination half-lives and in some parameters needed for physiologically-based pharmacokinetic modeling. *Risk Anal* 23: 117–142.

Heimann G. 1980. Enteral absorption and bioavailability in children in relation to age. *Eur J Clin Pharmacol* 18: 43–50.

Herngren L, Ehrenbo M, Boreus LO, 1983. Drug binding to plasma proteins during human pregnancy and in the perinatal period. Studies on cloxacillin and alprenolol. *Dev Pharmacol Ther* 6: 110–124.

Higuchi TT, Palmer JS, Gray LE Jr, Veeramachaneni DN. 2003. Effects of dibutyl phthalate in male rabbits following in utero, adolescent, or postpubertal exposure. *Toxicol Sci* 72: 301–313.

Hines RN, McCarver DG. 2002. The ontogeny of human drug-metabolizing enzymes: Phase I oxidative enzymes. *J Pharmacol Exp Ther* 300: 355–360.

Holladay SD, Smialowicz RJ. 2000. Development of the murine and human immune system: differential effects of immunotoxicants depend on time of exposure. *Environ Health Perspect* 108: 463–473.

Hoover S, Zeise L, Krowech G. 1991. *Exposure to environmental contaminants through breast milk*: In: *The Analysis, Communication and Perception of Risk*. Garrick BJ, Gekler WC, eds *Advances in Risk Analysis*, Plenum New York.

Hsu J-F, Guo YL, Liu C-H, Hu S-C, Wang J-N, Liao P-C. 2007. A comparison of PCDD/PCDFs exposure in infants via formula milk or breast milk feeding. *Chemosphere* 66 (2): 311–319.

Jacobs BL, van Praag H, Gage FH. 2000. Depression and the birth and death of brain cells. *Am Sci* 88: 340–345.

Jacobson JL, Jacobson SW. 1996. Intellectual impairment in children exposed to polychlorinated biphenyls in utero. *N Engl J Med* 335: 783–789.

Kearns GL, Abdel-Rahman SM, Alander SW, Blowey DL, Leeder JS, Kauffman RE. 2003. Developmental pharmacology: drug disposition, action, and therapy in infants and children. *N Engl J Med* 349: 1157–1167.

Kimmel CA. 2001. 1999 Warkany lecture: Improving the science for predicting risks to children's health. *Teratology* 63: 202–209.

Klaassen CD. 1973. Comparison of the toxicity of chemicals in newborn rats to bile duct-ligated and sham-operated rats and mice. *Toxicol Appl Pharmacol* 24: 37–44.

———. 1996. Absorption, distribution, and excretion of toxicants. In: *Casarett and Doull's Toxicology. The Basic Science of Poisons*. 5th ed. Klaassen CD, Amdur MO, Doull J, eds. McGraw-Hill, New York, pp. 99–101.

Klein RZ, Sargent JD, Larsen PR, Waisbren SE, Haddow JE, Mitchell ML. 2001. Relation of severity of maternal hypothyroidism to cognitive development of offspring. *J Med Screen* 8: 18–20.

Kushner BH, Heller G, Cheung N-KV, Wallner N, Kramer K, Bajorin D, Polyak T, Meyers PA. 1998. High risk of leukemia after short-term dose-intensive chemotherapy in young patients with solid tumors. *J Clin Oncol* 16 (9): 3016–3020.

Lacroix D, Sonnier M, Moncion A, Cheron G, Cresteil T. 1997. Expression of CYP3A in the human liver: evidence that the shift between CYP3A7 and CYP3A4 occurs immediately after birth. *Eur J Biochem* 247: 625–634.

Levy G, Khanna NN, Soda DM, Tsuzuki O, Stern L. 1975. Pharmacokinetics of acetaminophen in the human neonate: formation of acetaminophen glucuronide and sulfate in relation to plasma bilirubin concentration and *d*-glucaric acid excretion. *Pediatrics* 55: 818–825.

Maltoni C, Cotti G. 1988. Carcinogenicity of vinyl chloride in Sprague–Dawley rats after prenatal and postnatal exposure. *Ann NY Acad Sci* 534: 145–159.

McCarver DG, Hines RN. 2002. The ontogeny of human drug-metabolizing enzymes: phase II conjugation enzymes and regulatory mechanisms. *J Pharmacol Exp Ther* 300: 361–366.

McClain RM, Keller D, Casciano D, Fu P, MacDonald J, Popp J, Sagartz J. 2001. Neonatal mouse model: review of methods and results. *Toxicol Pathol* 29: 128–137.

McFadden SW, Haddow JE. 1969. Coma produced by topical application of isopropanol. *Pediatrics* 43: 622–623.

Miller RA. 1941. Observations on the gastric acidity during the first month of life. *Arch Dis Child* 16: 22–30.

Miller MD, Marty MA, Arcus A, Brown J, Morry D, Sandy M. 2002. Differences between children and adults: implications for risk assessment at California EPA. *Int J Toxicol* 21: 403–418.

Milsap RL, Jusko WJ. 1994. Pharmacokinetics in the infant. *Environ Health Perspect* 102 (suppl 11) 107–110

Morriss FH, Brewer ED, Spedale SB, Riddle L, Temple DM, Caprioli RM, West MS. 1986. Relationship of human milk pH during course of lactation to concentrations of citrate and fatty acids. *Pediatrics* 78: 458–464.

Morselli PL, Franco-Morselli R, Bossi L. 1980. Clinical pharmacokinetics in newborns and infants. Age-related differences and therapeutic implications. *Clin Pharmacokinet* 5: 484–527.

Muller G, Spassovski M, Henschler D. 1972. Trichloroethylene exposure and trichloroethylene metabolites in urine and blood. *Arch Toxicol* 29: 335–340.

Murphy GM, Signer E. 1974. Progress report: bile acid metabolism in infants and children. *Gut* 15: 151–163

Nagaraja TN, Desiraju T. 1994. Effects on operant learning and brain acetylcholine esterase activity in rats following chronic inorganic arsenic intake. *Hum Exp Toxicol* 13: 353–356.

Newbold RR, Jefferson WN, Padilla-Burgos E, Bullock BC. 1997. Uterine carcinoma in mice treated neonatally with tamoxifen. *Carcinogenesis* 18 (12): 2293–2298.

Newbold RR, Hanson RB, Jefferson WN, Bullock BC, Haseman J, McLachlan JA. 1998. Increased tumors but uncompromised fertility in the female descendants of mice exposed developmentally to diethylstilbestrol. *Carcinogenesis* 19 (9): 1655–1663.

NRC (National Research Council). 1977. *Drinking Water and Health*. Safe Drinking Water Committee. National Academies Press, Washington, DC.

______. 1986. *Drinking Water and Health*, Volume 6. Safe Drinking Water Committee. National Academies Press, Washington, DC.

______. 1990. *Health Effects of Exposure to Low Levels of Ionizing Radiation: BEIR V*. National Academy of Sciences, New York.

______. 1993. *Pesticides in the Diets of Infants and Children*. Committee on Pesticides in the Diets of Infants and Children. National Academies Press, Washington, DC.

______. 2005. *Health Implications of Perchlorate Ingestion*. Committee to Assess the Health Implications of Perchlorate Ingestion. National Academies Press, Washington, DC. Accessed at: www.nap.edu.

Olney JW, Farber NB, Wozniak DF, Jevtovic-Todorovic V, Ikonomidou C. 2000. Environmental agents that have the potential to trigger massive apoptotic neurodegeneration in the developing brain. *Environ Health Perspect* 108 (suppl 3): 383–388.

Oskarsson A, Schutz A, Skerfving S, Hallen IP, Ohlin B, Lagerkvist BJ. 1996. Total and inorganic mercury in breast milk and blood in relation to fish consumption and amalgam fillings in lactating women. *Arch Environ Health* 51: 234–241.

Patandin, S, Dagnelie PC, Mulder PG, Op de Coul E, van der Veen JE, Weisglas-Kuperus N, Sauer PJ. 1999. Dietary exposure to polychlorinated biphenyls and dioxins from infancy until adulthood: a comparison between breast-feeding, toddler, and long-term exposure. *Environ Health Perspect* 107: 45–51.

Paus T, Zijdenbos A, Worsley K, Collins DL, Blumenthal J, Giedd JN, Rapoport JL, Evans AC. 1999. Structural maturation of neural pathways in children and adolescents: in vivo study. *Science* 283: 1908–1911.

Perera F, Jedrychowski W, Ranh V, Whyatt RM. 1999. Molecular epidemiologic research on the effects of environmental pollutants on the fetus. *Environ Health Perspect* 107 (suppl 3): 451–460.

Plunkett LM, Turnbull D, Rodericks JV. 1992. Differences between adults and children affecting exposure assessment. In: *Similarities and Differences Between Children and Adults: Implications for Risk Assessment*. Guzelian PS, Henry CJ, Olin SS, eds. ILSI Press, Washington, DC, pp. 79–94.

Porterfield S. 2000. Thyroidal dysfunction and environmental chemicals: potential impact on brain development. *Environ Health Perspect* 108: 433–438.

Preston-Martin S. 1989. Epidemiological studies of perinatal carcinogenesis. *In: Perinatal and Multigenerational Carcinogenesis*, Napalokov NP, Rice JM, Tomatis L, Yamasaki H, eds. International Agency for Research on Cancer, Lyon, France, pp. 289–314.

Reed M. 1996. Principles of drug therapy. In: *Nelson Textbook of Pediatrics* 15th ed. Behrman RE, Kliegman RM, Arvin AM, Nelson WE, eds. W.B. Saunders, Philadelphia, PA.

Renwick AG. 1998. Toxicokinetics in infants and children in relation to the ADI and TDI. *Food Addit Contam* 15: 17–35.

Renwick AG, Dorne JL, Walton K. 2000. An analysis of the need for an additional uncertainty factor for infants and children. *Regul Toxicol Pharmacol* 31: 286–96.

Rice D, Barone S. 2000. Critical periods of vulnerability for the developing nervous system: evidence from humans and animals. *Environ Health Perspect* 108 (suppl 3): 511–533.

Robillard JE, Guillery EN, Petershack JA. 1999. Renal function during fetal life. In: *Pediatric Nephrology*. Barratt TM, Avner ED, Harmon WE, eds. Lippincott Williams &Wilkins, Baltimore, MD, pp.21–37.

Rodier P. 1994. Vulnerable periods and processes during central nervous system development. *Environ Health Perspect* 102 (suppl 2): 121–124.

______. 1995. Developing brain as a target of toxicity. *Environ Health Perspect* 103 (suppl 6): 73–76.

______. 2004. Environmental causes of central nervous system maldevelopment. *Pediatrics* 113: 1076–1083.

Rodier PM, Ingram JL, Tisdale B, Croog VJ. 1997. Linking etiologies in humans and animal models: studies of autism. *Reprod Toxicology* 11: 417–422.

Rogan W. 1995. Environmental poisoning of children: lessons from the past. *Environ Health Perspect* 103 (suppl 6): 19–23.

Rogan WJ, Gladen BC, McKinney JD, Cerreras, N, Hardy P, Thullen JD, Tingelstad J, Tully M. 1987. PCBs and DDE in human milk: effects on growth, morbidity and duration of lactation. *Am J Public Health* 77: 1294–1297.

Rozman KK, Klaassen CD. 1996. Absorption, distribution, and excretion of toxicants. In: *Casarett and Doull's Toxicology. The Basic Science of Poisons*, 5th ed. Klaassen CD, Amdur MO, Doull J, eds. McGraw-Hill, New York.

Schardein JL. 1993. Principles of teratogenesis applicable to drug and chemical exposure. In: *Chemically Induced Birth Defects*, 2nd ed. Marcel Decker, New York.

Shimada T, Yamazaki H, Mimura M, Wakamiya N, Ueng F-Y, Guengerich FP, Inui Y. 1996. Characterization of microsomal cytochrome P450 enzymes involved in the oxidation of xenobiotic chemicals in human fetal liver and adult lungs. *Drug Metab Dispos* 24: 515–522.

Siegel M, Lebenthal E, Krantz B. 1984. Effect of caloric density on gastric emptying in premature infants. *J Pediatr* 104: 118–122.

Siegner E, Fridrich R. 1975. Gastric emptying in newborns and young infants. *Acta Paediatr Scand* 64: 525–530.

Smiley-Jewell SM, Nishio SJ, Weir A, Plopper CG. 1998. Neonatal Clara cell toxicity by 4-ipomeanol alters bronchiolar organization in adult rabbits. *Am J Physiol* 274: L485–L498.

Smiley-Jewell SM, Liu FJ, Weir A, Plopper CG. 2000. Acute injury to differentiating Clara cells in neonatal rabbits results in age-related failure of bronchiolar epithelial repair. *Toxicol Pathol* 28 (2): 267–276.

Smith AH, Marshall G, Yuan Y, Ferreccio C, Liaw J, von Ehrenstein O, Steinmaus C, Bates MN, Selvin S. 2006. Increased mortality from lung cancer and bronchiectasis in young adults after exposure to arsenic in utero and in early childhood. *Environ Health Perspect* 114 (8): 1293–1296.

Snodgrass WR. 1992. Physiological and biochemical differences between children and adults as determinants of toxic response to environmental pollutants. In: *Similarities and Differences between Children and Adults. Implications for Risk Assessment*. Guzelian PS, Henry CJ, Olin SS, eds. ILSI Press, Washington, DC, pp. 35–42.

Swan SH, Main KM, Liu F, Stewart SL, Kruse RL, Calafat AM, Mao CS, Redmon JB, Ternand CL, Sullivan S, Teague JL. 2005. Decrease in anogenital distance among male infants with prenatal phthalate exposure. *Environ Health Perspect* 113: 1056–1061.

Takeuchi T, Morikawa N. 1960. *Psychiatr, Neurol Japan*. 62: 1850, as cited in *Minamata*, Smith WE, Smith AM. Holt, Rinehart and Winston, New York, 1975.

Thompson DE, Mabuchi K, Ron E, Soda M, Tokunaga M, Ochikubo S, Sugimoto S, Ikeda T, Terasaki M, Izumi S, Preston DL. 1994. Cancer incidence in atomic bomb survivors, II: Solid tumors, 1958–1987. *Radiat Res* 137: S17–S67.

Treluyer JM, Jacqz-Aigrain E, Alvarez F, Cresteil T. 1991. Expression of CYP2D6 in developing liver. *Eur J Biochem* 202: 583–588.

Treluyer JM, Gueret G, Cheron G, Sonnier M, Cresteil T. 1997. Developmental expression of CYP2C and CYP2C-dependent activities in the human liver: in vivo/in vitro correlation and inducibility. *Pharmacogenetics* 7: 441–452.

Tsai SY, Chou HY, The HW, Chen CM, Chen CJ. 2003. The effect of chronic arsenic exposure from drinking water on the neurobehavioral development in adolescence. *NeuroToxicology* 24: 747–753.

Tyrala EE, Hillman LS, Hillman RE, Dodson WE. 1977. Clinical pharmacology of hexachlorophene in newborn infants. *J Pediatr* 91: 481–486.

U.S. EPA.(Environmental Protection Agency), 1986. National primary and secondary drinking water regulations; fluoride; final rule. 40 CFR Parts 141, 142, and 143; *Fed Reg* 51 (63): 11396–11412.

______. 1990. *Drinking Water Criteria Document on Nitrate/Nitrite*. PB-91-142836. Prepared by Life Systems, Inc., Cleveland, OH for the Criteria and Standards Division, Office of Drinking Water, U.S. EPA, Washington, DC.

______. 1991a. Maximum contaminant level goals and national primary drinking water regulation for lead and copper; final rule.40 CFR Parts 141 and 142. *Fed Reg* 56 (110); 26460–26469 (June7).

______. 1991b. National primary drinking water regulations: final rule. 40 CFR Parts 141, 142, and 143. *Fed Reg* 56 (20): 3526–3597.

______. 1992. *Dermal Exposure Assessment: Principles and Applications*. EPA/600/8-91/0118 Office of Health and Environmental Assessment, U.S. EPA, Washington, DC.

______. 1997. *Exposure Factors Handbook*. Vols I to III. EPA/600/P-95/002Fa,b,c. Office of Research and Development, U.S. EPA, Washington, DC.

______. 2000. *Toxicological Review of Vinyl Chloride in Support of Summary Information on the Integrated Risk Information System (IRIS)*. National Center for Environmental Assessment, U.S. EPA, Washington, DC. Accessed at: http://www.epa.gov/iris.

______. 2005. *Supplemental Guidance for Assessing Susceptibility from Early-Life Exposure to Carcinogens*. EPA/630/R-03/003F. Risk Assessment Forum, U.S. EPA, Washington DC, March.

______. 2006a. *A Framework for Assessing Health Risks of Environmental Exposures to Children*. External Review Draft.EPA/600/R-05/093A. National Center for Environmental Assessment, Office of Research and Development, U.S. EPA, Washington, DC, March.

______. 2006b. Methoxychlor. *Integrated Risk Information System (IRIS)*. National Center for Environmental Assessment, U.S. EPA, Washington, DC. Last revised December 1, 1993. accessed at: http://www.epa.gov/iris.

______. 2006c. *Lead and Compounds (Inorganic)*. Integrated Risk Information System (IRIS), National Center for Environmental Assessment, U.S. EPA, Washington, DC. Last revised July 8, 2004. Accessed at: http://www.epa.gov/iris.

______. Vesselinovitch SD, Koda M, Mihailovich N, Rao KVN. 1984. Carcinogenicity of diethylnitrosamine in newborn, infant, and adult mice. *J Cancer Res Clin Oncol* 108: 60–65.

Vieira I, Sonnier M, Cresteil T. 1996. Developmental expression of CYP2E1 in human liver: hypermethylation control of gene expression during the neonatal period. *Eur J Biochem* 238: 476–483.

Waalkes MP, Ward JM, Liu J, Diwan BA. 2003. Transplacental carcinogenicity of inorganic arsenic in the drinking water: induction of hepatic, ovarian, pulmonary, and adrenal tumors in mice. *Toxicol Appl Pharmacol* 186: 7–17.

Watzka M, Bidlingmaier F, Schramm J, Klingmuller D, Stoffel-Wagner B. 1999. Sex and age-specific differences in human brain CYP11A1 mRNA expression. *Neuroendocrinology* 11: 901–905.

Wiley JA, Robinson JP, Piazza T, Garett K, Cirkensa K, Cheng YT, Martin G. 1991a. *Activity Patterns of California Residents*. Final Report. Survey Research Center, University of California, Berkeley, CA. Prepared for the California Air Resources Board, Contract A6-177-31.

Wiley JA, Robinson JP, Cheng YT, Piazza T, Stork L, Pladsen K. 1991b. *Study of Children's Activity Patterns*. Contract A733-149. Final Report. Air Resources Board, California Environmental Protection Agency, Oakland, CA.

Wilson JG. 1977. Current status of teratology. general principles and mechanisms derived from animal studies. In: *Handbook of Teratology*. Vol 1, *General Principles and Etiology*. Wilson JG, Fraser FC, eds. Plenum Press, New York, pp. 47–74.

Wright RO, Amarasiriwardena C, Woolf AD, Jim R, Bellinder DC. 2006. Neuropsychological correlates of hair arsenic, manganese, and cadmium levels in school-age children residing near a hazardous waste site. *Neurotoxicology* 27: 210–216.

Zartarian VG, Ozhaynak H, Xue J. 2000. Modeling aggregate chlorpyrifos exposure and dose to children. Presented at the 10th Annual Meeting of the International Society of Exposure Analysis, Monterey, CA, November 24–27.

Zsarnovsky A, Le HH, Wang H-S, Belcher S. 2005. Ontogeny of rapid estrogen-mediated extracellular signaling in the rat cerebral cortex: potent nongenomic agonist and endocrine disrupting activity of the xenoestrogen bisphenol A. *Endocrinology* 146: 5388–5396.

8

RISK ASSESSMENT FOR ESSENTIAL NUTRIENTS

JOYCE MORRISSEY DONOHUE
U.S. Environmental Protection Agency, Washington, DC

The nutritionally active compounds present in water are varied. They encompass an assortment of compounds of biological origin (i.e., sugars, amino acids, metabolic intermediates) as well as mineral nutrients. Nutritionally active compounds introduced by decay of living organisms are joined by those of anthropogenic origin from food processing and service establishments as well as from human and industrial wastes.

To date there are a small group of nutritionally active materials that have generated regulatory interest as drinking water contaminants:

- Essential trace elements (e.g., Cu, Zn, Se, F, Cr, Mn, Fe)
- Nonessential intermediary metabolites that are disinfection by-products (formaldehyde, acetaldehyde, glyoxal, methylglyoxal)
- Selected alcohols, aldehydes and ketones (i.e., ethanol, acetone, or isopropyl alcohol) that are accommodated by human metabolic pathways and generate ATP

Interest in electrolytes other than sodium and sulfate, macrominerals such as calcium or magnesium, essential amino acids, Krebs cycle acids, and monosaccharides that are likely to be present in water at very low concentrations has been minimal due to their weak toxic potency. The lack of concern for energy-yielding intermediary metabolites such as acetone and acetaldehyde stems from the ease with which they are degraded to carbon dioxide and water by a variety of life forms found in the aquatic environment.

Risk Assessment for Chemicals in Drinking Water, Edited by Robert A. Howd and Anna M. Fan

Nutritionally activecompounds can be divided into two broad categories: essential and nonessential. *Essential nutrients* are those required to support growth and development such as vitamins, minerals, and essential amino acids. *Nonessential nutrients* are those that can be synthesized de novo or from the essential nutrients. They are not required for growth and development but can be metabolized to supply energy or reduce the need for essential nutrients. It is the essential nutrients that are of the greatest health interest because of both their benefits and potential for toxicity.

Dose–Response Properties The dose–response profile for an essential nutrient is generally represented as a U-shaped curve (Figure 1). The left side of the curve represents the improvement in health or function that occurs as dose increases from a deficiency state to the level adequate to support function. The plateau region of the curve represents doses that are fully adequate to support function yet low enough to be physiologically processed and excreted without adverse interactions with tissues, cells, and/or macromolecules. Toxicity is manifest once homeostatic mechanisms can no longer maintain the physiological concentrations of the nutrient below the levels that cause an adverse effect, giving rise to the decline in health or function represented by the right–hand portion of the curve. The decline in function can be due either to interference of the chemical with biological processes or its reaction with, and subsequent modification of, critical biomolecules such as DNA, RNA, proteins, or membrane phospholipids. The goal of a risk assessment for an essential nutrient is to identify those doses that provide benefit to the organism as well as those associated with toxicity, thereby identifying doses that are safe and adequate. The narrower the plateau region of the dose–response curve, the more difficult it is to achieve this goal.

At first glance, Figure 1 is intellectually satisfying because it conveys a simple concept: too much of a nutrient can be as harmful as too little or "more is not necessarily better." What it does not show is that the opposite points along the function axis differ with regard to the severity of the effect and the system

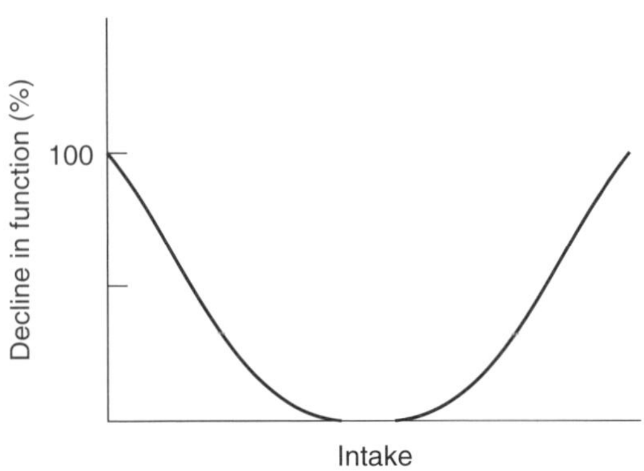

Figure 1. Generalized dose–response curve for an essential nutrient

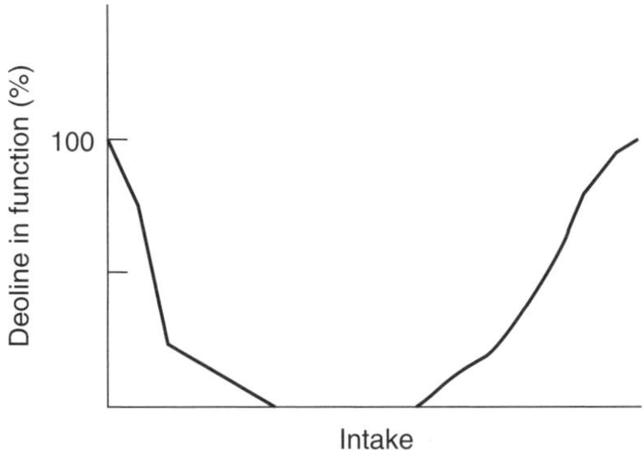

Figure 2. Conceptual asymmetric dose–response curve for an essential nutrient

involved, and the endpoints along the continuum are not necessarily symmetric. For example, the slope on the deficiency side of the curve may differ considerably from that of the toxicity side, and the progression from one measure of impaired function to the next may not be smooth. It may be better represented by several irregular transitions created when the dose–response curve for one adverse effect overlaps with that for another (Figure 2). A point on the portion of the curve that applies to nutritional inadequacy may represent a general feeling of malaise because supplies of a nutritional cofactor involved in oxidative metabolism are low, while the corresponding point on the toxicity side of the curve might indicate decreased food intake because of gastrointestinal irritation. Another point on the deficiency side of the curve could represent a facial dermatitis, while the corresponding point on the toxicity portion of the curve could indicate a birth defect such as a minor rib abnormality.

ASSESSMENT APPROACHES

Nutritional Guidelines

Development of reference values for essential nutrients for Canada and the United States is carried out by the Institute of Medicine (IOM, 1997) within the National Academy of Sciences. These guidelines are called dietary reference intake (DRI) values. They are an extension of the recommended dietary allowances set by the National Research Council (NRC) within NAS from 1941 to 1989, (IFIC, 1998). Similar guideline developing bodies exist within other countries and the World Health Organization (WHO, 2002). The terms used for the nutrient levels (doses) that fall on the essentiality (left) side of the plateau of the dose–response curve differ among the standard setting organizations but the methodologies are generally similar.

The IOM has established three guideline values that frame doses with associated nutritional benefits (1997):

1. *Adequate intakes* (AI): the recommended average daily dietary intake level for groups of healthy people within a designated lifestage and/or gender grouping. They are established when there are insufficient data to establish a recommended dietary allowance.
2. *Estimated average requirement* (EAR): the average daily nutrient intake value that is estimated to meet the requirement defined by a specified indicator of adequacy in 50% of healthy persons in a life-stage or gender grouping.
3. *Recommended dietary allowance* (RDA): the average daily dietary nutrient intake level that is sufficient to meet the nutrient requirements of nearly all (97 to 98%) healthy persons in a life-stage and gender grouping.

Separate values are established for different age and sex groupings and for pregnancy and lactation conditions. The AI values can be viewed as a reasonable guess of the dietary intakes consistent with maximum benefit. EAR and RDA are based to a greater extent on actual dose–response data for early biomarkers of deficiency.

Toxicity Guidelines

The IOM has recently begun to set nutritional guidelines for doses that fall at the far-right side of the plateau portion of the dose–response curve close to the point where toxicity might occur. This guideline is termed the tolerable upper intake level (UL). The UL is defined as the highest average daily dietary intake level that is likely to pose no risks of adverse health effects to almost all persons in the general population. It is an intake that can be tolerated biologically but is not a recommended dietary daily intake for healthy persons (IOM, 1997).

National toxicity guidelines applicable to water and environmental media are established by the U.S. Environmental Protection Agency (EPA). Guideline values are established for both cancer and noncancer effects. The *reference dose* (RfD), the oral exposure guideline for noncancer effects, is defined as an estimate (with uncertainty spanning perhaps an order of magnitude) of a daily oral exposure to the human population (including sensitive subgroups) that is likely to be without an appreciable risk of deleterious effects during a lifetime (U.S. EPA, 2005). It is derived from a lower confidence bound on a benchmark dose (BMDL), a no-observed-adverse-effect level (NOAEL), a lowest-observed-adverse-effectlevel (LOAEL), or another suitable point of departure. Uncertainty and variability factors are applied to reflect limitations of the data used (U.S. EPA, 2005). The RfD is similar to the WHO (2002) tolerable daily intake (TDI) and acceptable dietary intake (ADI). The RfD, TDI, and ADI guidelines apply to environmental contaminants (most of which lack nutritional benefits). All are categorized as estimates of the doses of environmental or additive chemicals that are unlikely to have adverse effects, most often for lifetime exposures.

The toxicity guideline applicable to cancer risk after oral exposure to a contaminant is the *unit risk*, defined as the upper-bound excess lifetime cancer risk estimated to result from continuous exposure to an agent at a concentration of 1 μg/L in water or 1 μg/m^3 in air (U.S. EPA, 2005). The unit risk is derived using the slope factor, representing the potency of the carcinogenic agent (based on the quantitative relationship between tumors and dose) for those chemicals with a genotoxic or undetermined mode of action. Carcinogenicity caused by a direct chemically induced modification of DNA is conceptually difficult when applied to nutrients because it implies that there is no exposure that is without risk, a dose–response concept that is difficult to apply to an essential nutrient.

The principles applied in the derivation of RfD, TDI, and ADI are similar to those used in establishing the UL values for essential nutrients. In each case, scientific studies are examined to identify dose–response properties for adverse effects. The *critical adverse effect* or group of effects is that demonstrated to occur at the lowest dose in the population studied (humans or laboratory animals). Adjustment factors are applied to account for intrahuman variability and other factors related to the applicability of the critical data to target human populations. Typically, adjustment factors (uncertainty or safety factors) are applied for intra- and interspecies extrapolation, adjustment for the duration of the study compared to the target exposure duration, and deficiencies in the data. In cases where the point of departure for the dose response calculation is a LOAEL rather than a NOAEL or its statistical equivalent, an additional uncertainty adjustment is made.

Commonalities Between Methodologies

There are a number of commonalities between the approaches used in setting DRI and RfD values. In both cases, experimental data are evaluated to select a point of departure for dose–response analysis. In both cases the preferred point of departure is a biomarker for adversity. For DRI values that apply to nutritional needs, the biomarker is an early sign of possible deficiency, whereas for the RfD and UL, the biomarker is an indicator of possible toxicity. In both cases, adjustments are made to the point of departure to allow for variability in the doses causing a human response and for uncertainties associated with the dose–response data and its applicability to humans. For DRI values the adjustment increases the guideline values to a dose above the point of departure so that it is protective for 50% (EAR), 98% (RDA), or $\geq$ 50% (AI) of the human population. For RfD values, the guideline value is a dose below the point of departure so that it protects most of the human population, including sensitive populations.

Differences Between Methodologies

There are also differences in the nutritional and toxicological approaches to deriving guidelines for nutrients. The direction of the adjustment to the point of departure discussed above is one major difference in the approaches. This narrows the plateau portion of the dose–response curve in trying to protect all target populations. The target populations also differ. Standard nutritional guidelines are directed toward a specific life stage and gender group and are designed for the healthy population. It is accepted within the nutrition community that there are health-related conditions that require that guideline values be increased or lowered from those applicable to the healthy population. On the other hand, RfD values apply to specific exposure durations rather than life stages or gender groupings and attempt to protect sensitive populations as well as the general population. RfD values are expressed as mg/kg·day doses, and DRIs are expressed in terms of daily intakes without the body weight unit. DRI values apply to foods as the vehicle of exposure; RfD values apply to oral exposures, including food, drinking water, ambient air, and other media. DRI values are established only for essential nutrients or groupings of nutrients, such as carbohydrates, fibers, and proteins, while RfD values can be estimated for any chemical that has been adequately tested for its toxicity in standard toxicological studies.

The *coefficient of variability* of a normal distribution plays an important role in the adjustments applied when deriving an RDA from an EAR. A 10% coefficient of variation is the default assumption (IOM, 1997) for situations where more quantitative data are not available. Derivation of an RfD relies on application of a series of uncertainty factors to the point of departure in increments of 1, 3, or 10 (derived from a logarithmic scale; U.S. EPA, 2000). The nutrition community, on the other hand, often applies uncertainty factors of 1, 1.5, or 2 as well as logarithmic factors when establishing an UL, estimates that are rarely applied in the typical RfD methodology. The richer human database that exists for most essential nutrients makes it easier to support lower uncertainty factors than is possible when relying on animal studies of toxic substances.

The methodological differences between the risk assessment policies of the nutrition and toxicology communities confound the process for establishing protective guidelines for essential nutrients. Standard application of RfD-type polices for toxic substances, when applied blindly to nutrients, would all too often result in the toxicity-based value being below or unreasonably close to the nutritionally derived RDA. In a similar fashion, RfD conventions applied to nonnutrient minerals present in foods may result in a guideline that is below the background levels in a typical diet.

COMPARISON OF GUIDELINE VALUES

Comparison of the RfD with the RDA or AI

Despite the methodological differences between the derivation of nutritional and toxicological guidelines, established values are not in conflict. Informed adjustments are made to the RfD methodology to avoid conflict with the well-established adult RDA. Table 1 compares the RfD with the RDA or AI for several trace mineral nutrients. To facilitate the comparison, the nutritional values have been converted

TABLE 1. Comparison of the Reference Dose with the Adult Recommended Dietary Allowance or Adequate Intake for Selected Trace Mineral Nutrients[a]

Nutrient	RfD (mg/kg·day)	RDA or AI (mg/kg·day)	
		Adult	Child (Age 1–3)
Boron	0.2	NE	NE
Chromium (III)	1.5	0.0005[b]	0.0009[b]
Fluoride	0.06	0.05[b]	0.07[b]
Manganese	0.14	0.03	0.1
Molybdenum	0.005	0.0006	0.02
Nickel	0.02	NE	NE
Selenium	0.005	0.0007	0.002
Zinc	0.3	0.2	0.2

[a]Body weights used for dose conversion: 76 kg for adults, 13 kg for a child. Values for the RDA and AI have been rounded to one significant figure. NE, none established.
[b]AI values.

to mg/kg·day doses by dividing the daily intake value by the standard body weight used by the IOM. Two groups were selected from the suite of RDA/AI values for this comparison:, adults (male, 19 to 50 years old; 76-kg body weight) and 1 to 3-year-old children 13-kg body weight). Children were selected over infants for the comparison because the RDAs for children rest on a more complete database than that of the AI values for infants. As illustrated in Table 1, there are relatively few trace nutrients for which the EPA has derived toxicological values. Minerals such as calcium, magnesium, and iron are present in both treated and untreated surface and ground waters, but their concentrations normally do not approach a level of toxicological concern.

There is no situation where the RfD value for a trace nutrient falls below the adult nutritional guidelines; however, the magnitude of the difference between the RDA and RfD varies and is sometimes very narrow. The difference between the RfD and adult male RDA for selenium is approximately 0.004 mg/kg·day. There is a similar difference between the RfD and adult AI for molybdenum. The difference between the RfD and adult AI for fluoride is 0.01 mg/kg·day and that for zinc and manganese is 0.1 mg/kg·day. The largest difference between the RDA and AI is that for chromium (III), where the RfD is three orders of magnitude higher than the AI. In almost all cases the difference between the RfD and RDA is smaller for children than for adults.

The RfDs for fluoride, manganese, molybdenum, and selenium are all based on human data. In each case, the uncertainty factors were adjusted so that the RfD would not be lower than the RDA/AI (Abernathy et al., 2004). In the case of manganese, the point of departure is a NOAEL from a human dietary study; no adjustment was made to this NOAEL. Had a standard 10-fold factor been applied to account for either intrahuman variability or the less than lifetime duration of the study, the RfD would have been lower than the RDA. For fluoride, the point of departure is the minimal LOAEL for dental fluorosis in children (a cosmetic effect) with no adjustment. Had a 30-fold uncertainty factor been applied for intrahuman variability (10) and a LOAEL/NOAEL adjustment (3) for use of a minimal LOAEL, the RfD for fluoride would have fallen below the level that makes the structure of the tooth and its enamel more resistant to dental carries (0.01 mg/kg·day) (IOM, 1997).

The RfD for zinc comes from several human short-term studies of adults that identified a minimal LOAEL for changes in copper homeostasis (Fischer et al., 1984; Yadrick et al., 1989; Davis et al., 2000; Milne et al., 2001). The LOAELs for these studies were averaged together to derive the LOAEL for the RfD calculation. Uncertainty factors of up to 300 could have been applied according to EPA policies (10 for an intraspecies adjustment, 10 because of the less than lifetime duration of the study and 3 for the use of a minimal LOAEL). However, in place of the possible 300-fold uncertainty factor, a simple threefold factor was applied so that the RfD would not fall below the RDA. In each of these cases, common sense took precedence over arbitrary application of uncertainty factor policies because the chemical is an essential nutrient.

Nickel and boron are interesting examples of toxicologically significant minerals that are widely distributed in foods but only classified as potentially essential by the nutrition community (IOM, 2001). No nutritional guidelines have been established for either mineral, but average concentration in the diet have been measured and are a fraction of the RfD. The dietary intake of boron ranges from 0.87 to 1.35 mg/day (0.01 to 0.019 mg/kg·day for adults), while that for nickel is 74 to 106 μg/day (1.1 to 5.8 μg/kg·day) (IOM, 2001).

Comparison of the RfD with the UL

As mentioned above, there are differences in uncertainty factor policies between the RfD and UL methodologies, which lead to some differences between the resulting recommendations. Table 2 compares the UL value established by the IOM for adult males and children ages 1 to 3 to the RfD value established by the EPA. As was the case with Table 1, the UL value was converted to units of mg/kg·day for this comparison. With few exceptions, the nutritional UL is the same or higher than the corresponding RfD, even when the same study is used as the basis of the derivation. The small differences in guideline values are primarily the result of differences in uncertainty policies.

With the exception of nickel, the UL values established by the IOM are all equal or slightly higher than the RfD values from EPA, even when the EPA RfD values were derived using a modified approach to avoid encroachment on the RDA. In a number of cases the same toxicological study or group of studies was used for both the EPA and IOM toxicity-based assessments. The use of uncertainty factors based on professional judgment by IOM rather than the logarithm-based factors applied by EPA accounts for the differences in values.

TABLE 2. Comparison of the RfD Values for Selected Nutrients with the Dietary UL[a]

Nutrient	RfD (mg/kg·day)	UL (mg/kg·day) Adult	UL (mg/kg·day) Child (Age 1–3)	Critical Effect
Boron	0.2	0.3	0.2	Decreased fetal weight; same study for EPA and IOM
Chromium (III)	1.5	NE	NE	No effect level in a chronic rat dietary study of Cr_2O_3 (EPA)
Fluoride	0.06	0.1	0.1	Skeletal fluorosis for adult; dental fluorosis for child (IOM). EPA RfD based on dental fluorosis in a child
Manganese	0.14	0.15	0.2	Upperlevel intakes from Western diets, same approach was used by IOM and EPA
Molybdenum	0.005	0.03	0.03	Adverse reproductive effects in female rats (IOM); increased serum uric acids in human population study (EPA)
Nickel (soluble)	0.02	0.01	0.015	Collective NOAEL from several rodent studies (IOM); decreased body and organ weights in a chronic oral study (EPA), which was one of the studies used by IOM
Selenium	0.005	0.005	0.008	Clinical selenosis in a human population; same studies for IOM and EPA
Zinc	0.3	0.5	0.6	Altered copper homeostasis; same studies for IOM and EPA

[a]Body weights used for dose conversion: 76 kg for adults, 13 kg for a child. Values for the RDA/AI have been rounded to one significant figure except for manganese (adults) and nickel (children). NE, none established.

As expressed by the IOM (1998), the rich database on the levels of nutrient in foods and in the diets of large human populations allows for the application of uncertainty factors that are lower than those that are typically applied to chemicals that lack nutritional benefits or are present in foods purely as contaminants. This is particularly true when the data used come from studies in humans, but is also the case when the toxicological data are derived from animal studies. Accordingly, the IOM uses uncertainty factor is of 1, 1.5, or 2, which are seldom applied by the EPA.

RISK ASSESSMENT RECOMMENDATIONS

The goal in establishing toxicity-based guidelines for essential trace nutrients is to determine a value that provides the dietary needs for most of the target population without increasing the risk for deficiency or toxicity in any segment of the population. Achieving this goal is complicated by the fact that nutritional needs during periods of growth and development, when expressed on a mg/kg·day basis, are generally greater than those associated with mature life stages, yet toxicity-based guidelines generally apply to the entire population: infants, children, the elderly, the sick, and the healthy.

The challenge in establishing toxicity-based standards for chemicals that are nonessential yet are generally present in the diet, such as ATP-generating, normal intermediary metabolites, is to set the toxicity threshold close to the level that can be accommodated metabolically by the exposed population without adverse effects. The pressure to harmonize nutritional and toxicological considerations for nonessential nutrient is less critical relative to health than it is for the essential nutrients. However, in terms of economics, it may be equally important. The costs associated with limiting discharge to the environment and/or cleaning up contaminated areas can be substantial for both essential and nonessential nutrients.

The following guidelines are recommended when conducting risk assessments for nutritionally active compounds:

- Recognize the biological function of the nutrient. Is it essential or nonessential?
- Determine the range of dietary intakes that can be accommodated by homeostatic controls.
- Utilize experimental data when deriving uncertainty factors rather than applying default conventions.
- Avoid conflict between the nutritional guideline value or normal dietary intake and the toxicity value. If there are age, sex, or life-stage groups to which the toxicity value should not apply, state the exceptions or consider using dietary intake data in determining uncertainty factors rather than applying default values.
- Recognize that even nutrients when taken in excess can have adverse health effects. Too many calories lead to obesity; too much cholesterol in the diet is a risk factor for cardiovascular disease.
- Differentiate between point-of-contact effects and dose. Some nutrients (e.g., copper, zinc) have adverse gastrointestinal effects that may occur at concentrations below those causing adverse systemic effects. Portal-of-entry effects should not automatically be considered to be related directly to dose. The concentration present at the portal surface may be a more important determinant of the effect than the mg/kg·day dose.
- Recognize that an RfD applies to total oral exposure. Adjustments are needed when applied to specific oral exposure routes (food, water, medicines).

- DRI guidelines should consider sources of exposure other than food and supplements as part of the risk characterization. When there is knowledge of the contribution of water to the total exposure from a controlled study, it should be quantified.
- Recognize that mineral nutrients in foods ultimately come from water and soils through uptake by plants and ingestion of feed and water by animals. The role of environmental media in providing mineral nutrients to the food supply should be recognized when setting limitations on the presence of essential minerals in soils and ambient water.
- Collaboration is important. Toxicologists and nutritionists should work together to ensure that guidelines provide for dietary needs and avoid adverse effects while recognizing the contribution of exposure route to the potential for risk.

Disclaimer

The opinions expressed in this chapter are those of the author and not necessarily those of the U.S. EPA.

REFERENCES

Abernathy CO, Donohue JM, Cicmanec J. 2004. Some comments of the selection of human uncertainty factors. *Hum Ecol Risk Assess* 10:29–37.

Davis CD, Milne DB, Nielsen FH. 2000. Changes in dietary zinc and copper affect zinc-status indicators of postmenopausal women, notably, extracellular superoxide dismutase and amyloid precursor proteins. *Am J Clin Nutr* 71:781–788.

Dourson M. 1994. Methods for establishing oral reference doses. In: *Risk Assessment of Essential Elements*. ILSI Press, Washington, DC, pp. 51–61.

Fischer PW, Giroux A, L'Abbe MR. 1984. Effect of zinc supplementation on copper status in adult man. *Am J Clin Nutr* 40:743–746.

Goyer RA. 1994. Biology and nutrition of essential elements. In: *Risk Assessment of Essential Elements*. ILSI Press, Washington, DC, pp. 13–19.

IFIC (International Food Information Council). 1998. Nutrition requirements get a make-over: the evolution of the recommended dietary allowances. *Food Insight News Letter*, September–October. Accessed at: www.ific.irg/foodinsight/1998/so/rdafi598.cfm.

IOM Institute of Medicine. 1997. *Dietary Reference Intakes for Calcium, Phosphorous, Magnesium, Vitamin D and fluoride*. A Report of the Panel on Micronutrient, IOM Food and Nutrition Board. National Academies Press, Washington, DC. Accessed at: http://www.nap.edu/openbook.

______. 1998. *Risk Assessment model for establishing Upper Intake Levels for Nutrients*. IOM Food and Nutrition Board. National Academies Press, Washingdon, DC. Accessed at: http://www.nap.edu/openbook.

______. 2000. *Dietary Reference Intakes for Vitamin C, Vitamin E, selenium and carotenoids*. IOM Food and Nutrition Board. National Academies Press, Washington, DC. Accessed at: http://www.nap.edu/openbook.

———. 2001. *Dietary reference intakes for vitamin A, vitamin K, arsenic, boron, chromium, copper, iodine, iron, manganese, molybdenum, nickel, silicon, vanadium and zinc*. Food and Nutrition Board, National Academies Press, Washington, DC. Accessed at: http://www.nap.edu/openbook.

Milne DB, Davis CD, Nielsen FH. 2001. Low dietary zinc alters indices of copper function and status in postmenopausal women. *Nutrition* 17:701–708.

Olin, SS. 1998. Between a rock and a hard place: methods for setting dietary allowances and exposure limits for essential minerals. *J Nutr* 128(2 suppl):364s–367s.

U.S. EPA (Environmental Protection Agency). 1985. *Integrated Risk Information Summary for Fluorine (Soluble Fluoride)*. National Center for Environmental Assessment, U.S. EPA, Cincinnati, OH. Accessed at: http://www.epa.gov/iris/subst/0053.htm.

———. 1987. *Integrated Risk Information Summary for Nickel Soluble Salts*. National Center for Environmental Assessment, U.S. EPA, Cincinnati, OH. Accessed at: http://www.epa.gov/iris/subst/0271.htm.

———. 1991a. *Integrated Risk Information Summary for Molybdenum*. National Center for Environmental Assessment, U.S. EPA, Cincinnati, OH. Accessed at: http://www.epa.gov/iris/subst/0425.htm.

———. 1991b. *Integrated Risk Information Summary for Selenium and Compounds*. National Center for Environmental Assessment, U.S. EPA, Cincinnati, OH. Accessed at: http://www.epa.gov/iris/subst/0472.htm.

———. 1995. *Integrated Risk Information Summary for Manganese*. National Center for Environmental Assessment, U.S. EPA, Cincinnati, OH. U.S. Accessed at: http://www.epa.gov/iris/subst/0373.htm.

———. 1998. *Integrated Risk Information Summary for Chromium III*. National Center for Environmental Assessment, U.S. EPA, Cincinnati, OH. Accessed at: http://www.epa.gov/iris/subst/0028.htm.

———. 2000. *Methodology for Deriving Ambient Water Quality Criteria for the Protection of Human Health*. Technical support document, Vol 1; *Risk Assessment*. U.S. EPA, Washington, DC, pp 3–6 to 3-9.

———. 2005. *Glossary of IRIS Terms*. Office of Research and Development, U.S. EPA, Washington, DC. Accessed at: http://www.epa.gov/iris/gloss8.htm.

WHO (World Health Organization). 2002. *WHO Guidelines for Drinking Water Quality: Policies and Procedures for Preparing and Updating of the WHO Guidelines for Drinking-Water Quality*. Geneva, Switzerland.

Yadrick MK, Kenney MA, Winterfeldt EA. 1989. Iron, copper, and zinc status: response to supplementation with zinc or zinc and iron in adult females. *Am J Clin Nutr* 49:145–50.

9

RISK ASSESSMENT FOR ARSENIC IN DRINKING WATER

Joseph P. Brown
California Environmental Protection Agency, Oakland, California

Arsenic is a naturally occurring element in Earth's crust and is very widely distributed in the environment. All humans are exposed to microgram quantities of arsenic largely from food (25 to 50 μg/day) and to a lesser degree from drinking water and air. Some seafood contains higher concentrations of arsenic, predominantly in less acutely toxic organic forms. Natural mineral deposits contain large quantities of arsenic in certain geographical areas, which may result in higher levels of arsenic in water. Waste chemical disposal sites may also be a source of arsenic contamination of water supplies. The main commercial use of arsenic in the United States is in pesticides, mostly herbicides and in wood preservatives. Misapplication or accidental spills of these materials could result in contamination of nearby water supplies. Burning of fossil fuels also produces low levels of arsenic emissions. Arsenic may also be found in low levels in tobacco smoke.

Most ingested arsenic is absorbed quickly through the gastrointestinal tract and will eventually be converted by the liver to a less toxic form and excreted in the urine. Arsenic does not have a tendency to accumulate in the body at low environmental exposure levels. In humans, while ingestion of large doses of arsenic may be lethal, lower exposure levels can cause a variety of systemic effects, including irritation of the digestive tract, nausea, vomiting, and diarrhea. Other effects of ingested arsenic include decreased production of erythrocytes and leukocytes, abnormal cardiac function, blood vessel damage, liver and/or kidney damage, and impaired nerve function in hands and feet (paresthesia). Characteristic skin abnormalities are also seen, appearing as dark or light spots

Risk Assessment for Chemicals in Drinking Water, Edited by Robert A. Howd and Anna M. Fan

on the skin and small "corns" on the palms, soles, and trunk. Some of the corns may ultimately progress to skin cancer. In addition, arsenic increases the risk of cancer at internal sites, especially lung, urinary bladder, kidney, and liver (Gosselin et al., 1984; ATSDR, 2000).

Arsenic contamination of groundwater in Bangladesh is associated with the largest poisoning of a population in history, with 35 to 77 million people at risk of adverse health effects (Smith et al., 2000; IARC, 2004). Of over 2000 water wells analyzed in 1998, 35% had arsenic concentrations above 50 μg/L and 8.4% had arsenic concentrations above 300 μg/L. In 1997, 200 villages were surveyed and 1802 of 469, 424 people were found to have arsenic-induced skin lesions. A more detailed analysis of four villages gave 430 of 1481 with skin lesions. Due to the latency for effects from chronic arsenic exposure, other more serious health effects are expected to occur in the exposed population in the future, including skin and internal cancers, neurological effects, hypertension and cardiovascular disease, peripheral vascular disease, and diabetes (Smith et al., 2000).

OCCURRENCE AND EXPOSURE

Arsenic ranks twentieth in abundance (0.0001%) among elements in Earth's crust but is widely distributed and commonly associated with ores of metals such as copper, lead, and gold (Cullen and Reimer, 1989; Oremland and Stoltz, 2003). According to the toxic chemical release inventory (NTIP, 1999), in 1997 the total releases of arsenic into the environment, including air, water, soil, and underground injection from 52 large industrial facilities in the United States, was 60,700 lb. In addition, another 989,000 lb of arsenic was transferred off-site (ATSDR, 2000).

The three largest sources of air emissions are fossil fuel combustion, pesticide use, and copper smelting. These sources accounted for 35, 26 and 19%, respectively, of total arsenic emissions and similar values for land deposition. The largest sources of arsenic in surface waters were urban runoff (37%), pesticide application (25%), and zinc production (20%). In comparison, the estimated total natural emission of arsenic in the northern hemisphere is 4×10^6 kg/yr, largely from volcanoes, forest fires, and weathering. Mean levels of arsenic in ambient air in the United States range from less than 1 to 3 ng/m^3 in remote areas and from 20 to 100 ng/m^3 in urban areas. This arsenic is usually a mixture of arsenite and arsenate except in areas where methylated arsenic pesticides are used (ATSDR, 1997). Urban areas often have higher airborne arsenic due to coal-fired power plants, but maximum concentrations are usually less than 100 ng/m^3.

Weathering of rocks and minerals appears to be a major source of arsenic found in soils and drinking water sources (U.S. EPA, 1987). Due to its ubiquitous nature, low concentrations of arsenic are present in almost all foods and drinking water, which are the primary sources of human exposure. Soil also receives arsenic from a variety of anthropogenic sources, including fly ash from power plants, smelting operations mining wastes, and municipal and industrial waste (ATSDR,

2000). Arsenic is released to soil from wood treated with chromated copper arsenate (CCA). Release of CCA from treated wood in playground equipment has long been a concern with respect to child exposures to arsenic (California DHS, 1987; CPSC, 1990), although it was discontinued for this and most other uses in 2003. Under normal environmental conditions, arsenate is the most stable form of arsenic and is therefore the main exposure form of inorganic arsenic. The average level of As in soil is about 5000 ppb (ATSDR, 1997). Arsenic residues in areas surrounding former copper smelters may be of concern with respect to child exposure (Hwang et al., 1997).

Water is the major means of transport of arsenic under natural conditions (Bencko, 1987). Sedimentation of arsenic in association with iron and aluminum represents a major factor in environmental transport and deposition of this element. It has generally been assumed that surface waters, including the sea, are "self-purifying" with respect to arsenic (i.e., that the arsenic is removed from solution by deposition with sediments) (Woolson, 1983). Water devoid of living organisms will very likely contain only inorganic arsenic in the form of arsenate and/or arsenite. Studies examining the form of arsenic in water supplies have largely reported only arsenate and arsenite in varying ratios (Irgolic et al., 1983; U.S. EPA, 1984).

The arsenate/arsenite ratio depends on the source of water and the redox conditions (Woolson, 1983). Pentavalent arsenic (As^{V}), which is the stable oxidation state in oxygen-containing waters, can be reduced to trivalent arsenic (As^{III}) in anoxic or reducing systems (Turner, 1987). Although As^{III} has been observed in estuarine waters and seawater, the proportion of As^{III} is low, and even in anoxic interstitial waters, complete reduction of arsenic to As^{III} has not been observed. As^{III} released to oxygenated waters can be reoxidized to As^{V} within a time scale of days. In most municipal water supplies the chief form of arsenic is As^{V}, due to aeration and chlorination (U.S. EPA, 1988). The major form of arsenic in well waters relatively rich in arsenic also appears to be As^{V} (U.S. EPA, 1984). In freshwater sources often more than 80% is As^{V}, while the remaining 20% or less is composed of As^{III}, monomethyl arsenate (MMA), and dimethyl arsenate (DMA) (Braman, 1983)

A recent survey by the EPA (EPA, 2000) compared arsenic occurrence data from a number of databases, including the Safe Drinking Water Information System (SDWIS), the National Arsenic Occurrence Survey (NAOS), the U.S. Geological Society (USGS) ambient groundwater arsenic databases, the National Inorganics and Radionuclides Survey (NIRS), and the Metropolitan Water District of Southern California Radionuclides Survey (MWDSC). The SDWIS is based on compliance monitoring data for ground and surface water community systems (CWS) and nontransient, noncommunity water supply systems (NTNCWS). According to the analysis of the SDWIS data, 11,873 groundwater CWS systems were estimated to have mean arsenic concentrations that exceeded 2 μg/L, 5252 systems that exceed 5 μg/L, and 2303 systems that exceed 10 μg/L. Arsenic concentrations were projected to be much lower in surface water systems (e.g., 1,052, 325, and 86 systems exceeding the 2-, 5-, and 10-μg/L levels, respectively).

For the MWDSC database the percent of groundwater systems exceeding 2, 5, 10, and 20 μg As/L were 19.2, 13.5, 5.8, and 1.9%, respectively. For surface water systems only 8% exceeded 2 μg As/L. A comparison of three other surveys of California arsenic occurrence (Tables 6–9ab in the EPA report) indicates somewhat higher levels in groundwater (3 to 9.5% exceedance at 20 μg/L; 0.6 to 1.9% 50 μg/L) and surface water (<1 to 3.5% at 20 μg/L; 0.5 to 2.0% at 50 μg/L).

Food arsenic values taken from U.S. Food and Drug Administration (FDA) surveys indicate an average daily dietary intake of approximately 50 μg As (Gartrell et al., 1985; U.S. EPA, 1988). The total diet study (1991–1997) (FNB, 2002) reported that individuals consumed an average of 37.9 μg As/day. The highest-consuming subgroup was males aged 51 to 70 years with a mean intake of 63.2 μg/day, followed by females of the same age with a mean intake of 54 μg/day. Meacher et al. 2002, using a Monte Carlo approach, estimated intakes of inorganic arsenic in the U.S. population from food, drinking water, air, and soil. The 90th percentile of total inorganic arsenic (As_i) intake was 11.4 μg/day for males and 9.4 μg/day for females, approximately 55% derived from food. Regional differences in inorganic As exposure were due mostly to exposure to arsenic in drinking water rather than to food preferences. For example, the mean intakes of As_i from all sources in the western U.S. region, with higher drinking water arsenic, were 10.6 μg/day in males and 9.3 μg/day in females, approximately 32% derived from food.

Most food contains low levels of arsenic, normally less than 0.25 mg/kg (Ishinishi et al., 1983). The ATSDR 1997 reports a typical range of As in food as 20 to 140 ppb. Generally, the meat, fish, and poultry group is the predominant source of arsenic intake for adults and has been estimated to account for about 80% of arsenic intake (U.S. EPA, 1988). Of this group, fish and seafood consistently contain the highest concentration of arsenic, generally one to two orders of magnitude higher than that in other foods. Freshwater fish contain much lower levels of arsenic than marine fish (Woolson, 1983). Approximately 5 to 10 percent of the arsenic in seafood is inorganic (GESAMP, 1986; Edmonds and Francesconi, 1993). The remainder is present in lipid- and water-soluble organic compounds (Pershagen, 1986). Arsenobetaine appears to be the major water-soluble organoarsenic compound in lobsters and shrimp. Rice was also relatively high in arsenic (Schoof et al., 1999; Lasky et al., 2004). Yost et al. (1998) estimated the intake of inorganic arsenic in North American diets at 8.3 to 14 μg/day in the United States and 4.8 to 12.7 μg/day in Canada for various age groups.

METABOLISM

Several comprehensive reviews of the absorption, distribution, metabolism, and elimination (ADME) of arsenic have been published (Vahter, 1983; Marcus and Rispin, 1988; U.S. EPA, 1988; Thompson, 1993). It has been suggested that the failure to demonstrate carcinogenicity of inorganic arsenic in experimental animals is possibly due to metabolic and/or distribution differences between humans and the animal models presently utilized. These differences are addressed below. Arsenic and arsenic compounds also cross the placenta. The kinetics of arsenic vary depending on the chemical form of arsenic and on the animal species. The following discussion is limited to the forms found in water and forms that are ingested via the aquatic food chain. These include the inorganic, soluble forms of arsenite (As^{III}) and arsenate (As^{V}) as well as organic methyl arsonate (MMA), dimethylarsinic acid (DMA), trimethylarsine (TMA), and arsenobetaine (in fish).

In general, only a small fraction of arsenic is excreted in the feces, so most kinetic and metabolic studies have focused on the urine. Soluble compounds of inorganic arsenic, whether in the trivalent or pentavalent form, are readily absorbed (80 to 90%) in most animal species following oral administration (Charbonneau et al., 1978; Vahter, 1981; Hughes et al. 1994); Freeman et al., 1995). Absorption of orally administered inorganic arsenic in humans has been shown to range between 54 and 80% (Tam et al., 1979; Buchet et al., 1981a,b; Kurttio et al., 1998).

Organic forms of arsenic are also absorbed extensively from the gastrointestinal tract. Human absorption of MMA, DMA, TMA, and arsenobetaine has been shown to be about 75 to 92%. At low-level exposures, excretion of arsenic and its metabolites seems to balance absorption of inorganic arsenic. With increasing arsenic intake, there is suggestive evidence that methylation appears less complete. In studies in mice and humans (Buchet et al., 1981a,b; Vahter, 1981), as the dose of inorganic arsenic increases, the percent of arsenic excreted as DMA decreases, accompanied by increased excretion in the percentage as inorganic arsenic. The percentage excreted as MMA remains virtually unchanged. In vitro studies have demonstrated that the liver is the site of arsenic methylating activity and that *S*-adenosylmethionine and reduced glutathione are required as methyl donors (Buchet and Lauwerys, 1985a,b).

Some potential for dermal absorption has also been reported. Rahman et al. 1994 conducted in vitro studies with clipped full-thickness mouse skin in a flow-through system, and sodium [^{74}As] arsenate applied as a solid, in an aqueous vehicle, or in soil. Absorption of arsenate increased linearly with applied dose from all vehicles. The maximum absorption of 62% of applied dose was obtained with the aqueous vehicle and the least (0.3%) with soil. Wester et al. (1993) evaluated the percutaneous absorption of [^{73}As]arsenate from soil or water in vivo in rhesus monkeys and in vitro in human cadaver skin. Water solutions of arsenate at low (0.024 ng/cm^2) or high (2.1 μg/cm^2) concentrations were compared. With 24-hour administration, in vivo absorption in the rhesus monkey was 6.4 ± 3.9[standard deviation(SD)] percent from the low dose and $2.0 \pm 1.2\%$ from the high dose. In vitro percutaneous absorption of the low dose from water in human skin was $0.93 \pm 1.1\%$ in receptor fluid and $0.98 \pm 0.96\%$ in the washed skin. Absorption from soil (0.4 ng/cm^2) was less, at 6.4 % in the monkey in vivo and 0.8% in human skin in vitro.

The retention and distribution patterns of arsenic are in part determined by its chemical properties. Arsenite (As^{III}) reacts and binds to sulfhydryl groups, while arsenate (As^{V}) has chemical properties similar to those of phosphate. As^{V} also has an affinity for sulfhydryl groups; however, its affinity is approximately 10-fold less than that if As^{III} (Jacobson-Kram and Montalbano, 1985). The distribution and retention patterns of As^{III} and As^{V} are also affected by species, dose level, methylation capacity, valence form, and route of administration.

Vahter et al. 1984 studied tissue distribution and retention of ^{74}As–DMA in mice and rats. About 80% of an oral dose of 0.4 mg As/kg was absorbed from the gastrointestinal tract. In mice more than 99% of the dose was excreted within 3 days compared to only 50% in rats, due largely to accumulation in blood. Tissue distribution in mice showed the highest initial (0.5 to 6 hours) concentrations in kidneys, lungs, intestinal mucosa, stomach, and testes. Tissues with the longest retention times were lungs, thyroid, intestinal walls, and lens.

The effect of dose on arsenate disposition was evaluated in adult female B6 $C3F_1$ mice dosed orally with 0.5 to 5000 μg/kg [^{73}As]arsenate in water (Hughes et al., 1994). The recovery of As-derived radioactivity in excreta and tissues ranged from 83.1 to 89.3% in 48 hours. As-derived radioactivity was detected in several tissues (urinary bladder, gallbladder, kidney, liver, lung), although the sum for each exposure level was very low (<0.5% of dose). The principal depot was the liver, followed by the kidneys. As the dose of arsenate increased there was a significant increase in the accumulation of radioactivity in the urinary bladder, kidney, liver, and lungs, with the greatest concentration in the urinary bladder.

An extensive review and analysis of the mammalian metabolic data on arsenic was conducted by Thompson 1993. The metabolism of arsenate can be viewed as a cascade of reductive and oxidative methylation steps leading successively to As^{III}, MMA^{V}, MMA^{III}, DMA^{V}, DMA^{III}, $TMAO^{V}$, and TMA (Jiang et al., 2003):

$$\underset{As^{V}}{As^{V}O(OH)_3} \xrightarrow{+2e} \underset{As^{III}}{As^{III}(OH)_3} \xrightarrow{+CH_3^+} \underset{MMA^{V}}{CH_3As^{V}O(OH)_2} \xrightarrow{+2e} \underset{MMA^{III}}{CH_3As^{III}(OH)_2} \xrightarrow{+CH_3^+}$$

$$\underset{DMA^{V}}{(CH_3)_2As^{V}O(OH)} \xrightarrow{+2e} \underset{DMA^{III}}{(CH_3)_2As^{III}OH} \xrightarrow{+CH_3^+} \underset{TMAO}{(CH_3)_3As^{V}O} \xrightarrow{+2e} \underset{TMA^{III}}{(CH_3)_3As^{III}}$$

MMA^{III} and DMA^{III} have only recently been detected as stable urinary metabolites in human subjects (Aposhian et al., 2000a,b; Le et al., 2000a,b), and trimethylarsine oxide (TMAO) and trimethylarsine (TMA) are rarely seen and are very minor metabolites in most mammals, if found at all. Few data are available on the tissue concentrations of trivalent methylated As species (Kitchin, 2001). Gregus et al. (2000) found that in bile duct-cannulated rats, As^{III} and its metabolites were preferentially excreted into bile (22%) versus 8% into urine in 2 hours. Arsenite appeared in bile rapidly and constituted the large majority in the first 20 minutes. Thereafter, As^{III} declined and MMA^{III} output gradually increased. From 40 minutes after intravenous As^{III} administration, MMA^{III} was the dominant form of biliary arsenic. Within 2 hours, 9.2% of the dose was excreted in the bile as MMA^{III}. Injection of arsenate produced a mixture of As^{V}, As^{III}, and MMA^{III} in the bile. Curiously, rats injected with MMA^{V} did not excrete MMA^{III}.

The metabolism results of Styblo et al. (1995) in rat liver cytosol in vitro seem to support the overall metabolic scheme noted above with MMA^{III} and MMA^{III}–diglutathione complex being more rapidly methylated to the dimethyl forms than MMA^{V}. Thompson also suggests that the data support the presence of two inhibitory loops: competitive inhibition by MMA^{III} of the $As^{III} \rightarrow MMA^{V}$ step catalyzed by MMTase, and possibly noncompetitive inhibition by As^{III} of the $MMA^{III} \rightarrow DMA^{V}$ step catalyzed by DMTase. Hayakawa et al. (2005) described a new metabolic pathway for arsenite mediated by human recombinant arsenic methyltransferase (Cyt19). In this scheme, arsenic triglutathione (ATG), generated nonenzymatically in the presence of 2 mM or higher glutathione, was converted to monomethylarsonic diglutathione (MADG) and dimethylarsenic glutathione (DMAG) conjugates. These glutathione conjugates undergo hydrolysis to MMA^{III} and DMA^{III}, respectively, providing a nonoxidative route to these trivalent methylated metabolites.

Styblo et al. (1996) observed 50 μM arsenite inhibition of DMA^{V} production in rat liver cytosol in vitro. Healy et al. (1998) studied the activity of MMTase in tissues of mice. The mean MMTase activities (units/mg ± SEM) measured in cytosol of mouse tissues were: liver, 0.40 ± 0.06; testis, 1.45 ± 0.08; kidney, 0.70 ± 0.06; and lung, 0.22 ± 0.01. Subchronic administration of arsenate to the mice in drinking water at 25 or 2500 μg As/L did not increase the MMTase activities. MMTases and DMTases have been partially purified from the livers of rabbits (Zakharyan et al., 1995), rhesus monkeys (Zakharyan et al., 1996), and hamsters (Wildfang et al., 1998). All of the enzyme preparations exhibited Michaelis–Menten enzyme kinetics with K_m values ranging from 8×10^{-4} M for hamster DMTase to 1.8×10^{-6} M for hamster MMTase. V_{max} values ranged from 0.007 pmol/mg protein/per hour for hamster DMTase to 39.6 pmol/mg protein/per hour for rabbit MMTase. Comparative studies have shown several species to be deficient in MTase activities, notably New World monkeys, marmosets, tamarin, squirrel, chimpanzee, and guinea pig (Vahter et al., 1995b; Aposhian, 1997).

While the reduction of arsenate and MMA^{V} can be accomplished nonenzymatically in vitro, and arsenate reduction by glutathione occurs in mammalian blood in vivo (Vahter and Envall, 1983; Winski and Carter, 1995), these reductive steps are probably enzymatically mediated in vivo. An arsenate reductase derived from human liver (Radabaugh and Aposhian, 2000) had an approximate molecular weight of 72,000, was specific for arsenite (i.e., did not reduce MMA^{V}), and exhibited substrate saturation at about 300 μM. The human arsenate reductase requires a thiol and a heat-stable cofactor and is apparently distinct from those isolated from bacteria (Gladysheva et al., 1992; Ji and Silver, 1992; Krafft and Macy, 1998).

Monomethyl arsonate (MMA^{V}) reductases have been isolated and described for rabbit (Zakharyan and Aposhian, 1999) and hamster (Sampayo-Reyes et al., 2000). In the latter study the distribution of MMA^{V} reductase activity ranged from brain (91.4 nmol MMA^{III}/mg protein per hour) and bladder (61.8 nmol MMA^{III}/mg·h) to skin > kidney > testis (all < 15 nmol/mg·h), with spleen > liver > lung > heart at intermediate levels. The high activity of MMA^{V} reductase in brain is curious and may help explain some of the neurotoxic effects of arsenic. Due to the relatively low affinity of the MMA^V reductase ($K_m = 2.2 \times 10^{-3}$ M) compared to the methyl transferases ($K_m = 5\text{–}9 \times 10^{-6}$ M), the MMA^V reduction is thought to be the rate-limiting step in arsenic metabolism (Zakharyan and Aposhian, 1999). The human liver MMA^V reductase appears to be identical to human glutathione *S*-transferase omega class hGSTO 1-1 (Zakharyan et al., 2001).

DMA is the main metabolite found in the tissues and urine of most experimental animals given inorganic arsenic. Humans are somewhat unique in that MMA is an important metabolite in addition to DMA. Human studies showed that within 4 to 7 days after oral As_i, 46 to 62% of the dose was excreted in the urine (Tam et al., 1979; Pomroy et al., 1980; Buchet et al., 1981a,b). Approximately 75% of the excreted arsenic is methylated, about one-third as MMA and two-thirds as DMA.

Apostoli et al. (1997) reported on the metabolism of arsine gas (AsH_3) in an occupationally exposed worker. Arsenic species most excreted in urine over 5 days postexposure were MMA, DMA, As^{III}, arsenobetaine (AsB), and to a lesser extent, As^V. The data indicate a capability to oxidize As^{III} to As^V species, probably via arsenite $As(OH)_3$. Arsenobetaine, an important form of arsenic in food, does not undergo subsequent biotransformation and is excreted via the urine.

The possibility of genetic polymorphism in arsenic metabolism has been suggested by Vahter et al. (1995a), who studied native Andean women in Argentina who were exposed to a wide range of As concentrations in drinking water (2.5 to 200 μg As/L). The women exposed to the highest As concentration in water exhibited surprisingly low levels of MMA in their urine (2.3%). The range of MMA in typical human urine is 12 to 20%. Chiou et al. (1997a) studied the relationships among arsenic methylation capacity, body retention, and genetic polymorphisms of glutathione-*S*-transferase (GST) M1 and T1 in 115 human subjects. Percentages of As species in urine (mean ± SE) were: As_i, 11.8 ± 1.0; MMA, 26.9 ± 1.2; and DMA, 61.3 ± 1.4. Genetic polymorphisms of GST M1 and T1 were significantly associated with As methylation. Subjects with the null genotype of GST M1 had an increased percentage of As_i in urine, whereas those with the null genotype GST T1 had elevated DMA in their urine samples.

Marnell et al. (2003) reported six polymorphisms in the MMA^V reductase hGSTO1 gene in DNA isolated from peripheral blood of 75 Mexican subjects. Two subjects with the same polymorphism showed five- to 10-fold higher concentrations (μg/g creatinine) of As_i in their urine than did other subjects. Yu et al. (2003) screened DNA of 22 subjects of European ancestry (EA) and 24 of indigenous American ancestry (IA) for polymorphisms in arsenate reductase and MMA^V reductase genes. For the arsenate reductase gene (hPNP), 48 polymorphic sites were identified while 33 were found in the MMA^V reductase gene (hGSTO1-1). For the EA individuals the MMA^V reductase gene showed greater polymorphism than the arsenate reductase gene, whereas the reverse was seen in the IA individuals. In the latter group only one polymorphism had a frequency of above 10%. Meza et al. (2005) screened 135 As-exposed subjects from Sonora, Mexico for polymorphisms in arsenic metabolism genes: arsenate reductase (hPNP), MMA^V reductase (hGSTO), and arsenic 3-methyltransferase (CYT19). The subjects were exposed to drinking water with 5.5 to 43.3 ppb arsenic. The screening was based on urinary DMA^V/MMA^V (D/M) ratios. The analysis revealed that all of the variation was due to a very strong association between CYT19 and D/M in children only (7 to 11 years of age). With children removed, no significant association was seen in adults (18 to 79 years). This developmentally regulated association between CYT19 and arsenic metabolism raises questions about the adequacy of arsenic risk assessment for children.

Several authors have studied the kinetics of As excretion in humans. Tam et al. (1979) administered ^{74}As arsenic acid (0.01 μg, ca. 6 μCi) to six adult males (age: 28 to 60; body weight: 64 to 84 kg) following an overnight fast. The urine was analyzed at 24-hour intervals for 5 days following As administration. In the first 24-hour period, As_i excretion exceeded that of the methylated metabolites, but thereafter the usual DMA > MMA > As_i pattern persisted, with DMA increasing as a percentage of cumulative excretion at the later times. A follow-up study (Pomroy et al., 1980) followed ^{74}As excretion for periods up to 103 days using a whole-body counter, with measurement of excreta for the first 7 days. These excretion data were best represented by a three-component exponential function. The coefficients for the pooled data accounted for 65.7% of the excretion with a half-life of 2.09 days, 30.4% with a half-life of 9.5 days, and 3.7% with a half-life of 38.4 days. A four-exponent function showed a better fit to one of the six subjects (half-lives: 0.017, 1.42, 7.70, and 44.1 days).

Physiologicallybased pharmacokinetic (PBPK) models utilize data for mathematical simulations of the uptake, distribution, metabolism, and excretion of toxic chemicals. Such models are used in risk assessment to estimate target tissue doses and to facilitate route-to-route and interspecies extrapolations. By contrast, pharmacodynamic (PD) models simulate biological responses to chemical exposures. A number of PBPK models for arsenic disposition and metabolism have been developed for experimental animals and humans (Mann et al., 1994, 1996a,b; Menzel et al., 1994; Yu, 1999; Gentry et al., 2004). Such models seem to do a fair job of predicting the overall disposition of arsenic in animals and humans, but biological response (PD) models are not yet available. Gentry et al. 2005 observed that pharmacodynamic changes occurred in mice without changes in PBPK-predicted arsenic tissue dosimetry.

HEALTH EFFECTS

Animal Studies

Acute and Subchronic Toxicity Acute effects seen in animals after oral exposure are similar to effects seen in humans (U.S. EPA, 1984), although these effects have not been documented as extensively as have the human effects (Gosselin et al., 1984; Marcus and Rispin, 1988; Lynn et al., 2000). Subacute and subchronic arsenic exposures generally affect many of the same organs or systems as those affected by acute arsenic exposure. The most-affected organs are those involved in absorption, accumulation, and/or excretion (i.e., the gastrointestinal tract, circulatory system, skin, liver, and kidney). However, other organs or systems that are particularly sensitive to the effects of arsenic, such as the nervous system and the heart, are also affected in the intermediate-term exposures (Squibb and Fowler, 1983).

Developmental and Reproductive Toxicity Information on the developmental and reproductive toxicity of inorganic arsenic is available mainly from animal studies using arsenite and arsenate salts and arsenic trioxide. Developmental effects observed in arsenic studies include malformation, growth retardation, and death in hamsters, mice, rats, and rabbits. A characteristic pattern of malformations is produced, with the specific developmental toxicity dependent on dose, exposure route and the point in the gestation period when exposure occurs; see OEHHA (1992), DeSesso et al. (1998), and Golub et al. (1998) for overviews. Machado et al. (1999) showed that a mutation in a single gene (the splotch mutation in C57BL/6J mice) can greatly alter sensitivity to As-induced birth defects.

Two fairly recent articles on arsenic developmental toxicity (Jacobsen et al., 1999; Holson et al., 2000) have concluded that based on the animal studies, environmental levels of arsenic exposure are unlikely to cause any developmental or reproductive effects in humans. However, a few epidemiological studies have indicated possible causal associations between arsenic exposures and fetal and infant mortality, neuropsychological development, and IQ (see "Human studies" below).

Genetic Toxicity Genetic toxicity data on arsenic are summarized in the genetic activity profiles (GAP) database for short-term tests. The GAP97WIN program and database is available online from the U.S. EPA (www.epa.gov). For trivalent arsenic (As^{III}), the GERMCELL and IARC databases list 11 positive findings in 25 nonhuman animal, plant, or microbial test systems. These include chromosomal aberrations in vitro and in vivo (3), micronuclei induction in mice in vivo (1), SCEs in mammalian cells (2), and cell transformation in vitro (3). For pentavalent arsenic (As^{V}), the IARC database lists 6 of 13 positive findings: chromosome aberrations in vitro (3); SCEs in vitro (2); and cell transformation in vitro (1). In general, the lowest effective doses (LEDs) for As^{III} in vitro were in the range 1 to 10 μM, whereas for As^{V} the LEDs were usually 10 to 50 μM.

Immunotoxicity Single inhalation exposures of mice to arsenic trioxide (0.94 mg As/m^3) led to increased susceptibility to respiratory bacterial pathogens, apparently via injury to alveolar macrophages (Aranyi et al., 1985). Sikorski et al. 1989 observed a decreased humoral response to antigens and decreases in several complement proteins in mice given 5.7 mg As/kg sodium arsenite intratracheally. No evidence of immunosuppression was detected in mice exposed orally to arsenate at doses up to 20 mg As/kg·day (Kerkvliet et al., 1980). Sakurai et al. (1998) observed that arsenite and arsenate were strongly toxic to mouse peritoneal or alveolar macrophages in vitro, with IC_{50} values of 5 and 650 μM, respectively. These inorganic arsenicals caused more necrotic death (80%) than apoptotic cell death (20%), and also induced a marked release of inflammatory cytokine, tumor necrosis factor α (TNFα), at cytotoxic doses. Cytotoxic effects of methylated arsenic compounds were less than those of the inorganic arsenicals. The IC_{50} value of DMA was about 5 mM, and MMA and TMAO had no toxicity at concentrations of 10 mM.

Neurotoxicity Neurological effects were not reported in chronic oral studies in dogs or monkeys exposed to arsenate or arsenite (Byron et al., 1967; Heywood and Sortwell, 1979). The NOAEL in the two-year dog study was 3.1 mg As/kg·day, and in the one-year monkey study, 2.8 mg As/kg·day. Some organic arsenicals such as phenyl arsenates may be neurotoxic at high doses. In pigs, subchronic oral exposure to roxarsone (0.87 to 5.8 mg As/kg·day for one month) caused muscle tremors, partial paralysis, and seizures (Rice et al., 1985; Edmonds and Baker, 1986). Time-dependent degeneration of myelin and axons in the spinal cord was also observed histologically (Kennedy et al., 1986). These signs were not seen in rats exposed to roxarsone, but hyperexcitability, ataxia, and trembling were observed at the highest dose of 11.4 mg As/kg·day (NTP, 1989).

Meija et al. (1997) studied the effects of combined exposure to lead and arsenic on central monoaminergic systems in the mouse brain. Lead acetate (116 mg/kg·day), sodium arsenite (11 or 13.8 mg/kg·day), a lead-arsenic mixture (116/13.8), or vehicle controls were administered to male BALB/c mice for 14 days. Regional brain concentrations of norepinephrine (NE), dopamine (DA), serotonin (5-HT), 3,4-dihydroxyphenylacetic acid (DOPAC), 5-hydroxyindole-3 acetic acid (5-HIAA), As, and Pb were measured. Arsenic alone caused regional increases in DOPAC, DA, 5-HT or their metabolites, and a decrease in NE. Arsenic combined with lead provoked significant changes in all three monoamines and their metabolites similar to those of As alone. The mixture also provoked a 38% decrease of NE in the hippocampus and increases of 5-HT in midbrain and frontal cortex (100 and 90%, respectively) over control values, alterations that were not elicited by either metal alone. Although the doses in this study were quite high, close to the LD_{10} for arsenite, this study demonstrates the interaction of two common environmental contaminants and the need for additional testing of chemical mixtures.

Chattopadhyay et al. (2002) observed arsenic-induced changes in growth, development, and apoptosis in neonatal and adult brain cells in vivo and in tissue culture. Sodium arsenite was administered in drinking water at 0, 0.03, 0.3, and 3.0 ppm for 20 days to pregnant rats (5 animals per dose group). The high dose level of arsenite led to a loss of gestation by 20% and to neonatal death of 25%. Both adults (postgestational) and neonatal rats were evaluated for spontaneous behavior, and both groups exhibited a dose-dependent reduction of activity at the highest dose of 50 and 70%, respectively. Postgestational and neonatal rat brain explants were cultured in medium with and without 0.3 ppm arsenite. The control explants showed signs of viability, outgrowth of cells, development of neuronal processes, and establishment of confluence and networking, whereas the treated explants showed reduced growth, loss of ground matrix, and inhibition of neural networking. The viability of cells was measured over 18 days in culture using Trypan Blue exclusion. The arsenite-exposed cells exhibited greater overall reductions in viability over the culture period.

Chronic Toxicity Most chronic animal studies have focused on the cancer endpoint. Byron et al. (1967) studied the chronic effects of inorganic arsenic in two-year studies in rats and dogs. Osborne-Mendel rats, 25 per sex per gender group, were fed sodium arsenite at 0, 15.63, 31.25, 62.5, 125, and 250 ppm As or sodium arsenate at 0, 31.25, 62.5, 125, 250, and 400 ppm As, both in commercial diet. Increased mortality was observed at one year at the high dose of both compounds: arsenite - 77% versus 27% in control; arsenate 8% versus 0% in controls. At concentrations of 31.25 ppm and above, body weight gain appeared to be depressed in females with either As treatment. With males, body weight gain was depressed at $\geq$62.5 ppm arsenite and at $\geq$125 ppm arsenate. Based on body weight gain depression, a chronic LOAEL for arsenate was 31.25 ppm (1.5 mg As^{V}/kg·day) and a chronic NOAEL for arsenite was 31.25 ppm (1.6 mg As^{III}/kg·day).

Beagle dogs, three per sex per dose group, were fed sodium arsenite or sodium arsenate at 0, 5, 25, 50, or 125 ppm in the diet for two years. At the highest sodium arsenite dose level, four dogs died after 3 to 9 months, one after 19 months, and the remaining dog was found moribund and was sacrificed at 8 months. All the high-dose dogs showed a weight loss of 44 to 61%. Anorexia and listlessness were the only clinical signs noted. Based on body weight depression and excess mortality in this study, a chronic NOAEL for arsenate was 50 ppm (1.25 mg As^{V}/kg) and for arsenite was 50 ppm (1.25 mg As^{III}/kg·day).

Schroeder and Balassa (1967) observed increased mortality in CD mice (54 per sex per dose group) administered 5 ppm arsenite in drinking water for 18 months. A companion study in Long-Evans rats (50 per sex per dose group), (Schroeder et al., 1968), also at 5 ppm arsenite in drinking water until natural death (52 months), showed no adverse effects.

Carcinogenicity A variety of arsenic compounds has been examined for carcinogenic activity: arsenic trioxide (As_2O_3), potassium arsenite ($KAsO_2$), sodium arsenite ($NaAsO_2$), sodium arsenate ($Na_2HAsO_4 \cdot 7H_2O$), and lead arsenate ($PbHAsO_4$). Two organic forms of arsenic have also been assessed: arsanilic acid [$C_6H_4NH_2AsO(OH_2)$] and dimethylarsinic acid ($C_2H_6AsO_2H$). Several of the arsenic compounds listed above have also been assessed in combination with various initiating and/or promoting agents. However, little carcinogenicity has been demonstrated in the animal bioassays (e.g., Boutwell, 1963).

Arsenic trioxide (As_2O_3) carcinogenicity has been evaluated in four studies in rats and mice. No increase in tumor incidence was seen by Hueper and Payne (1962) and Baroni et al. (1963), whether arsenic trioxide was given orally or applied dermally in combination with an initiator or promoter (i.e., DMBA, urethane, or croton oil). Knoth (1966) reported an increase in tumor incidence in treated animals [as reviewed by the IARC (1980) and the U.S. EPA (1984)]. Thirty mice were given orally 1 drop of a drug (Psor-Intern or Fowler's solution) containing arsenic trioxide once a week for five months. The author calculated the total dose as 7 mg As_2O_3 per animal. A higher incidence of adenomas of the skin, lung, peritoneum, and lymph nodes was reported at 14 months, than in concurrent controls. However, description of the study design, analysis, and results were very brief and incomplete, precluding critical assessment of this study. Rudnai and Borzsonyi (1981) administered arsenic trioxide by subcutaneous (s.c.) injection to pregnant mice and then postnatally to their offspring [as reviewed by the U.S. EPA (1984)]. Offspring were reported to have significantly increased incidence of lung tumors at 1 year of age. However, the study description was very limited, including no report of whether the tumors were malignant. Moreover, the s.c. injection route may not be relevant to humans exposed orally or by inhalation.

Sodium arsenite ($NaAsO_2$) has been the most extensively studied arsenic compound. Its carcinogenicity has been evaluated in rats, mice, and dogs (Byron et al., 1967; Kanisawa and Schroeder, 1967, 1969; Schrauzer and Ishmael, 1974; Schrauzer et al., 1978; Shirachi et al., 1983; Blakeley, 1987). Sodium arsenite did not exhibit carcinogenic activity when given in drinking water or in the diet at exposure levels ranging from 2 to 250 ppm arsenic. Tumor incidence actually decreased compared to controls in mice exposed to 5 to 10 ppm arsenic (Kanisawa and Schroeder, 1967; Schrauzer and Ishmael, 1974). Weight depression did occur at these levels (Kanisawa and Schroeder, 1967).

Waalkes et al. (2003) described a transplacental carcinogenicity assay of inorganic arsenic. Groups of 10 pregnant C3H mice received 0, 42.5 and 85 ppm arsenite ad libitum in drinking water from gestation day 8 to 18. The offspring received no additional arsenic treatment; they were weaned and put into gender-based groups ($n = 25$) according to maternal exposure. Male survival and body weights were affected by arsenic exposure and the study was limited to 74 weeks. Female mice were less affected and the study was carried out for the full 90-week period. At study termination, male offspring showed a marked dose-dependent increase in hepatocellular carcinoma (control, 12%; 42.5 ppm, 38%; 85 ppm, 61%; trend $p = 0.0006$) and in liver tumor multiplicity (tumors/liver, 5.6-fold over control at 85 ppm; trend $p < 0.0001$). A dose-dependent increase in adrenal tumor incidence and multiplicity (2.2-fold) was also seen (tumors: control, 38%; 42.5 ppm, 67%; 85 ppm, 91%; trend $p = 0.001$). In female offspring, dose-dependent increases occurred in ovarian tumors (control, 8%; 42.5 ppm, 26%; 85 ppm, 38%; trend $p = 0.015$) and in uterine proliferative lesions (hyperplasia + tumors; control, 16%, 42.5, 56%; 85 ppm, 62%; trend $p = 0.001$). Lung carcinoma was seen in female offspring (control, 0%; 42.5 ppm, 4%; 85 ppm, 21%; trend $p = 0.0086$). Oviduct proliferative lesions were seen in female offspring (4, 13, and 29%, respectively; trend $p = 0.0145$). Thus, inorganic arsenic exposure of pregnant mice during the later stage of gestation induced a variety of tumors, including aggressive epithelial malignancies of liver and lung, in their offspring, in a dose-dependent manner.

Rossman et al. (2002) have reported a UV radiation (UVR)-arsenite model in hairless mice. Two groups of 15 hairless but immunocompetent female Skh 1 mice were given 0 or 10 mg/L sodium arsenite (5.8 ppm arsenite) in drinking water and irradiated with a low (nonerythemic) 1.7 kJ/m^2 solar UVR dose three times per week. After 26 weeks the irradiated mice given arsenite had a 2.4-fold increase in skin tumors compared to the irradiated control mice (127 vs. 53 tumors, respectively, $p < 0.01$ by Fisher's exact test). The tumors were mostly squamous cell carcinomas, but those in the arsenite-treated mice were larger and more invasive than those seen in the controls (50% vs. 26%, respectively). The tumors appeared only in mice that received UVR, and only on the exposed areas (backs) of the animals. Times to first tumor ranged from about 55 to 130 days for the arsenite-treated mice versus. 85 to 175 days for the irradiated controls. These results are interesting but need to be confirmed in appropriate dose–response studies.

The carcinogenic activity of two forms of organic arsenic has been investigated. Arsanilic acid ($C_6H_8NH_2AsO(ON)_2$) has only been assessed in combination with DMBA and croton oil, in STS mice (Boutwell, 1963). The incidence of papillomas and carcinomas was not significantly different, regardless of whether or not the exposure regimen included arsanilic acid. The dimethylarsinic acid form ($C_2H_6AsO_2H$) of arsenic has also been assessed (Innes et al., 1969). Mice were exposed from 7 days to 18 months of age. No significant carcinogenic activity was observed.

Yamamoto et al. (1995) observed that dimethylarsinic acid (DMA) significantly enhanced tumor induction in the urinary bladder, kidney, liver, and thyroid in F 344/DuCrj rats pretreated with five carcinogens. Pretreatment with diethylnitrosamine (DEN), *N*-methyl-*N*-nitrosourea (MNU), *N*-butyl-*N*-(4-hydroxybutyl) nitrosamine (BBN), 1,2-dimethylhydrazine (DMH), and N-bis (2-hydroxypropyl) nitrosamine (DHPN) was administered for the first four weeks of the study. This was followed by no further treatment (control group), DMA administration at 50, 100, 200, or 400 ppm in drinking water on weeks 6 through 30, or DMA administration without pretreatment at 100 or 400 ppm in drinking water on weeks 6 through 30. Significant increases were observed in tumors of the bladder, kidney, liver, and thyroid, particularly at the highest doses. Urinary bladder carcinogenesis was strongly enhanced by DMA, even at the lowest dose level of 50 ppm. This study indicates that DMA may act as a carcinogen or promoter for urinary bladder, kidney, liver, and thyroid gland. However, due to the complexity of the protocol and high doses employed, the study results are difficult to interpret.

Wei et al. (1999, 2002) conducted a two-year bioassay with DMA in F344 rats. Male rats ($n = 36$ per group) were administered 0, 12.5, 50, or 200 ppm DMA in drinking water for 104 weeks. Significant increases were observed in urinary bladder tumors (0/36, 0/33, 8/31, and 12/31, respectively) and in preneoplastic lesions and papillary or nodular hyperplasias, at the middle and high dose. Various biochemical parameters were also evaluated (metabolites in urine,

BrdU labeling index in bladder epithelium, cancer-related genes in bladder tumors (H-ras, p53, K-ras, beta-catenin, p27), cyclin D1 activity, COX-2 expression, and 8-OHdG-formation, some of which had changes correlated with tumor development. The authors concluded that DMA is carcinogenic for the rat urinary bladder. The data also indicated that multiple genes were involved in stages of DMA-induced tumor development.

In general, inorganic or organic arsenic failed to exhibit significant carcinogenic activity when given orally to rodents. The only exception is the study by Knoth (1966). However, the description of the study design, analysis, and results was incomplete, precluding a critical assessment. Sodium arsenate or arsenic trioxide given by sub cutaneous injection resulted in insignificant carcinogenic activity. Although the applicability of these studies to human environmental exposure may be questionable, the production of leukemia or lymphomas in mice after parenterally administered arsenic cannot be discounted outright. Ishinishi et al. (1983) and Pershagen et al. (1984a,b) demonstrated increased incidences of lung tumors in hamsters given arsenic trioxide by intratracheal instillation, although the tumor incidences were low (e.g., 3/47). The observation of transplacental carcinogenicity of inorganic arsenic in mice (Waalkes et al., 2003) is particularly noteworthy. Huff et al. (2000) have noted 10 other human carcinogens with limited or no evidence of carcinogenicity in animal bioassays. These authors conclude that "while the collective evidence on the carcinogenicity of inorganic arsenic appears quite close to being considered sufficient evidence in experimental animals (IARC, 1980, 1987; Chan and Huff, 1997), an adequate and definitive long-term experiment on arsenic (and in particular arsenic trioxide) has not yet been done."

The studies of DMA-induced carcinogenicity following carcinogen pretreatment are difficult to interpret with respect to human risk, due largely to the high doses required to produce an effect. Also, there are significant differences in the metabolism of arsenic in the rat versus the human. Rats store arsenic in red blood cells, unlike humans, and the extent of methylation and dimethylation vary (Cohen et al., 2001). The findings of Wei et al. (1999) of direct carcinogenicity of DMA in rat urinary bladder appear to confirm the pretreatment studies.

The Waalkes et al. 2003 study of transplacental carcinogenesis in mice indicates that the gestational period is one of high sensitivity to the carcinogenic effects of arsenic. It also indicates that inorganic arsenic can be a complete carcinogen since its effects were seen long after exposure, did not require continued exposure, and were not reversible upon cessation of exposure as would be expected with a tumor promoter. The authors note that inorganic arsenic may act as a tumor progressor, affecting some pool of minimally neoplastic cells in the fetal target tissues. The findings of Rossman et al. (2001) that continuous exposure to inorganic arsenic in drinking water enhanced the aggressiveness of skin tumors in mice resulting from ultraviolet radiation tends to support the role of As as a tumor progressor.

Human Studies

Acute and Subchronic Toxicity The fatal dose of arsenic trioxide for humans is estimated to be between 70 and 180 mg (Vallee et al., 1960), although 120 mg appears to be the most commonly quoted minimum lethal dose. Some sources have quoted a lethal dose as low as 10 mg, while others have reported recovery from as much as 230 grains (15 grams) (Buchanan, 1962). The trivalent form of arsenic appears to be about four times as toxic as the pentavalent form. Victims of lethal oral arsenic poisoning generally followed one of two clinical patterns. *Acute massive intoxication* occurs when the victim takes a large dose of arsenic on an empty stomach, and may be fatal within a few hours as a consequence of cardiac failure (Jenkins, 1966). In the more typical cases involving the ingestion of a smaller amount of arsenic, the first sign of poisoning occurred from half an hour to several hours after ingestion. Initially, there is throat constriction, a metallic taste in the mouth, and a garlicky odor in the breath, followed by acute gastrointestinal effects, including severe abdominal pain, vomiting, and diarrhea, sometimes with muscular cramps and headache. Finally, 24 hours to several days after the initial exposure, there is a general vascular collapse, leading to shock, coma, and death.

At relatively low intake levels, arsenic may provoke mild gastrointestinal effects. Feinglass (1973) reported the gastrointestinal effects of acute and subchronic exposure to well water contaminated with 11,800 to 21,000 ppb of arsenic. Victims drinking between 10 and 85 cupfuls of such water over a 10-week period experienced nausea or vomiting, dryness or burning of the mouth and throat, abdominal pain, and diarrhea. One of the most common long-term indicators of acute arsenic exposure is Mees' lines, ridges that appear on the fingernails 6 to 8 weeks after exposure (Jenkins, 1966). General desquamation of the skin has also been seen several weeks after exposure (Zaloga et al., 1985).

Genetic Toxicity

In Vitro Studies Sodium arsenite has been evaluated for genetic toxicity in a number of test systems. In peripheral blood lymphocytes, effects included chromosomal aberrations and breaks and SCEs (Jha et al., 1992; Wiencke and Yager, 1992; Gebel et al., 1997; Rasmussen and Menzel, 1997), aneugenicity and mitotic arrest (Vega et al., 1997), micronuclei induction (Schaumloffel and Gebel, 1998), and positive Comet assay (Mass et al., 2001). Minimal effective concentrations were generally 1 μM or less. In fibroblasts, effects seen were micronuclei induction (Jha et al., 1992), DNA damage (Dong and Luo, 1994), gene mutation (Wiencke et al., 1997), and chromosomal aberrations and abnormal cells (Oya-Ohta et al., 1996), generally at 5 μM or less. In keratinocytes, effects were seen on interleukin-8 gene expression (Yen et al., 1996), altered β-adrenergic receptor density and affinity (Chang et al., 1998), growth factor secretion and transforming growth factor-α increases (Germolec et al., 1996, 1997), and gene amplification of SV 40 and DHFR sequences (Rossman and Wolosin, 1992). Minimal effective concentrations were 0.5 to 28 μM arsenite.

Mure et al. (2003) found that arsenite at low concentrations (0.0125 to 0.1 μM) transformed human osteosarcomas TE 85(HOS) cells to anchorage independence. Long-term exposure of HOS cells to 0.1 to 0.2 μM arsenite over 20 generations increased HPRT mutation frequency over 10-fold in a dose-dependent fashion. Arsenite also induced amplification of the DHFR gene over 10-fold in HOS cells over the arsenite range 0.0125 to 0.1 μM. Kligerman et al. (2003) found that MMA^{III} and DMA^{III} were the most clastogenic of six arsenicals tested in human lymphocytes. The dimethylated arsenicals were also spindle poisons, suggesting that they could induce aneuploidy. None of the six arsenicals tested induced gene mutations.

In Vivo Studies Chromosomal aberrations and SCE levels have been evaluated in humans with a history of arsenic exposure. Lerda 1994 studied human subjects exposed to drinking water containing 0.13 mg/L (130 ppb) arsenic for a period of at least 20 years. A control group of 155 people was exposed to less than 20 ppb arsenic in drinking water for more than 20 years. The exposed group had a significantly elevated blood lymphocyte SCE response of 10.46 mean SCE/cell $\pm$ 1.02 SD, versus 7.49 ± 0.97 for the control group (Student's t-test, $p < 0.001$). The urinary arsenic was also significantly higher in the exposed group, 0.16 mg/L versus 0.07 mg/L ($p < 0.001$).

Biggs et al. (1997) studied the occurrence of urinary bladder cell micronuclei in residents of northern Chile with low or high arsenic in their drinking water supplies. Urinary arsenic averaged 582 μg/L (range 61 to 1893) in the high-exposure group ($n = 124$) versus 59 μg/L (range 4 to 266) in the low-exposure group ($n = 108$). The groups were divided into quintiles based on urinary arsenic excretion (i.e., <54, 54 to 137, 138 to 415, 416 to 729, >729 μg/L). Each exposure quintile showed an increase in micronucleated cells (MNC) except the highest (i.e., 1.61, 3.39, 3.69, 4.77, and 1.52 MNC/1000 cells). Urinary arsenic was speciated to inorganic arsenic (As_i), MMA, and DMA. The strongest association was between the sum of species and the prevalence of bladder cell micronuclei.

Dulout et al. (1996) evaluated chromosomal aberrations in peripheral blood lymphocytes from 22 Andean women and children from Argentina exposed to arsenic in drinking water at about 200 μg/L. The genotoxicity endpoints studied were micronuclei in binucleated cells (MN), SCEs, and fluorescence in situ hybridization (FISH) with chromosome-specific DNA libraries. Compared to a control population exposed to very low arsenic in drinking water, the exposed group showed highly significant increases in the frequencies of micronuclei and of trisomy in lymphocytes. There were no notable effects on SCEs, specific translocations, or cell cycle progression. A portion of the micronuclei appears to originate from whole chromosome loss. The exposed children ($n = 10$) exhibited 35 ± 4.6 SEM MN per 1000 cells, and exposed women ($n = 12$), 41 ± 4.9 SEM MN per 1000 cells. A total of 22 control children and women exhibited 6.9 ± 1.7 SEM, indicating a highly statistically significant difference, $p < 0.001$.

Gonseblatt et al. (1997) evaluated two populations in Mexico for cytogenetic effects in blood lymphocytes associated with arsenic in their drinking water.

The groups of 30 to 35 residents were exposed to either 30 μg/L (range 7 to 62) or 408 μg/L (range 396 to 435) arsenic. Approximately one-third of each group was comprised of smokers. The incidence of chromosome aberrations was significantly higher in the high-exposure group: 7.12 ± 1.00 SEM percent versus 2.96 ± 0.54 SEM percent ($p < 0.05$ by t-test). Exposed individuals showed a significant increase in the frequency of chromatid and isochromatid deletions in lymphocytes and of MN in oral and urinary epithelial cells. Males were more affected than females, and a higher number of micronucleated oral cells were found among persons with skin lesions. The types of genetic damage observed provide additional evidence that arsenic is a clastogenic and aneugenic genotoxicant.

Maki-Paakkanen et al. (1998) described an association between structural chromosome aberrations (CAs) in peripheral blood lymphocytes and arsenic exposure via drinking water wells in 42 people in Finland. The median As concentration in well water was 410 μg/L, in urine, total As was 180 μg/L, and in hair, 1.3 μg/g. Eight control individuals were also analyzed who consumed water with low As (<1.0 μg/L). Increased As exposure indicated by increased concentrations of As species (As_i, MMA, DMA) in urine and cumulative arsenic dose (kg/lifetime) in crude and adjusted linear regression models was associated with increased frequency of CAs. An increased MMA/As total and decreased DMA/As total ratios were associated with increased CAs when all aberration types were considered. Current users of As-contaminated water showed stronger associations than all participants in the study.

The following general conclusions may be drawn from the genetic toxicity studies:

- Arsenic is a well-established genotoxicant in mammalian cells.
- Arsenic causes gene mutations in some systems but these are probably lethal in most and hence poorly recoverable.
- Arsenic does not appear to damage DNA directly except possibly at highly cytotoxic levels.
- Arsenic induces chromosomal aberrations (including micronuclei and aneuploidy) and SCEs.
- Arsenic enhances oxidative stress and influences the production of NO.
- Arsenic affects the methylation of DNA in tumor suppressor genes.
- Arsenic causes gene amplification.
- Arsenic inhibits DNA synthesis and repair.
- Arsenic acts as a co-mutagen.
- Methylated and dimethylated arsenic, although excreted more readily in vivo, also exhibit genotoxicity, albeit at higher exposure levels.
- Arsenic causes mitotic arrest, possibly by reaction with tubulin.

Relatively few studies have included MMA and/or DMA, and these have generally indicated either a lack of activity or activity at much higher effective concentrations than inorganic arsenic species. These results appear consistent with the in vivo initiation and promotion studies with DMA, which also required relatively high doses to achieve effects in rats. The one exception appears to be a study by Mass et al. (2001), which used trivalent and pentavalent arsenic species to assess DNA damage in human peripheral lymphocytes in vitro. By comparing the slopes of the regression lines relating the arsenical concentrations to the lengths of the induced tail moments in micrometers in the Comet assay (i. e., μm/μM), the authors were able to rank the "potencies" of the arsenicals as follows: $DMA^{III} > MMA^{III} \gg As^{III}\ As^{V} > MMA^{V}\ DMA^{V}$. No exogenous enzymatic activation was required for activity, and the trivalent arsenicals were considered to be direct-acting genotoxicants. This study indicates that inorganic arsenic is metabolized to highly reactive genotoxic trivalent methylated arsenic species. This work needs to be confirmed and extended to other tissues and model systems. It is also important to note, as the authors do, that MMA^{III} and DMA^{III} are not the only genotoxic species of arsenic that could exist.

Developmental and Reproductive Toxicity In an ecological study of a Hungarian population (n = 25,648), spontaneous abortion and stillbirth were examined with regard to arsenic exposure via drinking water (Borzsonyi et al., 1992). Data were collected over an 8-year period and compared with a population in a neighboring area with low arsenic. The arsenic-exposed population demonstrated increased incidence of hyperpigmentation and hyperkeratosis. There was some indication of an association of arsenic exposure with spontaneous abortion (RR = 1.36, 95% CI 1.1 to 1.6) and a stronger association with stillbirth (RR = 2.70, 95% CI 1.15 to 6.35); both effects were statistically significant.

Hopenhayn-Rich et al. (2000) conducted an ecologic retrospective study of chronic arsenic exposure and risk of infant mortality in two areas of Chile: Antofagasta, with a documented history of As-contaminated drinking water, and Valparaiso, a comparable low-exposure city. Between 1950 and 1996, infant and late fetal mortality rates declined markedly in both cities, as in other areas in Latin America. Despite the overall decline, rates for all mortality outcomes increased in Antofagasta during 1958–1961 and declined thereafter; the early increases coincide with the period of higher arsenic levels in the drinking water (860 μg As/L). Results of a Poisson regression analysis of the rates of late fetal, neonatal, and postneonatal mortality showed elevated relative risks for high arsenic exposure associated with each of the three mortality outcomes. The association between arsenic exposure and late fetal mortality was the strongest (RR = 1.72, 95% CI 1.54 to 1.93). Neonatal mortality (RR = 1.53, 95% CI 1.40 to 1.66) and postneonatal mortality (RR = 1.26, 95% CI 1.18 to 1.34) were also elevated. These findings provide suggestive evidence for arsenic-related human developmental toxicity.

Calderon et al. (2001) conducted a cross-sectional study on the effects of chronic exposure to lead (Pb), arsenic (As), and nutrition on the neuropsychological development of children. Two populations of children (n = 41, 39) with differing As exposure levels (63 vs. 40 μg/g) but similar Pb exposures (8.9 vs. 9.7 μg Pb/dL blood, respectively) were compared using the Wechsler Intelligence Scale for Children (WISC) Revised Version for Mexico. After controlling for significant potential confounders, verbal IQ was observed to decrease with increasing urinary arsenic concentration ($p < 0.01$). Language, verbal comprehension, and long-term memory also appeared to be affected adversely by increasing arsenic exposure. Blood lead was significantly associated with a decrease in attention (sequential factor). However, since blood lead is an imprecise measure of lead burden, there could be some residual confounding in this study.

The relationship between arsenic exposure via drinking water and neurological development as indicated by IQ was assessed in Thailand (Siripitayakunkit et al., 1999). A total of 529 children aged 6 to 9, selected randomly from 15 schools, were studied using a cross-sectional design. The male/female ratio was 1.08. The investigators stated that the subjects of the study were born in a period of chronic arsenic poisoning and that this cohort has been exposed continuously since birth, due to their nonmobility. Arsenic levels in hair were used to assess exposure, and the Wechsler Intelligence Scale Test for children was used to assess IQ. The mean hair arsenic was 3.52 μg/g (SD = 3.58); only 44(8.3%) had normal arsenic levels in hair (≤1 μg/g). The mean IQ of the population was 90.44 (range 54 to 123). The percentage of children in the average IQ group decreased significantly (57 to 40%) with increasing arsenic exposure, whereas the percentage in the lower IQ group increased with increasing As (23 to 38%) and in the very low IQ group (0 to 6%). In a comparison of IQ between children with As hair levels ≤2 ppm or >2 ppm, arsenic was found to explain 14% of the variance in IQ after controlling for father's occupation, mother's intelligence score, and family income. The study suffers from small numbers of children exposed to low arsenic (hair arsenic ≤ 1 ppm).

Wasserman et al. (2004) conducted a cross-sectional study of intellectual function in 201 As-exposed children 10 years of age in Bangladesh. Children's intellectual function was assessed with tests drawn from the Wechsler Intelligence Scale for Children version III, including the verbal, performance, and full-scale raw scores. Children provided urine for arsenic and creatinine and blood samples for blood lead and hemoglobin measurements. After adjustment for sociodemographic covariates and manganese in water, As was associated with reduced intellectual function in a dose-dependent manner. Children exposed to water arsenic of >50 μg/L had significantly lower performance and full-scale scores than did children with water As levels <5.5 μg/L. Using the full-scale raw score, As water concentrations of 10 and 50 μg/L were associated with intellectual function decrements of 3.8 and 6.4 points, respectively. Since there are no standardized measures of intelligence for use in Bangladesh, the decrements noted cannot be equated with IQ points. Nevertheless, the significant loss of intellectual function is clearly an adverse effect, especially when applied to a large population exposed to arsenic in drinking water.

Neurotoxicity Peripheral neuropathies, beginning with loss of sensation and developing into paralysis and muscle atrophy, frequently develop in patients 10 days to three weeks after acute exposure to arsenic (Chuttani et al., 1967; Hay and McCormack, 1987). These cases are frequently diagnosed as Guillain-Barre syndrome (Donofrio et al., 1987). Adults who have severe gastrointestinal reaction to arsenic rarely escape this complication (Jenkins, 1966). Six to 12 months after arsenic ingestion, these signs of poisoning may gradually disappear.

Studies in Japan and Czechoslovakia have reported hearing loss in children [studies described by Tabacova 1986]. In Japan, 12,000 infants were accidentally poisoned with dry milk contaminated with inorganic arsenic. Doses were estimated to be about 3.5 mg/day for 30 days. Anemia, kidney, and liver damage were seen, and there were 130 deaths (Hamamoto, 1955; Nakagawa and Ibuchi, 1970). Disturbances of CNS functions were reported in survivors 15 years after exposure, including severe hearing loss in 18% of 415 children studied, and electroencephalographic abnormalities (Ohira and Aoyama, 1972; Yamashita et al., 1972). Pathological eye effects were also seen, including a case of bilateral optic atrophy. Moderate hearing losses apparently due to inner ear damage were reported in children 10 years of age living near a coal-fired plant emitting large quantities of As (Bencko et al., 1977).

Hematotoxicity A number of arsenic compounds are toxic to blood cells. Exposure to arsenic can result in anemia and leukopenia, which may be because arsenic can cause bone marrow suppression. Acute exposures can produce decreased hematocrit and intravascular hemolysis. Winski and Carter (1998) evaluated arsenate toxicity in human erythrocytes and its possible role in vascular disease. Human erythrocytes were incubated in vitro with sodium arsenate (As^{V}) or sodium arsenite (As^{III}) and assessed for damage. After five hours of incubation with 10 mM As^{V} or As^{III}, significant cell death (hemolysis) occurred only in the As^{V}-treated cells. Morphologic changes were observed by scanning electron and light microscopy. As^{V} induced a concentration-dependent discocyte–echinocyte transformation extending to the formation of sphero-echinocytes. Significant echinocyte formation was seen at the lowest concentration of arsenate employed, 0.1 mM. Sphero-echinocytes were significantly increased at 5 mM and higher. Damaged cells exhibited depletion in cellular ATP, which became statistically significant at five-hour exposure to 0.01 mM arsenate. Treatment with 0.001 mM arsenate also showed depletion in ATP, and overall there was a clear doseresponse (percent of control ATP level vs. log As^{V} concentration). As^{V} was at least 1000 times more toxic than As^{III} based on ATP depletion. The consequences of ATP depletion for the red cell may be severe. ATP is used to maintain membrane shape, deformability, and osmotic stability. Depletion of ATP has been reported to decrease filterability and deformability and to increase blood viscosity (LaCelle, 1970; Rendell et al., 1992; Winski and Carter, 1998). Such changes may contribute to microvascular occlusion, local tissue ischemia, and consequent tissue damage (Weed et al., 1969; Somer and Meiselman, 1993). The occlusive nature of arsenic-induced circulatory disorders suggests that ATP depletion in red cells may play a role in the disease mechanism. Ma et al. (1997) reported that patients with arsenicism from Inner Mongolia, China, had circulating erythrocytes with abnormal shapes and damaged cellular membranes.

Vascular Disease Vascular diseases have long been noted to be associated with chronic arsenic exposures among German vineyard workers (Grobe, 1976) and inhabitants of Antofagasta, Chile (Borgono et al., 1977). Peripheral vascular diseases have been reported to be associated with the occurrence of arsenic in well waters in Taiwan (Chen and Wu, 1962; Chi and Blackwell, 1968; Tseng, 1977; Chen et al., 1988a). In an extensive review of the literature Engel and Smith (1994) concluded that there was good epidemiological evidence indicating that chronic arsenic consumption at high levels is a cause of severe vascular disease with resulting gangrene and limb amputations.

Wu et al. (1989) found significant trends of mortality rates from peripheral vascular diseases and cardiovascular diseases with concentrations of arsenic in well water. However, no significant association was observed for cerebrovascular accidents. Engel and Smith (1994) evaluated arsenic in drinking water and mortality from vascular disease in 30 U.S. counties from 1968 to 1984. Mean As levels in drinking water ranged from 5.4 to 91.5 μg/L. Standardized mortality ratios (SMRs) for diseases of arteries, arterioles, and capillaries (DAAC) for counties exceeding 20 μg/L were 1.9 (90% CI = 1.7 to 2.1) for females and 1.6 (90% CI = 1.5 to 1.8) for males. SMRs for three subgroups of DAAC including arteriosclerosis and aortic aneurysm were also elevated, as were congenital abnormalities of the heart and circulatory system.

Tseng et al. (1996) studied the dose relationship between peripheral vascular disease (PVD) and ingested inorganic arsenic in Taiwan villages where blackfoot disease is endemic. A total of 582 adults (263 men and 319 women) underwent Doppler ultrasound measurement of systolic pressures on bilateral ankle and brachial arteries for diagnosis of PVD, and estimation of long-term arsenic exposure. Multiple logistic regression analysis was used to assess the association between PVD and As exposure. A dose-response relationship was observed between the prevalence of PVD and long-term As exposure. The odds ratios (95% confidence intervals) after adjustment for age, sex, body mass index, cigarette smoking, serum cholesterol and triglyceride levels, diabetes mellitus, and hypertension were 2.77 (0.84 to 9.14), and 4.28 (1.26 to 14.54) for those who had cumulative As exposures of 0.1 to 19.9 and $\geq$20 (mg/L)·yr, respectively. A follow-up study (Tseng et al., 1977) indicated that PVD was correlated with ingested As and not with abnormal lipid profiles.

Chen et al. (1996) evaluated the dose-response relationship between ischemic heart disease (ISHD) mortality and long-term arsenic exposure. Mortality rates from ISHD among residents in 60 villages in an area of Taiwan with endemic arseniasis from 1973 through 1986 were analyzed for association with As concentrations in drinking water. Based on 1,355,915 person-years and 217 ISHD deaths, the cumulative ISHD mortalities from birth to age 79 years were 3.4, 3.5, 4.7, and 6.6% for the median As concentrations of <0.1, 0.1 to 0.34, 0.35 to 0.59, and ≥0.6 mg/L, respectively. Multivariate-adjusted relative risks [RRs (95% CI)] associated with cumulative arsenic exposure from well water were 2.46 (9.53 to 11.36), 3.97 (1.01 to 15.59), and 6.47 (1.88 to 22.24) for 0.1 to 9.9, 10.0 to 19.9, and 20 + (mg/L)·yr, respectively, compared with those without As exposure.

Chiou et al. (1997b) evaluated the dose-response relationship between prevalence of cerebrovascular disease and ingested arsenic among 8102 adults from 3901 households of the Lanyang Basin in northeastern Taiwan. Arsenic was assayed in well water of each household. Logistic regression analysis was used to estimate multivariate-adjusted odds ratios and 95% confidence intervals for various risk factors of cerebrovascular disease. A significant dose-response relationship was observed between As concentration in well water and prevalence of cerebrovascular disease after adjustment for age, sex, hypertension, diabetes mellitus, cigarette smoking, and alcohol consumption. The dose-response relationship was even more prominent for cerebral infarction, with multivariate-adjusted odds ratios (95% CI) of 1.0, 3.4 (1.6 to 7.3), 4.5 (2.0 to 9.9), and 6.9 (3.0 to 16), respectively, for those who consumed well water with an As concentration of 0, 0.1 to 50.0, 50.1 to 299.9, and >300 μg/L. For cumulative arsenic exposures of <0.1, 0.1 to 4.9, and ≥5.0 (mg/L)·yr, the odds ratios were 1.00, 2.26, and 2.69 for cerebrovascular disease, and 1.00, 2.66, and 3.39 for cerebral infarction, respectively. All of the values above for As-exposed groups were significantly greater than the values for unexposed groups at $p < 0.05$ or less.

Wang et al. (2002) studied the association between long-term arsenic exposure and carotid atherosclerosis (CA) in 463 residents of an arseniasis-endemic area in Taiwan. The extent of CA was measured by duplex ultrasonography; presence of plaque, and/or increase in intimal–medial thickness were used to assess the CA progression. Diabetes mellitus was assessed by oral glucose tolerance test and hypertension by mercury sphygmomanometers. Information on consumption of high-arsenic artesian well water, alcohol, and cigarette smoking was obtained by questionnaire interviews. Logistic regression was used to estimate the odds ratio and its 95% CI of CA for various risk factors. Three indices of long-term exposure to ingested arsenic, the duration of consuming artesian well water, the average concentration of arsenic in the water, and the cumulative arsenic exposure, were all significantly associated with prevalence of CA in a dose response-relationship. These relationships remained significant after adjustment for age, sex, hypertension, smoking, diabetes mellitus, alcohol consumption, waist-to-hip ratio, total serum cholesterol, and LDL cholesterol. The multivariate-adjusted odds ratio was 3.1 (95% CI 1.3 to 7.4) for those who had cumulative arsenic exposure of ≥20 (mg/L)·yr compared with those without exposure to arsenic from the well water. The authors conclude that chronic arsenic exposure is an independent risk factor for atherosclerosis and that CA is a novel biomarker for arseniasis. Although the effects in this study were subclinical, since they were observed before the development of events such as acute myocardial infarction and stroke, but still late in the atherosclerotic process, they may be considered "adverse" effects, due to the serious potentially fatal outcomes to which they may lead. Thus, the value noted above may indicate a chronic LOAEL for carotid atherosclerosis resulting from arsenic exposure via drinking water.

Chen et al. (1995) also investigated the association between long-term exposure to inorganic arsenic and the prevalence of hypertension. A total of 382 men and 516 women were studied in villages where arseniasis was hyperendemic. Hypertension was defined as a systolic blood pressure of 160 mmHg or greater, or a history of hypertension treated with antihypertensive drugs. The long-term arsenic exposure was calculated from the history of artesian well water consumption obtained through subject questionnaires and the measured arsenic concentration in well water. Residents in villages where long-term arseniasis was endemic had a 1.5-fold increase in age-and sex-adjusted prevalence of hypertension compared with residents in nonendemic areas. Duration of well water consumption, average As water concentration, and cumulative As exposure were all significantly associated with hypertension. For the cumulative As exposure in (mg/L)·yr the percent prevalence values were: 0, 5.0; 0.1 to 6.3, 4.9; 6.4 to 10.8, 12.8; 10.9 to 14.7, 22.1; 14.8 to 18.5, 26.5; $>$18.5 (mg/L)·yr, 29.2%.

Rahman et al. (1999) conducted a study of hypertension among 1595 subjects $\geq$30 years of age with and without exposure to arsenic via drinking water in Bangladesh; 114 of these were nonconsumers of well water, considered "unexposed." Time-weighted mean arsenic levels in mg/L and (mg/L)·yr exposures were estimated for each subject. Exposure categories were derived as $<$0.5, 0.5 to 1.0, and $>$1.0 mg/L and cumulative exposures as $<$1.0, 1.0 to 5.0, $>$5.0 to 10, and $>$10 (mg/L)·yr. Hypertension was defined as a systolic blood pressure $\geq$140 mmHg. Using "unexposed" subjects as the reference, the prevalence ratios (95% CI) for hypertension adjusted for age, sex, and body mass index (BMI) were 1.2 (0.6 to 2.3), 2.2 (1.1 to 4.3), and 2.5 (1.2 to 4.9) and 0.8 (0.3 to 1.7), 1.5 (0.7 to 2.9), 2.2 (1.1 to 4.4), and 3.0 (1.5 to 5.8) for the metrics of mg/L and (mg/L)·yr, respectively. Both metrics showed significant dose-response trends ($p \ll 0.001$) for crude and adjusted data sets.

Barchowsky et al. (1996) investigated the hypothesis that nonlethal levels of arsenic increase intracellular oxidant levels, promote nuclear translocation of *trans*-acting factors, and are mitogenic. Their results suggest that arsenite initiates vascular dysfunction by activating oxidant-sensitive endothelial cell signaling. Such dysfunction may induce an endothelial cell phenotype that is proinflammatory and retains monocytes in the vessel wall (Collins, 1993).

Lynn et al. (2000) studied arsenite-induced oxidative DNA damage in human vascular smooth muscle cells, concluding that arsenite activates NADH oxidase, producing superoxide and oxidative DNA damage in vascular smooth muscle cells. Such DNA-damaged cells may initiate an atherosclerotic plaque that may be considered a benign smooth muscle cell tumor. Alternatively, arsenic may act through alteration of cell signaling pathways. Arsenic causes blood vessel growth and remodeling in vivo and cell specific, dose-dependent induction of vascular endothelial growth factor-A (VEGF) in vitro (Soucy et al., 2004).

Diabetes Mellitus Chronic exposure to arsenic has been associated with late-onset or type 2 diabetes or diabetes mellitus in several studies. In a study related to those on vascular effects above, Lai et al. (1994) studied inorganic arsenic ingestion and the prevalence of diabetes mellitus in 891 adult residents of villages in southern Taiwan where arseniasis is hyperendemic. Diabetes status was determined by an oral glucose tolerance test and a history of diabetes regularly treated with sulfonylurea or insulin. Cumulative arsenic exposure in ppm·yr was determined from the detailed history of drinking artesian well water. There was a dose-response relation between cumulative arsenic exposure and prevalence of diabetes mellitus. The relation remained significant after adjustment for age, sex, body mass index, and activity level at work, by a multiple logistic regression analysis giving multivariate-adjusted odds ratios of 6.61 and 10.05, respectively, for exposures of 0.1 to 15 ppm·yr and >15.0 ppm·yr versus an unexposed group.

Rahman et al. (1998) assessed arsenic exposure as a risk factor for diabetes mellitus in western Bangladesh in a survey of 163 subjects with keratosis taken as exposed to arsenic and 854 unexposed persons. Diabetes mellitus was determined by a history of symptoms, previously diagnosed diabetes, glucosuria, and blood sugar level after glucose intake. Three time-weighted average exposure levels were derived: <0.5, 0.5 to 1.0, and >1.0 mg/L. For the unexposed and the three exposure levels the adjusted prevalence ratios (95% CI) were 1.0, 2.6 (1.2 to 5.7), 3.9 (1.5 to 8.2), and 8.0 (2.7 to 28.4), respectively. The chi-squared test for trend was very significant ($p < 0.001$). Although this study is somewhat weaker than the earlier study of Lai et al. (1994) in having smaller numbers and lack of comprehensive long-term well water analysis for arsenic, it does corroborate the earlier Taiwanese study.

Tseng et al. (2000) followed up the study of Lai et al. (1994) with a prospective cohort study of 446 nondiabetic residents of arseniasis-endemic villages in Taiwan, followed biannually by oral glucose tolerance test. Diabetes was defined as a fasting plasma glucose level ≥7.8 mmol/L and/or a 2-h postload glucose level of ≥11.1 mmol/L. During the follow-up period of 1500 person-years, 41 cases developed diabetes, with an overall incidence of 27.4 per 1000 persons per year. The incidence of diabetes correlated with age, BMI, and cumulative arsenic exposure (CAE). The multivariate adjusted risks were 1.6, 2.3, and 2.1 for greater versus less than 55 years, 25 kg/m^2, and 17 (mg As/L)·yr, respectively. The incidence rates (per 1000 persons per year) were 18.9 for CAE < 17 (mg/L)·yr and 47.6 for CAE ≥ 17 (mg/L)·yr. The crude relative risk (95% CI) was 2.5 (1.4 to 4.7) and the adjusted relative risk was 2.1 (1.1 to 4.2) for higher vs. lower CAE. The results support the earlier finding of a dose-dependent association between long-term arsenic exposure and diabetes mellitus.

Skin Effects Tseng et al. (1968) examined 40,421 inhabitants of 37 villages in southwestern Taiwan where artesian well water with a high arsenic concentration (mostly 0.4 to 0.6 ppm, but ranging from 0.01 to 1.82 ppm) had been used for more than 45 years. The examination paid particular attention to skin lesions, peripheral vascular disorders, and cancers. Well water samples were collected from most of the villages where such water was still being used, and villages were designated into "low," "mid," and "high" groups based on arsenic concentrations of <0.3, 0.3 to 0.6, and >0.6 ppm, respectively. Overall, there were 7418 cases of hyperpigmentation, 2868 of keratosis, 428 of skin cancer, and 360 of blackfoot disease. In a control population of 7500 persons exposed to <0.017 ppm arsenic in the well water, no cases of any of the these disorders were found.

Mazumder et al. (1998) investigated arsenic-associated skin lesions of keratosis and hyperpigmentation in 7683 exposed subjects in West Bengal, India. Although water arsenic concentrations ranged up to 3400 μg/L, over 80% of the subjects consumed water with <500 μg/L. The age-adjusted prevalence of keratosis was strongly related to water As concentration, rising from zero with <50 μg/L to 8.3% at >800 μg As/L for females and 0.2 to 10.7% in males, respectively. A similar dose-response was observed for hyperpigmentation: 0.3 to 11.5% for females and 0.4 to 22.7% for males. Overall, males had two to three times the prevalence of both keratosis and hyperpigmentation than females apparently ingesting the same doses of arsenic per body weight. Subjects that were more than 20% below standard body weight for their age and sex had a 1.6-fold increase in the prevalence of keratoses, suggesting that malnutrition may play a role in increasing susceptibility.

Ahsan et al. (2000) studied associations between drinking water and urinary arsenic levels and occurrence of skin lesions in Bangladesh. The survey included 167 residents of three contiguous villages, of which 27 (16.2%) had keratosis, 34 (20.4%) had melanosis, and 36 (21.6%) had either keratosis and/or melanosis. Subjects with skin lesions were more likely to have a higher level of arsenic in their drinking water or urine. Also subjects with skin lesions were more likely to have a higher cumulative arsenic exposure. Significantly, a sizable proportion of subjects with skin lesions was seen at the lowest As levels: 13/36 (36.1%) drank water with <50 μg As/L, and 5/36 (13.9%) with <10 μg As/L. Overall there was more than a threefold elevated risk of skin lesions for those subjects who had the highest levels of urinary arsenic.

Respiratory Disease Noncancer lung disease has also been associated with ingestion of inorganic arsenic. Studies in Chile, India, and Bangladesh support this association. A 1976 cross-sectional survey in Antofagasta, Chile examined 144 schoolchildren with arsenic-induced skin lesions and reported that bronchopulmonary disease occurred 2.5 times more often in these children (15.9%) than in children with normal skin (6.9%) (Borgono et al., 1977). In data collected between 1968 and 1972 in Antofagasta, Chile the prevalence of cough and/or dyspnea among 398 children correlated with mean drinking water arsenic concentrations (Zaldivar and Ghai, 1980). The prevalence of bronchiectasis was 23 times higher and recurrent bronchopneumonia was 3.44 times higher in children with chronic arsenical dermatosis than in the general population of Chilean children (Zaldivar, 1980). Finally, over a three-year period following installation of a water arsenic treatment plant in Antofagasta, the prevalence of cough and/or dyspnea dropped from 38% to 7% ($p < 0.001$), a rate similar to that found in a nonexposed region of Chile.

In West Bengal, India, symptoms of cough were reported by 89 of 156 patients with arsenic-associated skin lesions, and 17 of these patients showed evidence of restrictive disease on pulmonary function testing (Mazumder et al., 1998). Following up on this, the authors analyzed data from a cross-sectional survey of 6864 nonsmokers who were clinically examined and interviewed, and the arsenic content in their current primary drinking water source was measured (Mazumder et al., 2000). Subjects included those who had arsenic-associated skin lesions, such as hyperpigmentation and hyperkeratosis and who were also highly exposed at the time of the survey (arsenic water concentration ≥ 500 μg/L). Individuals with normal skin and low arsenic water concentration (<50 μg/L) were used as the referent group. In females, shortness of breath was 23-fold greater among the exposed subjects with skin lesions than in the referent group [age-adjusted prevalence odds ratio (POR) = 23.3, 95% CI 5.8 to 92.8]. The POR for cough was 7.8 (95% CI = 3.1 to 19.5) and for chest sounds was 9.6 (95% CI 4.0 to 22.9). In males, the age-adjusted POR for cough was 5.0 (95% CI 2.6 to 9.9), and chest sounds, 6.9 (95% CI 3.1 to 15.6). This study, while not blinded, had a large sample population, high odds ratios, and a positive trend for respiratory effects and arsenic exposure (NRC, 2001).

Milton et al. (2001) reported an association between chronic ingestion of inorganic arsenic and chronic bronchitis in a small study in Bangladesh of 94 subjects with arsenic-related skin lesions. The mean drinking water arsenic concentration was 614 μg/L (range 136 to 1000 μg/L). The study included 124 nonexposed controls. All subjects were nonsmokers. Chronic bronchitis was present in 14/40 exposed males, 11/50 nonexposed males, 15/54 exposed females, and 2/74 nonexposed females. The crude prevalence ratios were 1.6 (95% CI 0.8 to 3.1) and 10.3 (95% CI 2.4 to 43.1) for males and females, respectively (NRC, 2001). After Mantel–Haenszel adjustment the prevalence ratio was 3.0 (95% CI 1.6 to 5.3) (see NRC, 2001, for additional discussion).

All-Cause Mortality Tsai et al. (1999) compared mortality due to all causes in areas of Taiwan with high levels of arsenic in drinking water. Standardized mortality ratios (SMRs) for noncancer and cancer diseases, by sex, from 1971 to 1994 were calculated both with local and national reference groups. Arsenic levels in the study group drinking water ranged from 0.25 to 1.14 ppm (median = 0.78 ppm). The local study area reported 11,193 male and 8874 female deaths compared to 113,576 and 80,350 in the local reference and 1,290,606 and 836,203 in the national reference groups, respectively.

For males with the local reference, significant SMRs (95% CI) were seen for diabetes mellitus 1.35 (1.16 to 1.55), ischemic heart disease 1.75 (1.59 to 1.92), cerebrovascular disease 1.14 (1.08 to 1.21), vascular disease 3.56 (2.91 to 4.30), bronchitis 1.48 (1.25 to 1.73), asthma 1.18 (1.08 to 1.31), liver cirrhosis 1.17 (1.02 to 1.34), and nephritis 1.16 (1.01 to 1.39). Comparisons with the national reference group gave significant SMRs for ischemic heart disease 1.50 (1.36 to 1.64), heart disease 1.17 (1.08 to 1.28), cerebrovascular disease 1.09 (1.03 to 1.15), vascular disease 3.09 (2.53 to 3.73), bronchitis 1.87 (1.59 to 2.18), liver cirrhosis 1.17 (1.08 to 1.28), and nephritis 1.23 (1.07 to 1.41). For females with the local reference, significant SMRs (95% CI) were seen for diabetes mellitus 1.55 (1.39 to 1.72), hypertension 1.20 (1.06 to 1.37), ischemic heart disease 1.44 (1.27 to 1.61), cerebrovascular disease 1.24 (1.18 to 1.31), vascular disease 2.30 (1.78 to 2.93), bronchitis 1.53 (1.30 to 1.80), and nephritis 1.16 (1.01 to 1.39). Similar values were seen with the national reference group, except that hypertension and nephritis were no longer significant.

SMRs for all malignant cancers in males were 2.19 (2.11 to 2.28) for local and 1.94 (1.87 to 2.01) for national reference. Significant male SMRs for local and national reference groups included intestine 2.10 (1.20 to 1.83, local), lung 2.46 (1.77 to 3.34, local), skin 5.97 (4.62 to 7.60, national), prostate 2.52 (1.86 to 3.34, local), urinary bladder 10.50 (9.37 to 11.73, national), kidney 6.80 (5.49 to 8.32, national), and lymphoma 1.63 (1.23-2.11, local). The higher of the two reference values is given in each case. For females, SMRs for all malignant cancers were 2.40 (2.30 to 2.51) and 2.05 (1.96 to 2.14) for local and national, respectively. Significant individual cancer SMRs with both reference groups included pharyngeal 2.36 (1.13 to 4.34, local), rectum 1.87 (1.64 to 2.14, national), lung 4.13 (3.77 to 4.52, local), skin 6.81 (5.29 to 8.63, national), kidney 10.49 (8.75 to 12.47), bladder 17.65 (5.70 to 19.79), and lymphoma 1.70 (1.18 to 2.37). This study compares mortality; however, since not all diseases are fatal, the figures tend to underestimate the risks of serious adverse health effects. Also, it is important to note that the noncancer SMRs are not much lower than SMRs due to malignant cancers. Hence, the noncancer endpoints described here need to be taken as seriously as the cancer endpoints.

Fortunately, several studies appear to provide suitable data for the quantitative risk assessment of several significant noncancer disease endpoints, including cerebrovascular and cardiovascular disease, hypertension, diabetes mellitus, and skin keratosis. Table 1 provides a summary of the available data showing that exposure to arsenic via drinking water is associated with a number of serious health effects, often in a dose-related manner.

Carcinogenicity Arsenic has long been recognized as a human carcinogen. Hutchinson (1887, 1888) reported skin cancer in patients treated with arsenical medications. Several epidemiologic studies have since confirmed that ingested arsenic can cause skin cancer and inhaled arsenic lung cancer (IARC, 1980, 1987). Early evidence that ingestion of arsenic is a cause of various cancers other than skin cancer came mainly from studies in Taiwan (Chen et al., 1988a,b; Wu et al., 1989) and to a lesser extent from two studies in Japan (Tsuda et al., 1990). Since then several studies have provided strong additional evidence that arsenic ingestion causes internal cancers, in particular cancers of the urinary bladder and lung (Chiou et al., 1995, 2001; Tsuda et al., 1995; Hopenhayn-Rich et al., 1996a; Guo et al., 1997; Smith et al., 1998).

Tseng et al. (1968) and Tseng (1977) correlated risk of skin cancer with arsenic exposures in a large population (>100,000 persons) in southwestern Taiwan. Levels of arsenic in their artesian drinking water wells ranged from 0.001 to

TABLE 1. Summary of Arsenic-Induced Noncancer Toxicity Observed In Vivo

Toxic Endpoint Category	Population Studied, Effect[a]	Study
Genetic toxicity	22, Argentina; MN	Dulout et al., 1996
	30–35, Mexico; CAs	Gonseblatt et al., 1997
	42, Finland; CAs	Maki-Paakkanen et al., 1998
Developmental and reproductive toxicity	25,648, Hungary; SA, SB, skin effects	Borzsonyi et al., 1992
	Chile; infant mortality	Hopenhayn-Rich et al., 2000
	80, Mexico children; IQ	Calderon et al., 2001
	529, Thailand children; IQ, visual perception	Siripitayakunkit et al., 1999, 2001
	201, Bangladesh children; intellectual function	Wasserman et al. 2004
	109, Taiwan children; cognitive function	Tsai et al., 2003
Neurotoxicity	Czechoslovakia; 415, Japan children; hearing loss	Yamashita et al., 1972; Bencko et al., 1977; Tabacova, 1986
Vascular disease	582, Taiwan; PVD	Tseng et al. 1996
	217, Taiwan; ISHD	Chen et al., 1966
	8102, Taiwan; CI	Chiou et al., 1997b
	463, Taiwan; carotid atherosclerosis	Wang et al., 2002
	898, Taiwan; HT	Chen et al., 1995
	1595, Bangladesh; HT	Rahman et al., 1999
Diabetes mellitus	891, Taiwan; DM	Lai et al., 1994
	1017, Bangladesh; DM	Rahman et al., 1998
	446, Taiwan; DM	Tseng et al., 2000
Skin effects	40,421, Taiwan	Tseng et al., 1968
	7683, West Bengal, India; keratosis, hyperpigmentation	Mazumder et al., 1998
	167, Bangladesh; keratosis, melanosis	Ahsan et al., 2000
Respiratory disease	144, Chile; bronchopulmonary disease in children	Borgono et al., 1977
	398, Chile; cough, dyspnea in children	
	6864, West Bengal, India; shortness of breath, cough, chest sounds	Zaldivar and Ghai, 1980 (see also Smith et al., 1998); Mazumder et al., 1998, 2000
	218, Bangladesh; chronic bronchitis	Milton et al., 2001

[a]MN, micronucleus test; CAs, chromosome aberrations; SA, spontaneous abortion; SB, stillbirth; IQ, intelligence quotient; PVD, peripheral vascular disease; ISHD, ischemic heart disease; CI, cerebrovascular infarct; HT, hypertension; DM, diabetes mellitus.

1.82 mg/L, with average levels of around 0.4 to 0.6 mg/L. A house-to-house medical survey of 40,421 exposed persons demonstrated a dose-response relationship between prevalence of skin cancer, the arsenic concentration of drinking water, and duration of water intake. Overall prevalence rates for skin cancer, hyperkeratosis, and hyperpigmentation were 10.6, 71.0, and 183.5 per 1000, respectively. The youngest patient with hyperpigmentation was 3 years old, the youngest with hyperkeratosis 4 years old, and the youngest with skin cancer 24 years old. Ninety-nine percent of those with skin cancer had multiple skin cancers. Overall, skin cancers are not as great a concern as internal cancers, such as lung or bladder cancer, because internal cancers are often life threatening, whereas most skin cancers are not (NRC, 2001). Therefore, skin cancer is not covered further here and the reader is directed to other studies and analyses (Buchet and Lison, 1998; Tsai et al., 1998; Ahmad et al., 1999; Hinwood et al., 1999; Ma et al., 1999; Karagas et al., 2001; IARC, 2004).

Chen et al. (1988b) briefly described a steep dose-response relationship between arsenic levels in artesian well water in 42 villages of the blackfoot disease endemic area of southwestern Taiwan and rates of bladder, lung, kidney, and skin cancer, as well as prostate cancer for men. The study period (1973–1986) covered 899,811 person years of observation. Exposure was grouped in three categories based on arsenic levels from a 1962 to 1964 survey of over 83,656 wells in Taiwan. The exposure categories were <0.3 ppm, 0.3 to 0.59 ppm, and ≥0.6 ppm arsenic in drinking water. Mortality rates were age adjusted using the working population in 1976 as the standard. For all cancer sites in males the standardized mortality per 100,000 persons was: 128 (control); 154, <0.3 ppm; 258.9, 0.3 to 0.59 ppm; and 434.7, ≥0.6 ppm. For females the values were 85.5, 113.3, 182.6, and 369.4, respectively.

Chen and Wang (1990) investigated cancer mortality rates in the arseniasis-endemic areas of Taiwan compared to other areas of the country, extending the analysis of Chen et al. (1988b). Drinking water arsenic analyses conducted from 1974 to 1976 showed that among 83,656 wells tested, 18.7% had arsenic concentrations ≥50 μg/L and 2.7% had arsenic concentrations ≥350 μg/L. Urbanization and industrialization indices were included in the analysis to adjust for possible confounding by socioeconomic status. Mortality data from 1972 to 1983 were used to evaluate 21 malignant neoplasms, using correlation analysis weighted by person-years at risk. Seven cancer sites were significantly associated with arsenic levels in the water: urinary bladder, lung, liver, kidney, skin, and nasal cavity for both men and women, and prostate cancer for men. Multivariate analysis indicated an increase in mortality per 100,000 of up to 0.4 for nasal cancer in females and 6.8 for liver cancer in males for every 0.1 mg/L increase in arsenic concentration in water. The main limitation of this study is its ecological design, in which groups are compared rather than individuals, with their corresponding mean exposure levels and cancer rates. Nevertheless, the results are consistent with those of other studies in the area (Chen et al., , 1988a,b; Wu et al., 1989).

Chiou et al. (1995) investigated the relationship between internal cancers and arsenic in relation to blackfoot disease. Patients ($n = 263$) and 2293 healthy controls, all residents of the arsenic endemic area of southwestern Taiwan, were followed for seven years. After controlling for the effects of age, sex, and smoking in the regression analysis, a dose-response relationship was observed between arsenic exposure from drinking well water and the incidence of bladder and lung cancer. Blackfoot disease patients were found to be at a significantly increased risk compared to controls after adjustment for cumulative arsenic exposure.

Chiou et al. (2001) studied the incidence of urinary tract cancers among 8102 residents of an arseniasis-endemic area in northeastern Taiwan. Arsenic levels in the drinking water ranged from less than 0.15 μg/L (undetectable) to 3590 μg/L. Exposure was estimated from arsenic concentrations in the well associated with each subject *at one point in time only*, although most households had used their current wells for at least 10 years (Chen and Chiou, 2001). Each home was said to have its own well and that they had been in use for more than 50 years. The incidence of urinary tract cancers (kidney and bladder) was significantly increased in the cohort relative to the general population of Taiwan (SIR = 2.05, 95% CI 1.22 to 3.24). The SIR for bladder cancer was 1.96 (95% CI 0.94 to 3.61) and for kidney cancer was 2.82 (95% CI 1.29 to 5.36), based on nine subjects with bladder cancer, eight with kidney cancer, and one with both. A significant dose-response relationship was observed between urinary tract cancers, particularly transitional cell carcinoma (TCC) after adjusting for age, sex, and smoking. The relative risks of developing TCC were 1.9, 8.2, and 15.3 for arsenic concentrations of 10.1 to 50.0, 50.1 to 100, and greater than 100 μg/L, respectively, based on one, two, and six diagnoses of TCC in the low-, mid-, and high-dose groups, respectively. The confidence intervals are huge because of these small numbers. Conclusions regarding a dose-response relationship are difficult for this reason and because exposure was assessed at only one point in time (Cantor, 2001).

Smith et al. (1998) investigated cancer mortality in a population of around 400,000 people in a region of northern Chile (region II) which had been exposed to high arsenic levels in drinking water. Arsenic concentrations from 1950 to the present were obtained. Population-weighted average arsenic levels reached 570 μg/L from 1955 to 1969, and decreased to less than 100 μg/L by 1980. Cancer mortality for the years 1989 through 1993 in region II was compared to the rest of Chile. The results indicated marked increases for bladder, kidney, lung, and skin cancer mortality in region II, with corresponding SMRs of 6.0, 1.6, 3.8, and 7.7 for men, and 8.2, 2.7, 3.1, and 3.2 for women. All results were significant at the 95% confidence level, except liver cancer, with an SMR of 1.1 for both men and women, respectively. This study showed considerably elevated rates for the same cancers found to be consistently elevated in the Taiwanese studies except for liver cancer. Smoking survey data and mortality rates from chronic obstructive pulmonary disease (COPD) provided evidence that smoking did not contribute to the increased mortality from these cancers. These findings provide additional evidence that ingestion of inorganic arsenic in drinking water is indeed a cause of bladder and lung cancer. It was estimated that arsenic might account for 7% of all deaths among those aged 30 and over. If so, the impact of arsenic on the population mortality in region II of Chile is greater than any other reported to date from environmental exposure to a carcinogen in a major population.

Ferreccio et al. (2000) studied the relationship between lung cancer and arsenic in drinking water in northern Chile in a case–control study involving patients diagnosed between 1994 and 1996 and frequency-matched hospital controls. To avoid the problem of matching on exposure, eligible controls included all patients admitted to any public hospital in the entire study area. Each lung cancer case

TABLE 2. Summary of Arsenic-Induced Internal Cancers Observed in Drinking Water Studies

Tumor Sites	Population Studied	Study
Bladder, kidney, lung	1001, Taiwan	Chen et al., 1985
Bladder, lung, liver	368, Taiwan	Chen et al., 1986
Bladder, kidney, lung, liver, skin, prostate (male)	899,811 person-years, Taiwan	Chen et al., 1988b
Bladder, kidney, lung, liver, skin	898,806 person-years, Taiwan	Wu et al., 1989
Bladder, kidney, lung, liver	898,806 person-years, Taiwan	Chen et al., 1992
Bladder, kidney, lung	Taiwan	Chen and Wang, 1990
Bladder	383, Utah, U.S.	Bates et al., 1995
Bladder, lung	263 experimental, 2293 controls, Taiwan	Chiou et al., 1995
	855, Chile	Smith et al., 1998
Lung, liver, bladder, kidney	1999, Taiwan	Tsai et al., 1999
Urinary organs, transitional cell carcinoma	8102, Taiwan	Chiou et al., 2001
Lung, urinary cancer	113, Japan	Tsuda et al., 1995
Bladder, kidney, lung	3647, Argentina	Hopenhayn-Rich et al., 1996, 1998
Lung	151 experimental, 419 controls, Chile	Ferreccio et al., 2000
Respiratory, bladder/other urinary	4058, Utah, U.S.	Lewis et al., 1999
Bladder, kidney	275, Finland	Kurttio et al., 1999
Lung	814, Japan	Nakadaira et al., 2002
	10,591, Taiwan	Chen et al., 2004
Bladder	509, Nevada, California, U.S.	Steinmaus et al., 2003

was matched to both a cancer and noncancer control. The study area included regions I, II, and III, the three northernmost provinces. The population in region II experienced high exposure to inorganic arsenic in past years from natural contamination of water originating in the Andes Mountains, whereas water sources in regions I and III contained relatively little arsenic. The study involved 152 lung cancer cases and 419 controls, who were interviewed regarding drinking water sources, cigarette smoking, socioeconomic status, lifetime residential history, and occupation. Using lifetime residential histories, each participant was assigned the average water arsenic concentration for the county in which they resided for each year. Counties generally have just one important supply of water in this extremely dry desert part of Chile. Average arsenic water concentrations were calculated from 1930 to the present. The lowest exposure categories were used as a reference to calculate ORs. Logistic regression analysis revealed a clear trend in lung cancer odds ratios (95% CIs) with increasing concentration of arsenic in drinking water: 1, 0.3 (0.1 to 1.2), 1.8 (0.5 to 6.9), 4.1 (1.8 to 9.6), 2.7 (1.0 to 7.1), 4.7 (2.0 to 11.0), 5.7 (1.9 to 16.9), 7.1 (3.4 to 14.8) for arsenic concentrations ranging from less than 10 μg/L to 990 μg/L. Despite some limitations this is "the only study available for risk assessment that has individual estimates of exposure on all subjects for more than 40 years, well beyond the latency for lung cancer" (NRC, 2001). A summary of these and a few other important studies linking arsenic consumption via drinking water and internal cancers is provided in Table 2.

RISK ASSESSMENT

Noncancer Effects

The California OEHHA (2004) applied risk assessment criteria to a number of noncancer toxic endpoints. The results of the assessment are summarized in Table 3. Appropriate uncertainty factors are shown for endpoints based on a NOAEL, LOAEL, or benchmark dose (lowest effect dose for 5% or 1% effect level, based on the day quality, expressed as LED_{05} or LED_{01}).

The human studies analyzed by benchmark response (BMR) methodology appear to give the best basis for determining a health-protective concentration for arsenic in drinking water. This is because they involve a dose-response methodology that utilizes more of the available data and the health effects analyzed in these studies are of public health concern. We concluded that the study by Chiou et al. (1997b) was the most robust of the studies evaluated. The single most representative value from the study was for CVD by the cumulative exposure metric, 0.86 μg/L. A higher value could be derived from the data of Lai et al. (1994) and Rahman et al. (1998) on diabetes mellitus (2.5 to 4.2 ppb, μg/L). Calculated values for peripheral vascular disease and skin effects were somewhat higher at about 6 ppb. Thus, based on this analysis, a level of 0.0009 mg/L, or 0.9 ppb, based on cerebrovascular disease (Chiou et al., 1997b) would represent a suitable health-protective value for noncancer adverse health effects due to chronic intake of arsenic in drinking water.

TABLE 3. Estimation of Health Protective Drinking Water Arsenic Concentrations Based on Noncancer Toxicity

Species and Study	Toxicity Endpoint	Dose–Response Criterion	Dose	UF/Risk[a]	Health Protective Concentration (mg/L)
Hamster					
Hanlon and Ferm, 1986	Fetal malformation	LOAEL, PBPK adjusted	2.8 mg/kg·day	1000	0.017
Dog					
Byron et al., 1967	Death, anorexia, listlessness, weight loss, slight to moderate anemia	NOAEL	1.25 mg/kg·day	100	0.088
Rat					
Byron et al., 1967	Decreased survival, weight loss, bile duct enlargement	NOAEL	3.12 mg/kg·day	100	0.22
Rhesus monkey					
Heywood and Sortwell, 1979	Sudden death without other clear clinical signs, possible CNS effects	LOAEL	2.8 mg/kg·	1000	0.02
Human					
Tseng et al., 1968; Tseng, 1977	Skin hyperpigmentation and keratoses	RfD U.S. EPA	0.3 μg/kg·day	3	0.0021
Mazumder et al., 1998	Skin keratosis	LED_{01}	50 μg/L	10	0.0010
				1×10^{-4}	0.0005
		LED_{05}	9.4 μg/kg·day	10	0.0066
				1×10^{-4}	0.0007
Chiou et al., 1997	Cerebral infarct	LED_{01}	166 μg/L	30	0.011
				1×10^{-4}	0.0016
		LED_{01}	3.5 (mg/L)·yr	30	0.00033
				1×10^{-4}	0.0024

	Cerebrovascular disease	LED_{01}	189 μg/L	10	0.0038
				1×10^{-4}	0.0019
		LED_{01}	3.0 (mg/L)·yr	10	0.00086
				1×10^{-4}	0.0021
Chen et al., 1996	Ischemic heart disease mortality	LED_{01}	5.5 (mg/L)·yr	30	0.00052
				1×10^{-4}	0.00078
Chen et al., 1995	Hypertension	LED_{10}	7.2 (mg/L)·yr	10	0.0020
				1×10^{-4}	0.00010
Rahman et al., 1999	Hypertension	LED_{10}	5.8 (mg/L)·yr	10	0.0016
				1×10^{-4}	0.00008
Lai et al., 1994	Diabetes mellitus	LED_{05}	8.8 (mg/L)·yr	10	0.0025
				1×10^{-4}	0.00025
Rahman et al., 1998	Diabetes mellitus	LED_{05}	0.21 mg/L	10	0.0042
				1×10^{-4}	0.00042
Wang et al., 2002	Carotid atherosclerosis (subclinical)	LOAEL estimated	20 (mg/L)·yr	10	0.006
Tseng et al., 1996	Peripheral vascular disease	LOAEL estimated	20 (mg/L)·yr	10	0.006
Siripitayakunkit et al., 1999	Developmental neurotoxicity, IQ deficit in children	LOAEL estimated and LED_{05}	30 μg/L per IQ point loss	30	0.0002
				1×10^{-2}	0.0003
Tsai et al., 2003	Developmental neurotoxicity, switching attention in children	LED_{05}	7 μg/day	30	0.0002
Wasserman et al., 2004	Developmental neurotoxicity, intellectual function deficit in children raw full-scale score vs. μg As/L	LOAEL estimated,	10 μg/L	30	0.0003
		fitted slope, exact fit $r^2 = 1$	−0.44 μg/L	30	7.6×10^{-5}

[a]Risk calculations assume low dose linearity, where x is the point of departure fraction (e.g., 0.01 for 1%, and risk is the criterion, e.g., risk*(70 yr)/[(x/LEDx)(1000)] for the cumulative dose metric and risk/x/LEDx for the water concentration metric).

Cancer Effects

The mechanisms for arsenic carcinogenicity are unknown. Because arsenic does not cause point mutations in experimental systems, some investigators have postulated that the results are consistent with theories of sublinearity for arsenic dose–response relationships. However, inference of sublinearity from toxicological considerations is at best speculative without support from more mechanistic or empirical data. Several mechanisms may be involved, including multiple interactions with other factors, both extrinsic and intrinsic, and available theories are inadequate to predict the dose–response relationship for this multifactorial, multisite carcinogen. In addition, no information has been produced to identify the range of arsenic exposures in which meaningful sublinearity might occur for any postulated theoretical mechanisms.

As with other major causes of human cancer, it is not likely that mechanisms allowing for valid predictions of dose–response relationships for low levels of arsenic will be identified in the near future. Indeed, mechanistic theories to date do not even predict why such high rates of bladder and lung cancer are observed in humans exposed to arsenic at levels not much higher than the current drinking water standards. Until they do, it is futile to use such theories to postulate what might be happening below the as-yet detectable effect levels in humans. This is not to say that mechanistic research is not important; however, our current understanding of arsenic toxic mechanisms appears rather primitive. It is also noteworthy that for many established causes of human cancer, the dose–response relationships found in epidemiological studies are linear, whether or not the particular agents involved cause point mutations [e.g., asbestos, chromium (VI), beryllium, and nickel subsulfide;] (California OEHHA, 1999).

A full discussion of potential modes of action of arsenic-induced carcinogenicity is beyond the scope of this chapter; however, an expert panel convened by the U.S. EPA (1997) identified the most plausible MOAs as increased chromosomal abnormalities; reduced DNA repair, altered DNA methylation, oxidative stress, reactive oxygen species damaging DNA, cell proliferation, and co-carcinogenicity. Arsenic is also known to affect a number of signal transduction pathways in human target cells, thereby altering gene expression (Porter et al., 1999; Hu et al. 2002; Rea, 2003).

Recent quantitative risk assessments of arsenic-induced cancer via drinking water ingestion have been based on lung and bladder cancer data from Taiwan or a combination of data from Taiwan, Chile, and Argentina (Table 4). The California OEHHA (2004) risk assessment derived an arsenic negligible risk concentration of 4 ppt in drinking water based on a unit risk value of 2.7×10^{-4} $(\mu g/L)^{-1}$ and a theoretical lifetime cancer risk level of 1×10^{-6}. The unit risk was based on linear regression analysis of lung and urinary bladder cancer mortality data in epidemiological studies in Taiwan, Chile, and Argentina and background mortality rates for these cancers in the United States. Other estimates of unit risks include 2.6×10^{-4} $(\mu g/L)^{-1}$ based on California mortality rates; 3.1×10^{-4} $(\mu g/L)^{-1}$ based on the sum of lung, bladder, skin, and kidney cancer mortality; and 5.9×10^{-4} $(\mu g/L)^{-1}$ based on lung and bladder cancer incidences rather than

TABLE 4. Summary of Quantitative Cancer Risk Assessment of Arsenic in Drinking Water

Risk Assessment	Methodology	Unit Risk, Health Protective Concentration
California OEHHA, 2004	Linear regression of lung data from Taiwan, Chile, and Argentina, value adjusted to include bladder cancer	2.7×10^{-4} (μg/L)$^{-1}$, 4.0 ppt (ng/L) (range 1.7–3.8)
NRC, 2001	Additive Poisson model Bladder cancer: male, female Lung cancer: male, female Taiwan data	2.3×10^{-4}, 1.2×10^{-4} (μg/L)$^{-1}$ 1.4×10^{-4}, 1.8×10^{-4} (μg/L)$^{-1}$ 1.5 ppt
U.S. EPA, 2001; Morales et al., 2000	Various dose-response models, ED_{01} based; lung and bladder cancer Taiwan data	2.41×10^{-5} to 6.05×10^{-5} (μg/L)$^{-1}$, 16–40 ppt; 0.2–11 ppt with Taiwanese comparison population

mortality. Unit risk estimates based on a transplacental carcinogenicity assay in mice were generally in the 1×10^{-4} to 1×10^{-3} (μg/L)$^{-1}$ range for various tumors and dose averaging methods. Thus the range of plausible negligible-risk values based on these unit risks is 1.7 to 3.8 ppt. The latter figure rounded to one significant figure is considered the most robust estimate in this assessment.

The National Research Council in their Arsenic in Drinking Water: 2001 Update (NRC, 2001) concluded that fitting the additive Poisson model with a linear term for dose to the lung and urinary bladder cancer incidence data from southwestern Taiwan to estimate ED_{01} values at specific As levels of interest (i.e., 3, 5, 10, and 20 μg/L) was the preferred analytical approach to assessing human cancer risk. They estimated excess lifetime risks per 10,000 people exposed to 10 μg As/L of 14 to 18 for lung cancer and 23 to 12 for bladder cancer in males and females, respectively. Assuming linear low dose extrapolation, these values would correspond to unit risks of 1.2×10^{-4} and 2.3×10^{-4} (μg/L)$^{-1}$ for bladder cancer in females and males and 1.4×10^{-4} and 1.8×10^{-4} (μg/L)$^{-1}$ for lung cancer in males and females.

The EPA (U.S. EPA, 2001) in their final rule on arsenic in drinking water assumes an average water EPA consumption of 1.0 and 1.2 L/d for tap and total water and 90th percentile values of 2.1 and 2.3 L/d, respectively. For cancer risks the agency has essentially used risk estimates for lung and bladder from Morales et al. (2000). That assessment is similar in approach to the NRC assessment described above and is based entirely on data from Taiwan. At 10 μg As/L, EPA estimates the mean population cancer risk as 2.41×10^{-4} to 2.99×10^{-4} and the 90th percentile upper bounds as 5.23×10^{-4} to 6.09×10^{-4}. Assuming low dose linearity of dose-response, these values would correspond to unit risks of 2.4×10^{-5} to 6.1×10^{-5} (μg/L)$^{-1}$, equivalent to negligible risk drinking water concentrations of 16 to 40 ppt. Thus, the EPA risk estimates are about four to 10 times less than those of the California OEHHA or the NRC. Morales et al. (2000) estimated a broad range of risks depending on the mathematical model that was fit to the data sets and what comparison population was used. For lung cancer without a comparison population, Morales et al. (2000) estimated LED_{01}s (lower bounds on the ED_{01}s) of 213 to 396 μg/L. With a Taiwanese comparison population these values were 6 to 196 μg/L, and with a southwestern Taiwanese comparison population, 8 to 181 μg/L. The EPA chose the highest of these analytical options (i.e., those without a comparison population). The combined tumor LED_{01} estimates with a Taiwanese comparison population ranged from 2 to 106 μg/L, values, which if extrapolated to negligible risk levels would bracket the negligible risk concentrations from the California OEHHA and NRC assessments (i.e., 0.2 to 11 ppt).

CONCLUSIONS

Recent studies add to the evidence that ingestion of inorganic arsenic increases the risk of lung cancer. Clear increased risks were found in ecological studies in both Argentina and Chile. Confounding due to smoking could be excluded as the explanation in both populations. Increased lung cancer risks have been reported in a small study in Japan involving drinking water and a case-control study with individual exposure data from Chile. The biological plausibility that ingestion of arsenic might increase lung cancer risks is strengthened by the fact that it is a confirmed lung carcinogen by inhalation. Taking this into account, there is now sufficient evidence to conclude that ingestion of inorganic arsenic is a cause of human lung cancer.

There is also sufficient evidence from studies in several countries to conclude that ingestion of arsenic is a cause of human bladder cancer. Beyond the findings in Taiwan, the strongest evidence comes from large population studies in Chile and Argentina, both of which found that the highest relative risks for internal cancer mortality associated with arsenic exposure were for bladder cancer. These ecological studies are supplemented by studies with individual data, in particular in Taiwan and in the Fowler's solution study in England. There is therefore ample evidence to conclude that inorganic arsenic ingestion is a cause of human bladder cancer.

Although recent studies add to the existing evidence and make it probable that ingestion of arsenic can cause kidney cancer, the findings are not as strong as for bladder and lung cancer. The evidence concerning liver cancer has actually been weakened by recent studies, especially the lack of increased risks of liver cancer in region II of Chile in the presence of dramatic increases in bladder and lung cancer mortality. The IARC (2004) concluded that "there is sufficient evidence in humans that arsenic in drinking water causes cancer of the urinary bladder, lung and skin."

The major single source of uncertainty in quantitative cancer risk estimation is the shape of the dose–response curve (U.S. EPA, 2005). It has been argued that the dose-response relationship between ingested arsenic and cancer may not be linear, but rather, a threshold or sublinear response may exist (IARC, 1987; Lu, 1990; Carlson-Lynch et al., 1994). If this were the case, the assumption of linearity would overestimate risk at low dose levels. Currently, evidence is limited that a threshold or significant sublinear dose-response mechanism exists for ingested arsenic and cancer. The study with the best exposure data, that of Ferreccio et al. (2000), suggests supralinearity at low dose. Nevertheless, the confidence intervals are broad, and therefore this study does not establish supralinearity as the shape of the dose–response relation.

Inorganic arsenic in drinking water represents possibly the highest potential risk of any waterborne chemical, with the possible exception of radon. It is interesting to consider why effective regulation of this chemical was delayed for so many years. Among the main reasons we ought to include the absence of any workable animal model(s) of carcinogenicity. The IARC 2004 concluded that "the studies on inorganic arsenic provide *limited evidence* for carcinogenicity in experimental animals." Many risk assessors are toxicologists familiar with analyzing animal toxicity data but less comfortable with assessing epidemiological data, which are almost always deficient in accurate exposure information. Also, the fact that early data indicated skin cancer, which was often not fatal or life-threatening, allowed risk managers to discount the risks to some extent. Arguments about nutritional deficiency of Taiwanese cancer subjects and a possible (though unproven) human nutritional role for arsenic added to the discount. It was not until serious internal cancers were apparent and later quantified that more stringent regulation was proposed (California OEHHA, 1992; Brown and Fan, 1994) and eventually became unavoidable. The lesson for risk assessors is not to ignore or overly discount human data despite its many limitations. This is particularly true for arsenic and other environmental toxicants (e.g., lead and methylmercury) with serious noncancer health effects, such as heart attack, stroke, and neurotoxicity. There is a need for robust quantitative risk assessment methodology for noncancer health effects that complements cancer risk assessment (Clewell and Crump, 2005).

Acknowledgments

This chapter was based in part on the 2001 report "Cancer Risk Assessment of Inorganic Arsenic in Drinking Water" prepared by Drs. Allan Smith and Peggy Lopipero of the University of California-Berkeley under contract to the Office of Environmental Health Hazard Assessment, California Environmental Protection Agency, Sacramento, CA.

Disclaimer

The opinions exressed in this paper are those of the author and not necessarily those of the Office of Environmental Health Hazard Assessment or the California Environmental Protection Agency.

REFERENCES

Ahmad SA, Sayed MHSU, Hadi SA, Faruquee MH, Kahn MH, Jalil MA, Ahmed R, Kahn AW. 1999. Arsenicosis in a village in Bangladesh. *Int J Environ Health Res* 9: 187–195.

Ahsan H, Perrin M, Rahman A, Faruque P, Stute M, Zheng Y, Milton H, Abul MBBS, Brant-Rauf P, van Geen A, Graziano J. 2000. Associations between drinking water and urinary arsenic levels and skin lesions in Bangladesh. *J Occup Environ Med* 42: 1195–1201.

Aposhian HV. 1997. Enzymatic methylation of arsenic species and other new approaches to arsenic toxicity. *Anu Rev Pharmacol Toxicol* 37: 397–419.

Aposhian HV, Gurzau ES, Le XC, Gurzau A, Healy SM, Lu X, Ma M, Yip L, Zakharyan RA, Maiorino RM, Dart RC, Tircus MG, Gonzalez-Ramirez D, Morgan DL, Avram D, Aposhian MM. 2000a. Occurrence of monomethylarsonous acid in urine of humans exposed to inorganic arsenic. *Chem Res Toxicol* 13: 693–697.

Aposhian HV, Zheng B, Aposhian MM, Le XC, Cebrian ME, Cullen W, Zakharyan RA, Ma M, Dart RC, Cheng Z, Andrewes P, Yip L, O'Malley GF, Maiorino RM, Van Voorhies W, Healy SM, Titcomb A 2000b. DMPS-arsenic challenge test, II: Modulation of arsenic species, including monomethylarsonous acid (MMA^{III}), excreted in human urine. *Toxicol Appl Pharmacol* 165: 74–83.

Apostoli P, Alessio L, Romeo L, Buchet JP, Leone R. 1997. Metabolism of arsenic after acute occupational arsine intoxication. *J Toxicol Environ Health* 52: 331–342.

Aranyi C, Bradof JN, O' Shea WJ, Graham JA, Miller FJ. 1985. Effects of arsenic trioxide inhalation exposure on preliminary antibacterial defenses in mice. *Arch Environ Health* 41: 171–177.

ATSDR 1997. Agency for Toxic Substances and Disease Registry). *Arsenic. Toxicological Profile*. Public Health Service, U.S. Department of Health and Human Services, Atlanta, GA.

———. 2000. *Arsenic. Toxicological Profile* (Update). Public Health Service, U.S. Department of Health and Human Services, Atlanta, GA.

Barchowsky A, Dudek EJ, Treadwell MD, Wetterhahn K. 1996. Arsenic induces oxidant stress and NF-κB activation in cultured aortic endothelial cells. *Free Radic Biol Med* 21: 783–790.

Baroni C, Van Esch GJ, Saffiotti U. 1963. Carcinogenesis tests of two inorganic arsenicals. *Arch Environ Health* 7: 668–674.

Bates MN, Smith AH, Cantor KP. 1995. Case–control study of bladder cancer and arsenic in drinking water. *Am J Epidemiol* 141: 523–530.

Bencko V. 1987. *Arsenic*. In: *Genotoxic and Carcinogenic Metals: Environmental and Occupational Occurrence and Exposure*, Fishbein L, Furst A, Mehlman MA, eds. Princeton Scientific Publishing Co., Princeton, NJ, pp. 1–30.

Bencko V, Symon K, Chladek V, Pihrt J. 1977. Health aspects of burning coal with a high arsenic content. II. Hearing changes in exposed children. *Environ Res* 13: 386–395.

Biggs ML, Kalman DA, Moore LE, Hopenhayn-Rich C, Smith MT, Smith AH. (1997). Relationship of urinary arsenic to intake estimates and a biomarker of effect, bladder cell micronuclei. *Mutat Res* 386: 185–195.

Blakeley BR. 1987. The effect of arsenic on urethane-induced adenoma formation in Swiss mice. *Can J Vet Res* 51: 240–243.

Borgono JM, Vicent P, Venturino H, Infante A. 1977. Arsenic in the drinking water of the city of Antofagasta: epidemiological and clinical study before and after the installation of a treatment plant. *Environ Health Perspect* 19: 103–105.

Borzsonyi M, Bereczky A, Rudnai P, Csnady M, Horvath A. 1992. Epidemiologic studies on human subjects exposed to arsenic in drinking water in southeast Hungary. *Arch Toxicol* 66: 77–78.

Boutwell RK. 1963. A carcinogenicity evaluation of potassium arsenite and arsanilic acid. *Agric Food Chem* 11: 381–385.

Braman RS. 1983. *Environmental reaction and analysis methods*. In: *Biological and Environmental Effects of Arsenic*, Fowler BA, ed. Elsevier, Amsterdam, The Netherlands, pp. 141–154.

Brown JP, Fan AM. 1994. Arsenic: risk assessment for California drinking water standards. *J Haz and Matter* 39: 149–159.

Buchanan WD. 1962. *Toxicity of Arsenic Compounds*. Elsevier, New York.

Buchet JP, Lauwerys R. 1985a. Study of inorganic arsenic methylation by rat liver invitro: relevance for the interpretation of observations in man. *Arch Toxicol* 57: 125–129.

______. 1985b. Study of factors influencing the invivo methylation of inorganic arsenic in rats. *Toxicol Appl Pharmacol* 91: 65–74.

Buchet JP, Lison D. 1998. Mortality by cancer in groups of the Belgian population with a moderately increased intake of arsenic. *Int Arch Occup Environ Health* 71: 125–130.

Buchet JP, Lauwerys R, Roels H. 1981a. Urinary excretion of inorganic arsenic and its metabolites after repeated ingestion of sodium meta arsenite by volunteers. *Int Arch Occup Environ Health* 48: 111–118.

______. 1981b. Comparison of the urinary excretion of arsenic metabolites after a single oral dose of sodium arsenite, monomethylarsonate, or dimethylarsinate in man. *Int Arch Occup Environ Health* 48: 71–79.

Byron WR, Bierbower GW, Brouwer JB, Hansen WH. 1967. Pathologic changes in rats and dogs from two-year feeding of sodium arsenite or sodium arsenate. *Toxicol Appl Pharmacol* 10: 132–147.

Calderon J, Navarro ME, Jimenez-Capdeville ME, Santos-Diaz MA, Golden A, Rodriguez-Leyva I, Borja-Aburto V, Diaz-Barriga F. 2001. Exposure to arsenic and lead and neuropsychological development in Mexican children. *Environ Res A* 85: 69–76.

California DHS (Department of Health Services). 1987. Evaluation of hazards posed by the use of wood preservatives on playground equipment, Report to the Legislature. Office of Environmental Health Hazard Assessment, California DHS, Sacramento, CA.

California OEHHA (Office of Environmental Health Hazard Assessment). 1992. *Arsenic Recommended Public Health Level for Drinking Water. Draft*. California Environmental Protection Agency, Berkeley, CA.

______. 1999. *Air Toxics Hot Spots Program Risk Assessment Guidelines. Part II,* Technical Support Document for Describing Available Cancer Potency Factors. California Environmental Protection Agency, Sacramento, CA.

______. 2004. *Public Health Goal for Arsenic in Drinking Water*. California Environmental Protection Agency, Sacramento, CA.

Cantor KP. 2001. Invited commentary: Arsenic and cancer of the urinary tract. *Am J Epidemiol* 153: 419–421.

Carlson-Lynch H, Beck BD, Boardman PD. 1994. Arsenic risk assessment. *Environ Health Perspect* 102: 354–356.

Chan P, Huff, JE. 1997. Arsenic carcinogenesis in animals and in humans: mechanistic, experimental, and epidemiological evidence. *Environ Carcin Ecotox Rev* C 15: 83–122.

Chang HR, Yu HS, Chai CY, Lin YR, Yu CL. 1998. Arsenic induces decreased expression of β2-adrenergic receptors in cultured keratinocytes. *Arch Dermatol Res* 290: 402–404.

Charbonneau SM, Spencer K, Bryce F, Sandi E. 1978. Arsenic excretion by monkeys dosed with arsenic-containing fish or with inorganic arsenic. *Bull Environ Contam Toxicol* 20: 470–477.

Chattopadhyay S, Bhaumik S, Chaudhury AN, Gupta SD. 2002. Arsenic induced changes in growth development and apoptosis in neonatal and adult brain cells in vivo and in tissue culture. *Toxicol Lett* 128: 73–84.

Chen CJ, Chiou H. 2001. Chen and Chiou respond to "Arsenic and cancer of the urinary tract" by Cantor. *Am J Epidemiol* 153: 422–423.

Chen CJ, Wang CJ. 1990. Ecological correlation between arsenic level in well water and age-adjusted mortality from malignant neoplasms. *Cancer Res* 50: 5470–5474.

Chen KP, Wu HY. 1962. Epidemiologic studies on blackfoot disease, II: A study of source of drinking water in relation to the disease. *J Formosan Med Assoc* 61: 61–618.

Chen CJ, Chuang YC, Lin TM, Wu HY. 1985. Malignant neoplasms among residents of a blackfoot disease-endemic area in Taiwan. *Br J Cancer* 53: 399–405.

Chen CJ, Chuang YC, You SL, Lin HY. 1986. A retrospective study on malignant neoplasms of bladder, lung and liver in blackfoot disease endemic area in Taiwan. *Br J Cancer* 53: 399–405.

Chen CJ, Wu MM, Lee SS, Wang JD, Cheng SH, Wu HY. 1988a. Atherogenicity and carcinogenicity of high-arsenic artesian well water: multiple risk factors and related malignant neoplasms of blackfoot disease. *Arteriosclerosis* 8: 452–460.

Chen CJ, Kuo TL, Wu MM. 1988b. Arsenic and cancers. *Lancet* 1:414.

Chen CJ, Chen CW, Wu MM, Kuo TL. 1992. Cancer potential in liver, lung, bladder and kidney due to ingested inorganic arsenic in drinking water. *Br J Cancer* 66: 888–892.

Chen CJ, Hsueh YM, Lai MS, Shyu MP, Chen SY, Wu MM, Kuo TL, Tai TY. 1995. Increased prevalence of hypertension and long-term arsenic exposure. *Hypertension* 25: 53–60.

Chen CJ, Chiou HY, Chiang MH, Lin LJ, Tai TY. 1996. Dose–response relationship between ischemic heart disease mortality and long-term arsenic exposure. *Atheroscler Thromb Vasc Biol* 16: 504–510.

Chen CL, Hsu LI, Chiou HY, Hsueh YM, Chen SY, Wu MM, Chen CJ. 2004. Ingested arsenic, cigarette smoking, and lung cancer risk: a follow-up study in arseniasis-endemic areas of Taiwan. *JAMA* 292: 2984–2990.

Chi IC, Blackwell RQ. 1968. A controlled retrospective study of blackfoot disease, an endemic peripheral gangrene disease in Taiwan. *Am J Epidemiol* 88: 7–24.

Chiou HY, Hsueh YM, Liaw KF, Horng SF, Chiang MH, Pu YS, Lin JS, Huang CH, Chen CJ. 1995. Incidence of internal cancers and ingested inorganic arsenic: a seven-year follow-up study in Taiwan. *Cancer Res* 55: 1296–1300.

Chiou HY, Hsueh YM, Hsieh LL, Hsu LI, Hsu YH, Hsieh FI, Wei ML, Chen HC, Yang HT, Leu LC, Chu TH, Chen-Wu C, Yang MH, Chen CJ. 1997a. Arsenic methylation capacity, body retention, and null genotypes of glutathione S-transferase M1 and T1 among current arsenic-exposed residents in Taiwan. *Mutat Res* 386: 197–297.

Chiou HY, Huang YI, Su CL, Chang SF, et al. 1997b. Dose-response relationship between prevalence of cerebrovascular disease and ingested inorganic arsenic. *Stroke* 28: 1717–1723.

Chiou H-Y, Chiou S-T, Hsu Y-H, Chou Y-L, Tseng CH, Wei M-L, Chen C-J. 2001. Incidence of transitional cell carcinoma and arsenic in drinking water: a follow-up study of 8102 residents in an arseniasis-endemic area in northeastern Taiwan. *Am J Epidemiol* 153: 411–418.

Chuttani PN, Chawla LS, Sharma TD. 1967. Arsenical neuropathy. *Neurology* 17: 269–274.

Clewell HJ, Crump KS. 2005. Quantitative estimates of risk for noncancer endpoints. *Risk Anal* 25: 285–289.

Cohen SM, Yamamoto S, Cano M, Arnold LL. 2001. Urothelial cytotoxicity and regeneration induced by dimethylarsinic acid in rats. *Toxicol Sci* 59: 68–74.

Collins T. 1993. Endothelial nuclear factor-kappa B activation and the initiation of the atherosclerotic lesion. *Lab Invest* 68: 499–508.

CPSC (Consumer Product Safety Commission). 1990. Project on Playground Equipment: Transmittal of estimate of risk of skin cancer for dislodgeable arsenic on pressure treated wood playground equipment. Memorandum to the Commission from SE Dunn (Secretary) containing four separate reports and executive summary. August 2. CPSC, Washington, DC.

Cullen WR, Reimer KJ. 1989. Arsenic speciation in the environment. *Chem Rev* 89: 713–764.

DeSesso JM, Jacobson CF, Scalli AR, Farr CH, Holson JF. 1998. An assessment of the developmental toxicity of inorganic arsenic. *Reprod Toxicol* 12: 385–433.

Dong J-T, Luo X-M. 1994. Effects of arsenic on DNA damage and repair in human fetal lung fibroblasts. *Mutat Res* 315: 11–15.

Donofrio PD, Wilbourn AJ, Albers JW, Rogers L, Salanga V, Greenberg HS. 1987. Acute arsenic intoxication presenting as Guillain-Barré-like syndrome. *Muscle Nerve* 10: 114–120.

Dulout FN, Grillo CA, Seoane AI, Maderna CR, Nilsson R, Vahter M, Darroudi F, Natarajan AT. 1996. Chromosomal aberrations in peripheral blood lymphocytes from native Andean women and children from northwestern Argentina exposed to arsenic in drinking water. *Mutat Res* 370: 151–158.

Edmonds MS, Baker DH. 1986. Toxic effects of supplemental copper and roxarsone when fed alone or in combination to young pigs. *J Anim Sci* 63: 533–537.

Edmonds JS, Francesconi KA. 1993. Arsenic in seafoods: human health aspects and regulations. *Mar Poll Bull* 26: 665–674.

Engel RR, Smith AH. 1994. Arsenic in drinking water and mortality from vascular disease: an ecologic analysis in 30 U.S. counties. *Arch Environ Health* 49: 418–427.

Feinglass EJ. 1973. Arsenic intoxication from well water in the United States. *N Engl J Med* 288: 828–830.

Ferreccio C, Gonzalez C, Milosavjlevic V, Marshall G, Sancha AM, Smith AH. 2000. Lung cancer case control study in Chile identifies a strong exposure-response trend with arsenic concentrations in drinking water. *Epidemiology* 11: 673–679.

FNB (Food and Nutrition Board). 2002. *Dietary Reference Intakes for Vitamin A, Vitamin K, Arsenic, Boron, Chromium, Copper, Iodine, Iron, Manganese, Molybdenum, Nickel, Silicon, Vanadium, and Zinc*. Appendix E. Mean and Percentiles for Usual Intake of Arsenic (μg/day) from Food, Total Diet Study (1991–1997). Institute of Medicine, Washington, DC, pp. 658–659. Accessed at: www.nap.edu/openbook/0309072794/html/505.html.

Freeman GB, Schoof RA, Ruby MV, Davis AO, Dill JA, Liao SC, Lapin CA, Bergstrom PD. 1995. Bioavailability of arsenic in soil and house dust impacted by smelter activities following oral administration in *Cynomolgus* monkeys. *Fundam Appl Toxicol* 28: 215–222.

Gartrell MJ, Craun JC, Podrebarac DS, Gunderson EL. 1985. Pesticides, selected elements, and other chemicals in adult total diet samples, October 1979–September 1980. *J Assoc Off Anal Chem* 68: 1184–1197.

Gebel T, Christensen S, Dunkelberg H. 1997. Comparative and environmental genotoxicity of antimony and arsenic. *Anticancer Res* 17: 2603–2608.

Gentry PR, Covington TR, Mann S, Shipp AM, Yager JW, Clewell HJ III. 2004. Physiologically based pharmacokinetic modeling of arsenic in the mouse. *J Toxicol Environ Health A* 67: 43–71.

Gentry PR, Covington TR, Lawrence G, McDonald T, Snow ET, Germolec D, Moser G, Yager JW, Clewell HJ III. 2005. Comparison of tissue dosimetry in the mouse following chronic exposure to arsenic compounds. *J Toxicol Environ Health A* 68: 329–351.

Germolec DR, Yoshida T, Gaido K, Wilmer JL, Simeonova PP, Kayama F, Burleson F, Dong W, Lange RW, Luster MI. 1996. Arsenic induces overexpression of growth factors in human keratinocytes. *Toxicol Appl Pharmacol* 141: 308–318.

Germolec DR, Spalding J, Boorman GA, Wilmer JL, Yoshida T, Simeonova PP, Bruccoleri A, Kayama F, Gaido K, Tennant R, Burleson F, Dong W, Lang RW, Luster MI. 1997. Arsenic can mediate skin neoplasia by chronic stimulation of keratinocyte-derived growth factors. *Mutat Res* 386: 209–218.

GESAMP. 1986. Review of Potentially Harmful Substances: Arsenic, Mercury and Selenium. Reports and Studies. GESAMP (28), joint group of experts on the scientific aspects of marine pollution. IMO/FAO/UNESCO/WMO/WHO/IAEA/UN/UNEP.

Gladysheva TB, Oden KL, Rosen BP. 1992. Properties of the arsenate reductase of plasmid R733. *Biochemistry* 33: 7288–7293.

Golub MS, Macintosh MS, Baumrind N. 1998. Developmental and reproductive toxicity of inorganic arsenic: animal studies and human concerns. *J Toxicol Environ Health B* 1: 199–241.

Gonseblatt ME, Vega L, Salazar AM, Montero R, Guzman P, Blas J, Del Razo LM, Gaecia-Vargas G, Albores A, Cebrian ME, Kelsh M, Ostrosky-Wegman P. 1997. Cytogenetic effects in human exposure to arsenic. *Mutat Res* 386: 210–228.

Gosselin RE, Smith RP, Hodge HC, Braddock, JE. 1984. In: *Clinical Toxicology of Commercial Products*. 5th ed. Williams & Wilkins, Baltimore, MD, p. III-42.

Gregus Z, Gyurasics A, Csanaky I. 2000. Biliary and urinary excretion of inorganic arsenic: monomethylarsonous acid as a major biliary metabolite in rats. *Toxicol Sci* 56: 18–25.

Grobe JW. 1976. Periphere Durchblutungsst Horungen und Akrocyanose bei Arsengeschadigten Moselwintzern [Peripheral circulatory disorders and acrocyanosis in Moselle valley vineyard workers with arsenic poisoning]. *Berufsdermatosen* 24: 78–84.

Guo H, Chiang H, Hu H, Lipsitz SR, Monson RR. 1997. Arsenic in drinking water and incidence of urinary cancers. *Epidemiology* 8: 545–550.

Hamamoto E. 1955. Infant arsenic poisoning by powdered milk. *Nihon Iji Shimpo* 1649: 3–12.

Hanlon DP, Ferm VH. 1986. Teratogen concentration changes as the basis of the heat stress enhancement of arsenate teratogenesis in hamsters. *Teratology* 34: 189–193.

Hay R, McCormack JG. 1987. Arsenic poisoning and peripheral neuropathy. *Aust Fam Phys* 16: 287–289.

Hayakawa T, Kobayashi Y, Cui X, Hirano S. 2005. A new metabolic pathway of arsenite-glutathione complexes are substrates for human arsenic methyltransferase Cyt19. *Arch Toxicol* 79: 183–191.

Healy SM, Casarez EA, Ayala-Fierro F, Aposhian HV. 1998. Enzymatic methylation of arsenic compounds, V: Arsenite methyltransferase activity in tissues of mice. *Toxicol Appl Pharmacol* 148: 65–70.

Heywood R, Sortwell RJ. 1979. Arsenic intoxication in the rhesus monkey. *Toxicol Lett* 3: 137–144.

Hinwood AL, Jolley DJ, Sim MR. 1999. Cancer incidence and high environmental arsenic concentrations in rural populations: results of an ecological study. *Int J Environ Health Res* 9: 131–141.

Holson JF, Stump DG, Clevidence KJ, Knapp JF, Farr CH. 2000. Evaluation of the prenatal toxicity of orally administered arsenic trioxide in rats. *Food Chem Toxicol* 38: 459–466.

Hopenhayn-Rich C, Biggs ML, Smith AH, Kalman DA, Moore LE. 1996. Methylation study of a population environmentally exposed to arsenic in drinking water. *Environ Health Perspect* 104: 620–628.

Hopenhayn-Rich C, Biggs ML, Smith AH. 1998. Lung and kidney cancer mortality associated with arsenic in drinking water in Cordoba, Argentina. *Int J Epidemiol* 27: 561–569.

Hopenhayn-Rich C, Browning SR, Hertz-Picciotto I, Ferreccio C, Peralia C, Gibb H. 2000. Chronic arsenic exposure and risk of infant mortality in two areas of Chile. *Environ Health Perspect* 108: 667–673.

Hu Y, Jin X, Snow ET. 2002. Effect of arsenic on transcription factor AP-1 and NF-κB DNA binding activity and related gene expression. *Toxicol Lett* 133: 33–45.

Hueper WC, Payne WW. 1962. Experimental studies in metal carcinogenesis. *Arch Environ Health* 5: 445–462.

Huff J, Chan P, Nyska A. 2000. Is the human carcinogen arsenic carcinogenic to laboratory animals? *Toxicol Sci* 55: 17–23.

Hughes MF, Menache M, Thompson DJ. 1994. Dose-dependent disposition of sodium arsenate in mice following acute oral exposure. *Fund am Appl Toxicol* 22: 80–89.

Hutchinson J. 1887. Arsenic cancer. *Br Med J* 2: 1280–1281.

_____. 1888. On some examples of arsenic-keratosis of the skin and of arsenic-cancer. *Trans Pathol Soc Lond* 39: 352–363.

Hwang YH, Bornschein RL, Grote J, Menrath W, Roda S. 1997. Environmental arsenic exposure of children around a former copper smelter site. *Environ Res* 72: 72–81.

IARC International Agency for Research on Cancer. 1980. *IARC Monographs on the Evaluation of the Carcinogenic Risk of Chemicals to Man: Some Metals and Metallic Compounds*, Vol 23. IARC Lyon, France.

______. 1987. *IARC Monographs on the evaluation of carcinogenic risks to humans. Overall Evaluations of Carcinogenicity—An Updating of the IARC Monographs,* Vols 1 to 42, Suppl 7. IARC, Lyon, France.

______. 2004. *IARC Monographs on the Evaluation of Carcinogenic Risks to Humans. Vol 84, Some Drinking-Water Disinfectants and Contaminants, Including Arsenic*. IARC Lyon, France pp 41–267.

Innes JRM, Ulland BM, Valerio MG, Petrucelli L, Fishbein L, Hart ER, Pallotta AJ, Bates RR, Falk HL, Gart JJ, Klein M, Mitchell I, Peters J. 1969. Bioassay of pesticides and industrial chemicals for tumorigenicity in mice: a preliminary note. *J Natl Cancer Inst* 42: 1101–1114.

Irgolic KJ, Stockton RA, Chakraborti D. 1983. Determination of arsenic and arsenic compounds in water supplies. In: *Arsenic: Industrial, Biomedical, Environmental Perspectives*, Lederer WH, Fensterheim RJ, eds. Van Nostrand Reinhold, New York, pp. 282–308.

Ishinishi N, Yamamoto A, Hisanaga A, Imanasu T. 1983. Tumorigenicity of arsenic trioxide to the lung in Syrian golden hamsters by intermittent instillations. *Cancer Lett* 21: 141–147.

Ishinishi N, Tsuchiya K, Vahter M, Fowler BA. 1986. Arsenic. In: *Handbook on the Toxicology of Metals*, 2nd ed, Friberg L, Nordberg GF, Vouk V, eds. Elsevier Science, New york, pp. 43–83.

Jacobson CF, Stump DG, Nemec MD, Holson JF, Desesso JM. 1999. Appropriate exposure routes and doses in studies designed to assess developmental toxicity: a case study of inorganic arsenic. *Int J Toxicol* 18: 361–368.

Jacobson-Kram D, Montalbano D. 1985. The reproductive effects assessment group′s report on the mutagenicity of inorganic arsenic. *Environ Mutagen* 7: 787–804.

Jenkins RB. 1966. Inorganic arsenic and the nervous system. *Brain* 89: 479–498.

Jha AN, Noditi M, Nilsson R, Natarajan AT. 1992. Genotoxic effects of sodium arsenite on human cells. *Mutat Res* 284: 215–221.

Ji G, Silver S. 1992. Reduction of arsenate to arsenite by the ArsC protein of the arsenic resistance operon of Staphylococcus aureus plasmid pI258. *Proc Natl Acad Sci USA* 89: 9474–9478.

Jiang G, Gong Z, Li XF, Cullen WR, Le XC. 2003. Interaction of trivalent arsenicals with metallothionein. *Chem Res Toxicol* 16: 873–880.

Kanisawa M, Schroeder HA. 1967. Life term studies on the effects of arsenic, germanium, tin, and vanadium on spontaneous tumors in mice. *Cancer Res* 27: 1192–1195.

______. 1969. Life term studies on the effect of trace elements on spontaneous tumors in mice and rats. *Cancer Res* 29: 892–895.

Karagas MR, Stukel TA, Morris JS, Tosteson TD, Weiss JE, Spencer SK, Greenberg ER. 2001. Skin cancer risk in relation to toenail arsenic concentrations in a US population-based case-control study. *Am J Epidemiol* 153: 559–565.

Kennedy S, Rice DA, Cush PF. 1986. Neuropathology of experimental 3-nitro-4-hydroxyphenylarsonic acid toxicosis in pigs. *Vet Pathol* 23: 454–461.

Kerkvliet NI, Steppan LB, Koller LD, Exon JH. 1980. Immunotoxicology studies of sodium arsenate: effects of exposure on tumor growth and cell-mediated tumor immunity. *J Environ Pathol Toxicol* 4: 65–79.

Kitchin KT. 2001. Recent advances in arsenic carcinogenesis: modes of action, animal model systems, and methylated arsenic metabolites. *Toxicol Appl Pharmacol* 172: 249–261.

Kligerman AD, Doerr CL, Tennant AH, Harrington-Brock K, Allen JW, Winkfield E, Poorman-Allen P, Kundu B, Funasaka K, Roop BC, Mass MJ, DeMarini DM. 2003. Methylated trivalent arsenicals as candidate ultimate genotoxic forms of arsenic: induction of chromosomal mutations but not gene mutations. *Environ Mol Mutagen* 42: 192–205.

Knoth W. 1966. Arsenic treatment. *Arch Klin Exp Dermatol* 227: 228–234.

Krafft T, Macy JM. 1998. Purification and characterization of the respiratory arsenate reductase of *Chrysiogenes arsenatis*. *Eur J Biochem* 255: 647–653.

Kurttio P, Komulainen H, Hakala E, Kahelin H, Pekkanen J. 1998. Urinary excretion of arsenic species after exposure to arsenic present in drinking water. *Arch Environ Contam Toxicol* 34: 297–305.

Kurttio P, Pukkala E, Kahelin H, Auvinen A, Pekkanen J. 1999. Arsenic concentrations in well water and risk of bladder and kidney cancer in Finland. *Environ Health Perspect* 107: 705–729.

LaCelle PL. 1970. Alteration in membrane deformability in hemolytic anemias. *Semin Hematol* 7: 355–371.

Lai MS, Hsueh YM, Chen CJ, Shyu MP, Chen SY, Kuo TL, Wu TL, Wu MM, Tai TY. 1994. Ingested inorganic arsenic and prevalence of diabetes mellitus. *Am J Epidemiol* 139: 484–492.

Lasky T, Sun W, Kadry A, Hoffman M. 2004. Mean total arsenic concentrations in chicken 1989–2000 and estimated exposures for consumers of chicken. *Environ Health Perspect* 112: 18–21.

Le XC, Lu X, Ma M, Cullen WR, Aposhian HV, Zheng B. 2000a. Speciation of key arsenic metabolic intermediates in human urine. *Anal Chem* 72: 5172–5177.

______. 2000b. Determination of monomethylarsonous acid, a key arsenic methylation intermediate, in human urine. *Environ Health Perspect* 108: 1015–1018.

Lerda D. 1994. Sister-chromatid exchange (SCE) among individuals chronically exposed to arsenic in drinking water. *Mutat Res* 312: 111–120.

Lewis RD, Southwick JW, Ouellet-Hellstrom R, Rench J, Calderon RL. 1999. Drinking water arsenic in Utah: A cohort mortality study. *Environ Health Perspect* 107: 359–365.

Lu FJ. 1990. Blackfoot disease: Arsenic or humic acid? *Lancet* 336: 115–116.

Lynn S, Gurr JR, Lai HT, Jan KY. 2000. NADH oxidase activation is involved in arsenite-induced oxidative damage in human vascular smooth muscle cells. *Circ Res* 86: 514–519.

Ma HZ, Sun TZ, Xia YJ, Yu GJ. 1997. Endemic arsenic poisoning in Inner Mongolia, China (abstr). *Toxicologist* 36: 239.

Ma HZ, Xia YJ, Wu KG, Sun TZ, Mumford JL. 1999. Human exposure to arsenic and health effects in Bayingnormen, Inner Mongolia. In: *Arsenic Exposure and Health*

Effects. Chappell WR, Abernathy CO, Calderon RL, eds. Elsevier Science, Amsterdam, The Netherlands, pp. 127–131.

Machado AF, Hovland DN, Pilafas S, Collins MD. 1999. Teratogenic response to arsenite during neurulation: relative sensitivities of C57BL/6 J and SWV/Fnn mice and impact of the splotch allele. *Toxicol Sci* 51: 98–107.

Maki-Paakkanen J, Kurttio P, Paldy A, Pekkanen J. 1998. Association between the clastogenic effect in peripheral lymphocytes and human exposure to arsenic through drinking water. *Environ Mol Mutagen* 32: 301–313.

Mann S, Droz PO, Vahter M. 1994. A physiologically based pharmacokinetic model for the four major arsenic species in mammals. In: *Arsenic Exposure and Health*. Chappell AWR, Abernathy, CO, Cothern CR, eds. Science and Technology Letters, Chicago, IL, pp. 219–231.

_____. 1996a. A physiologically based pharmacokinetic model for arsenic exposure, I: Development in hamsters and rabbits. *Toxicol Appl Pharmacol* 137: 8–22.

_____. 1996b. A physiologically based pharmacokinetic model for arsenic exposure II. Validation and application in humans. *Toxicol Appl Pharmacol* 140: 471–486.

Marcus WL, Rispin AS. 1988. *Threshold carcinogenicity using arsenic as an example*. In: *Advances in Modern Environmental Toxicology: Risk Assessment and Risk Management of Industrial and Environmental Chemicals*, Vol 15. Cothern CR, Mehlman MA, eds. Princeton Publishing, Princeton, NJ, pp. 133–158.

Marnell LL, Garcia-Vargas GGG, Chowdhury UK, Zakharyan RA, Walsh B, Avram MD, Kopplin MJ, Cebrian ME, Silbergeld EK, Aposhian HV. 2003. Polymorphisms in the human monomethylarsonic acid (MMA^{V}) reductase/hGSTO1 gene and changes in urinary arsenic profiles. *Chem Res Toxicol* 16: 1507–1513.

Mass MJ, Tennant A, Roop BC, Cullen WR, Styblo M, Thomas DJ, Kligerman AD. 2001. Methylated trivalent arsenic species are genotoxic. *Chem Res Toxicol* 14: 355–361.

Mazumder DNG, Haque R, Ghosh N, De BK, Santra A, Chakraborty D, Smith AH. 1998. Arsenic levels in drinking water and the prevalence of skin lesions in West Bengal, India. *Int J Epidemiol* 27: 871–877.

_____. 2000. Arsenic levels in drinking water and the prevalence of respiratory effects in West Bengal, India. *Int J Epidemiol* 29: 1047–1052.

Meacher DM, Menzel DB, Dillencourt MD, Bic LF, Schoof RA, Yost LJ, Eickhoff JC, Farr CH. 2002. Estimation of multimedia inorganic arsenic intake in the U.S. population. *Hum Ecol Risk Assess* 8: 1697–1721.

Meija JJ, Diaz-Barriga F, Calderon J, Rios C, Jimenez-Capdeville ME. 1997. Effects of lead-arsenic combined exposure on central monoaminergic systems. *Neurotoxicol Teratol* 19: 489–497.

Menzel DB, Ross M, Oddo SV, Bergstrom P, Greene H, Roth RN. 1994. A physiologically based pharmacokinetic model for ingested arsenic. In: *Arsenic Exposure and Health*. Chappell WR, Abernathy CO, Cothern CR, eds. Science and Technology Letters, Chicago, IL, pp. 209–218.

Meza MM, Yu L, Rodriguez YY, Guild M, Thompson D, Gandolfi AJ, Klimecki WT. 2005. Developmentally restricted genetic determinants of human arsenic metabolism: association between urinary methylated arsenic and CYT19 polymorphisms in children. *Environ Health Perspect* Doi:10.1289/ehp.7780. Accessed at: http://dx.doi.org/.

Milton AH, Hasan Z, Rahman A, Rahman M. 2001. Chronic arsenic poisoning and respiratory effects in Bangladesh. *J Occup Health* 43: 136–140.

Morales KH, Ryan L, Kuo TL, Wu MM, Chen CJ. 2000. Risk of internal cancers from arsenic in drinking water. *Environ Health Perspect* 108: 655–661.

Mure K, Uddin AN, Lopez LC, Styblo M, Rossman TG. 2003. Arsenite induces delayed mutagenesis and transformation in human osteosarcoma cells at extremely low concentrations. *Environ Mol Mutagen* 41: 322–331.

Nakadaira H, Endoh K, Katagiri M, Yamamoto M. 2002. Elevated mortality from lung cancer associated with arsenic exposure for a limited duration. *J Occup Environ Med* 44: 291–299.

Nakagawa Y, Ibuchi Y. 1970. On the follow-up investigation of Morinaga milk arsenic poisoning. *Igaku no Ayumi* 74: 1–3.

NRC (National Research Council). 2001. *Arsenic in Drinking Water: 2001. Update*. National Academies Press, Washington, DC.

NTIP National Toxicology Information Program. 1999. *Toxic Chemical Release Inventory*. TRI97 National Library of Medicine, Bethesda, MD.

NTP National Toxicology Program. 1989. *Toxicology and Carcinogenesis Studies of Roxarsone (CAS No. 121-19-7) in F344/N Rats and B6 C3F1 Mice (Feed Studies)*. Technical Report Series 345. NTP, Research Triangle Park, NC.

Ohira M, Aoyama H. 1972. Epidemiological studies on the Morinaga powdered milk-poisoning incident. *Jpn J Hyg* 27: 500–531.

Oremland RS, Stolz JF. 2003. The ecology of arsenic. *Science* 300: 939–944.

Oya-Ohta Y, Kaise T, Ochi T. 1996. Induction of chromosomal aberrations in cultured human fibroblasts by inorganic and organic arsenic compounds and the different roles of glutathione in such induction. *Mutat Res* 357: 123–129.

Pershagen G. 1986. Sources of exposure and biological effects of arsenic. In: *Environmental Carcinogens Selected Methods of Analysis,* Vol 8, *Some Metals: As, Be, Cd, Cr, Ni, Pb, Se, Zn*. IARC Publication 71, O'Neill IK, Schuller P, Fishbein L, eds. International Agency for Research on Cancer, Lyon, France, pp. 45–61.

Pershagen G, Nordberg G, Bjorklund NE. 1984a. Carcinomas of the respiratory tract in hamsters given arsenic trioxide and/or benzo[a]pyrene by the pulmonary route. *Environ Res* 34: 227–241.

______. 1984b. Experimental evidence on the pulmonary carcinogenicity of arsenic trioxide. *Arch Toxicol Suppl* 7: 403–404.

Pomroy C, Charbonneau SM, McCullough RS, Tam GKH. 1980. Human retention studies with 74-As. *Toxicol Appl Pharmacol* 53: 550–556.

Porter AC, Fanger GR, Vaillancourt RR. 1999. Signal transduction pathways regulated by arsenate and arsenite. *Oncogene* 18: 7794–7802.

Radabaugh TR, Aposhian HV. 2000. Enzymatic reduction of arsenic compounds in mammalian systems: Reduction of arsenate to arsenite by human liver arsenate reductase. *Chem Res Toxicol* 13: 26–30.

Rahman MS, Hall LL, Hughes MF. 1994. In vitro percutaneous absorption of sodium arsenate in B6 C3F1 mice. *Toxicol in Vitro* 8: 441–448.

Rahman M, Tondel M, Ahmad SA, Axelson O. 1998. Diabetes mellitus associated with arsenic exposure in Bangladesh. *Am J Epidemiol* 148: 198–203.

Rahman M, Tondel M, Ahmad SA, Chowdhury IA, Faruquee MH, Axelson O. 1999. Hypertension and arsenic exposure in Bangladesh. *Hypertension* 33: 74–78.

Rasmussen RE, Menzel DB. 1997. Variation in arsenic-induced sister chromatid exchange in human lymphocytes and lymphoblastoid cell lines. *Mutat Res* 386: 299–306.

Rea M, Gregg J, Qin Q, Phillips MA, Rice RH. 2003. Global alteration of gene expression in human keratinocytes by inorganic arsenic. *Carcinogenesis* 24: 747–756.

Rendell M, Luu T, Quinlan E, Knox S, Fox M, Kelly S, Kahler K. 1992. Red cell filterability determined using the cell transit time analyzer (CTTA): effects of ATP depletion and changes in calcium concentration. *Biochim Biophys Acta* 1133: 293–300.

Rice DA, Kennedy S, McMurray CH, Blanchflower WJ. 1985. Experimental 3-nitro-4-hydroxyphenylarsonic acid toxicosis in pigs. *Res Vet Sci* 39: 47–51.

Rossman TG, Wolosin D. 1992. Differential susceptibility to carcinogen-induced amplification of SV40 and dhfr sequences in SV40-transformed human keratinocytes. *Mol Carcinogen* 6: 203–213.

Rossman TG, Uddin AN, Burns FJ, Bosland MC. 2001. Arsenite is a cocarcinogen with solar ultraviolet radiation for mouse skin: an animal model for arsenic carcinogenesis. *Toxicol Appl Pharmacol* 176: 64–71.

Rossman TG, Uddin AN, Burns FJ, Bosland MC. 2002. Arsenite cocarcinogenesis: an animal model derived from genetic toxicology studies. *Environ Health Perspect* 110 (suppl 5): 749–752.

Rudnai P, Borzsonyi M. 1981. Tumor inducing effect of arsenic trioxide treatment in CFLP mice. *Magyar Onkol* 25: 73–77.

Sakurai T, Kaise T, Matsubara C. 1998. Inorganic and methylated arsenic compounds induce cell death in murine macrophages via different mechanisms. *Chem Res Toxicol* 11: 273–283.

Sampayo-Reyes A, Zakharyan RA, Healy SM, Aposhian HV. 2000. Monomethylarsonic acid reductase and monomethylarsonous acid in hamster tissue. *Chem Res Toxicol* 13: 1181–1186.

Schaumloffel N, Gebel T. 1998. Heterogeneity of the DNA damage provoked by antimony and arsenic. *Mutagenesis* 13: 281–286.

Schoof RA, Yost LJ, Eickhoff J, Crecelius EA, Cragin DW, Meacher DM, Menzel DB. 1999. A market basket survey of inorganic arsenic in food. *Food Chem Toxicol* 37: 839–846.

Schrauzer GN, Ishmael D. 1974. Effects of selenium and of arsenic on the genesis of spontaneous mammary tumors in inbred C3H mice. *Ann Clin Lab Sci* 4: 441–447.

Schrauzer GN, White DA, McGinness JE, Schneider CJ, Bell LJ. 1978. Arsenic and cancer: effects of joint administration of arsenite and selenite on the genesis of mammary adenocarcinoma in inbred female C3H/St mice. *Bioinorg Chem* 9: 245–253.

Schroeder HA, Balassa JJ. 1967. Arsenic, germanium, tin, and vanadium in mice: effects on growth, survival and tissue levels. *J Nutr* 92: 245–252.

Schroeder HA, Kanisawa M, Frost DV, Mitchener M. 1968. Germanium, tin and arsenic in rats: effects on growth, survival, pathological lesions and life span. *J Nutr* 96: 37–45.

Shirachi DY, Johansen MG, McGowan JP, Tu SH. 1983. Tumorigenic effect of sodium arsenite in rat kidney. *Proc West Pharmacol Soc* 26: 413–415.

Sikorski EE, McKay JA, White KL Jr, Bradley SG, Munson AE. 1989. Immunotoxicity of the semiconductor gallium arsenide in female B6 C3F1 mice. *Fundam Appl Toxicol* 13: 843–858.

Siripitayakunkit U, Visudhiphan P, Pradipasen M, Vorapongsathro T. 1999. Association between chronic arsenic exposure and children's intelligence in Thailand. In: *Arsenic Exposure and Health Effects*. Chappell WR, Abernathy CO, Calderon RL, eds. Elsevier Science, Amsterdam, The Netherlands, pp. 141–149.

Siripitayakunkit U, Lue S, Choprapawon CT. 2001. *Possible effects of arsenic on visual perception and visual-motor integration of children in Thailand*. In: *Arsenic Exposure and Health Effects IV*. Chappell WR, Abernathy CO, Calderon RL, eds. Elsevier Science, Amsterdam, The Netherlands, pp. 165–172.

Smith AH, Goycolea M, Haque R, Biggs ML. 1998. Marked increase in bladder and lung cancer mortality in a region of northern Chile due to arsenic in drinking water. *Am J Epidemiol* 147: 660–669.

Smith AH, Lingas EO, Rahman M. 2000. Contamination of drinking-water by arsenic in Bangladesh: a public health emergency. *Bull WHO* 78; 1093–1103.

Somer T, Meiselman HJ. 1993. Disorders of blood viscosity. *Trends Mol Med* 25: 31–39.

Soucy NV, Klei LR, Mayka DD, Barchowsky A. 2004. Signaling pathways for arsenic-stimulated vascular endothelial growth factor-A expression in primary vascular smooth muscle cells. *Chem Res Toxicol* 17: 555–563.

Squibb KS, Fowler BA. 1983. The toxicity of arsenic and its compounds. In: *Biological and Environmental Effects of Arsenic*. Fowler BA, ed., Elsevier Science, Amsterdam, The Netherlands, pp. 233–269.

Steinmaus C, Yuan Y, Bates MN, Smith AH. 2003. Case-control study of bladder cancer and drinking water arsenic in the western United States. *Am J Epidemiol* 158: 1193–1201.

Styblo M, Yamauchi H, Thomas DJ. 1995. Comparative in vitro methylation of trivalent and pentavalent arsenicals. *Toxicol Appl Pharmacol* 135: 172–178.

Styblo M, Delnomdedieu M, Thomas DJ. 1996. Mono- and dimethylation of arsenic in rat liver cytosol in vitro. *Chem-Biol Interact* 99: 147–164.

Tabacova S. 1986. Maternal exposure to environmental chemicals. *NeuroToxicology* 7: 421–440.

Tam GKH, Charbonneau SM, Bryce F, Pomroy C, Sandi E. 1979. Metabolism of inorganic arsenic (^{74}As) in humans following oral ingestion. *Toxicol Appl Pharmacol* 50: 319–322.

Thompson DJ. 1993. A chemical hypothesis for arsenic methylation in mammals. *Chem-Biol Interact* 88: 89–114.

Tsai SM, Wang TN, Ko YC. 1998. Cancer mortality trends in a blackfoot disease endemic community of Taiwan following water source replacement. *J Toxicol Environ Health A* 55: 389–404.

_____. 1999. Mortality for certain diseases in areas with high levels of arsenic in drinking water. *Arch Environ Health* 54: 186–193.

Tsai SY, Chou HY, The HW, Chen CM, Chen CJ. 2003. The effects of chronic arsenic exposure from drinking water on the neurobehavioral development in adolescence. *NeuroToxicology* 24: 747–753.

Tseng WP. 1977. Effects and dose-response relationships of skin cancer and blackfoot disease with arsenic. *Environ Health Perspect* 19: 109–119.

Tseng WP, Chu HM, How SW, Fong JM, Lin CS, Yeh S. 1968. Prevalence of skin cancer in an endemic area of chronic arsenicism in Taiwan. *J Natl Cancer Inst* 40: 453–463.

Tseng CH, Chong CK, Chen CJ, Tai TY. 1996. Dose-response relationship between peripheral vascular disease and ingested inorganic arsenic among residents in blackfoot disease endemic villages in Taiwan. *Atherosclerosis* 120: 125–133.

———. 1997. Lipid profile and peripheral vascular disease in arseniasis-hyperendemic villages in Taiwan. *Angiology* 48: 321–335.

Tseng CH, Tai TY, Chong CK, Tseng CP, Lai MS, Lin BJ, Chiou HY, Hsueh YM, Hsu KH, Chen CJ. 2000. Long-term arsenic exposure and incidence of non-insulin-dependent diabetes mellitus: a cohort study in arseniasis-hyperendemic villages in Taiwan. *Environ Health Perspect* 108: 847–851.

Tsuda T, Nagira T, Yamamoto M, Kume Y. 1990. An epidemiological study on cancer in certified arsenic poisoning patients in Toroku. *Ind Health* 28: 53–62.

Tsuda T, Babazono A, Yamamoto E, Kurumatani N, Mino Y, Ogawa T, Kishi Y, Aoyama H. 1995. Ingested arsenic and internal cancer: a historical cohort study followed for 33 years. *Am J Epidemiol* 141: 198–209.

Turner DR. 1987. Speciation and cycling of arsenic, cadmium, lead and mercury in natural waters. In: *Lead, Mercury, Cadmium and Arsenic in the Environment*. Hutchinson TC, Meema KM, eds. Wiley, New York, pp. 175–186.

U.S. EPA (Environmental Protection Agency). 1984. *Health Assessment Document for Inorganic Arsenic*. EPA/600/8-83/021F. U.S. EPA. Washington, DC.

———. 1987. *The Risk Assessment Guidelines of 1986*. EPA/600/8-87/045. Office of Health and Environmental Assessment, U.S. EPA, Washington, DC.

———. 1988. *Special Report on Ingested Arsenic: Skin Cancer; Nutritional Essentiality*. EPA/625/3-87/013. U.S. EPA, Washington, DC.

———. 1997. *Report on the Expert Panel on Arsenic Carcinogenicity: Review and Workshop*. Prepared by Eastern Research Group, Inc. for the National Center for Environmental Assessment, U.S. EPA, Washington, DC.

———. 2000. *Arsenic Occurrence in Public Drinking Water Supplies*. EPA/815/R-00/023. Office of Water, U.S. EPA, Washington, DC.

———. 2001. National primary drinking water regulations; arsenic and clarifications to compliance and new source contaminants monitoring: final rule. fed reg 66:6976–7066, January 22.

———. 2005. *Guidelines for Carcinogen Risk Assessment*. EPA/630/P-03/001B. Risk Assessment Forum, U.S. EPA, Washington, DC.

Vahter M. 1981. Biotransformation of trivalent and pentavalent inorganic arsenic in mice and rats. *Environ Res* 25: 286–293.

———. 1983. Metabolism of arsenic. In: *Biological and Environmental Effects of Arsenic*. Fowler, BA, ed. Elsevier, Amsterdam, The Netherlands, pp. 171–198.

Vahter M, Envall J. 1983. In vivo reduction of arsenate in mice and rabbits. *Environ Res* 32: 14–24.

Vahter M, Marafante E, Dencker L. 1984. Tissue distribution and retention of ^{74}As-dimethylarsinic acid in mice and rats. *Arch Environ Contam Toxicol* 13: 259–264.

Vahter M, Concha G, Nermell B, Nilsson R, et al. 1995a. A unique metabolism of inorganic arsenic in native Andean women. *Eur J Pharmacol* 293: 455–462.

Vahter M, Couch R, Nermell B, Nilsson R. 1995b. Lack of methylation of inorganic arsenic in the chimpanzee. *Toxicol Appl Pharmacol* 133: 262–268.

Vallee BL, Ulmer DD, Wacker WEC. 1960. Arsenic toxicology and biochemistry. *Arch Ind Health* 21: 132–151.

Vega L, Gonsebatt ME, Ostrosky-Wegman P. 1997. Aneugenic effect of sodium arsenite on human lymphocytes in vitro: an individual susceptibility effect detected. *Mutat Res* 334: 365–373.

Waalkes MP, Ward JM, Liu J, Diwan BA. 2003. Transplacental carcinogenicity of inorganic arsenic in the drinking water: induction of hepatic, ovarian, pulmonary, and adrenal tumors in mice. *Toxicol Appl Pharmacol* 186: 7–17.

Wang CH, Jeng JS, Yip PK, Chen CL, Hsu LI, Hsueh YM, Chiou HY, Wu MM, Chen CJ. 2002. Biological gradient between long-term arsenic exposure and carotid atherosclerosis. *Circulation* 105: 1804–1809.

Wasserman GA, Liu X, Parvez F, Ahsan H, Factor-Litvak P, van Geen A, Slavkovich V, Lolacono NJ, Cheng Z, Hussain I, Momotaj H, Graziano JH. 2004. Water arsenic exposure and children's intellectual function in Araihazar, Bangladesh. *Environ Health Perspect* 112: 1329–1333.

Weed RI, LaCelle PL, Merrill EW. 1969. Metabolic dependence of red cell deformability. *J Clin Invest* 48: 795–809.

Wei M, Wanibuchi H, Yamamoto S, Li W, Fukushima S. 1999. Urinary bladder carcinogenicity of dimethylarsenic acid in male F344 rats. *Carcinogenesis* 20: 1873–1876.

Wei M, Wanibuchi H, Morimura K, Iwai S, Yoshida K, Endo G, Nakae D, Fukushima S. 2002. Carcinogenicity of dimethylarsinic acid in male F344 rats and genetic alterations in induced urinary bladder tumors. *Carcinogenesis* 23: 1387–1397.

Wester RC, Maibach HI, Sedik L, Melendres J, Wade M. 1993. In vivo and in vitro percutaneous absorption and skin decontamination of arsenic from water and soil. *Fundam Appl Toxicol* 20: 336–340.

Wiencke JK, Yager JW. 1992. Specificity of arsenite in potentiating cytogenetic damage by the DNA crosslinking agent diepoxybutane. *Environ Mol Mutagen* 19: 195–200.

Wiencke JK, Yager JW, Varonyi A, Hultner M, Lutze LH. 1997. Study of arsenic mutagenesis using the plasmid shuttle vector pZ189 propagated in DNA repair proficient human cells. *Mutat Res* 386: 335–344.

Wildfang E, Zakharyan RA, Aposhian HV. 1998. Enzymatic methylation of arsenic compounds, VI: Characterization of hamster liver arsenite and methylarsonic acid methyltransferase activities in vitro. *Toxicol Appl Pharmacol* 152: 366–375.

Winski SL, Carter DE. 1995. Interactions of the rat blood cell sulfhydryls with arsenate and arsenite. *J Toxicol Environ Health* 46: 379–397.

______. 1998. Arsenate toxicity in human erythrocytes: characterization of morphologic changes and determination of the mechanism of damage. *J Toxicol Environ Health A* 53: 345–355.

Woolson EA. 1983. Emissions, cycling and effects of arsenic in soil ecosystems. In: *Biological and Environmental Effects of Arsenic*. Fowler BA, ed. Elsevier, Amsterdam, The Netherlands, pp 51–139.

Wu MM, Kuo TL, Hwang YH, Chen CJ. 1989. Dose-response relation between arsenic concentration in well water and mortality from cancers and vascular diseases. *Am J Epidemiol* 130: 1123–1132.

Yamamoto S, Konishi Y, Matsuda T, Murai T, Shibata MA, Matsui-Yuasa I, Otani S, Kuroda K, Endo G, Fukushima S. 1995. Cancer induction by an organic arsenic compound, dimethylarsenic acid (cacodylic acid), in F344/DuCrj rats after pretreatment with five carcinogens. *Cancer Res* 55: 1271–1286.

Yamashita N, Doi M, Nishio M, Hoj H, Tanuka M. 1972. Current state of Kyoto children poisoned by arsenic tainted Morinaga dry milk. *Jpn J Hyg* 28: 364–399.

Yen HT, Chiang LC, Wen KH, Chang SF, Tsai CC, Yu CL, Yu HS. 1996. Arsenic induces interleukin-8 expression in cultured keratinocytes. *Arch Dermatol Res* 288: 716–717.

Yost LJ, Schoof RA, Aucoin R. 1998. Intake of inorganic arsenic in the North American diet. *Hum Ecol Risk Assess* 4: 137–152.

Yu D. 1999. A physiologically based pharmacokinetic model of inorganic arsenic. *Regul Toxicol Pharmacol* 29: 128–141.

Yu L, Kalla K, Guthrie E, Vidrine A, Klimecki WT. 2003. Genetic variation in genes associated with arsenic metabolism: Glutathione-*S*-transferase omega 1-1 and purine nucleoside phosphorylase polymorphisms in European and indigenous Americans. *Toxicogenomics* 111: 1421–1427.

Zakharyan RA, Aposhian HV. 1999. Enzymatic reduction of arsenic compounds in mammalian systems: the rate-limiting enzyme of rabbit liver arsenic biotransformation is MMA^V reductase. *Chem Res Toxicol* 12: 1278–1283.

Zakharyan R, Wu Y, Bogdan GM, Aposhian HV. 1995. Enzymatic methylation of arsenic compounds: assay, partial purification, and properties of arsenite methyltransferase and monomethylarsonic acid methyltransferase of rabbit liver. *Chem Res Toxicol* 8: 1029–1038.

Zakharyan R, Wildfang E, Aposhian HV. 1996. Enzymatic methylation of arsenic compounds, III: The marmoset and tamarin, but not the rhesus, monkeys are deficient in methyltransferases that methylate inorganic arsenic. *Toxicol Appl Pharmacol* 140: 77–84.

Zakharyan RA, Sampayo-Reyes A, Healy SM, Tsaprailis G, Board PG, Liebler DC, Aposhian HV. 2001. Human monomethylarsonic acid (MMA^V) reductase is a member of the glutathione-S transferase superfamily. *Chem Res Toxicol* 14: 1051–1057.

Zaldivar R. 1980. A morbid condition involving cardiovascular, bronchopulmonary, digestive and neural lesions in children and young adults after dietary arsenic exposure. *Zentbl Bakteriol Abt Orig B Hyg Krankenhhyg, Betrhyg, Praev Med* 170: 44–56.

Zaldivar R, Ghai GL. 1980. Clinical epidemiological studies on endemic chronic arsenic poisoning in children and adults, including observations on children with high and low-intake of dietary arsenic. *Zentrbl Bakteriol. Abt Orig B Hyg, Krankenhhyg, Betrhyg, Praev Med* 170: 409–421.

Zaloga GP, Deal J, Spurling T, Richter J, Chernow B. 1985. Case report: unusual manifestations of arsenic intoxication. *Am J Med Sci* 289: 210–214.

10

RISK ASSESSMENT FOR CHLOROFORM, RECONSIDERED

RICHARD SEDMAN
California Environmental Protection Agency, Oakland, California

Disinfection by-products are formed when water purification chemicals such as chlorine, chloramine, ozone, or chloride dioxide react with natural organic and inorganic matter in water. The trihalomethanes (THMs) are a prominent group of disinfection by-products, including chloroform, bromoform, bromodichloromethane, and chlorodibromomethane. Another important class of disinfection by-products are the haloacetic acids, which include dichloroacetic acid and dichloroacetate. Health Canada (1996) estimates that THMs comprise up to 50% by weight of all disinfection by-products in drinking water. Among the trihalomethanes, the most common, based on frequency of detection and concentration in drinking water, is chloroform, followed by bromodichloromethane, then chlorodibromomethane, and bromoform (WHO, 1994; U.S. EPA, 1998b; Health Canada, 2006).

Following the publication of new guidelines for carcinogen risk assessment by the EPA in 1996 (U.S. EPA, 1996), an expert panel was formed that included representatives from government, industry, and academia for the purpose of applying the new guidelines (ILSI, 1997). Chloroform and dichloroacetate, two disinfection by-products, were selected as case studies (ILSI, 1997). The findings of the panel and EPA's independent analysis (U.S. EPA, 2001) using the new guidelines, markedly altered the approach used to assess risk associated with exposure to chloroform and the approach ultimately proposed to establish the maximum contaminant level goal (MCLG). The revised risk assessment for chloroform, which was based on information about toxic and cancer mechanism, is well

Risk Assessment for Chemicals in Drinking Water, Edited by Robert A. Howd and Anna M. Fan

documented and needs no new summary (ILSI, 1997; U.S. EPA, 2001). Given the risk assessment for chloroform inaugurated the use of new cancer guidelines, the intent of this chapter is to reexamine the chloroform case study and the new approach to promulgate the proposed MCLG. The case study for chloroform illustrates some of the difficulties and uncertainties in conducting a risk assessment using the new guidelines.

CARCINOGENIC EFFECTS

Chloroform is a well-studied THM because it occurs in high levels in drinking water and because administration of chloroform has resulted in statistically significant increases in tumors in rodents. The administration of chloroform to rats and mice yielded statistically significant increases in liver and kidney tumors (Table 1). Increased tumors in the liver occurred in female Strain A mice that received 30 doses of 600 or 1200 mg/kg in olive oil over a four-month period (Eschenbrenner and Miller, 1945), in male and female B6 $C3F_1$ mice treated with 138 to 477 mg/kg·day by gavage in corn oil for 78 weeks (NCI, 1976), in female Wistar rats exposed to 240 mg/kg·day in the drinking water (Tumasonis et al., 1985, 1987), and in male Wistar rats exposed to 180 mg/kg·day in the drinking water for life (Tumasonis et al., 1985, 1987).

Kidney tumors, tubular cell adenomas, or carcinomas were observed in male BDF1 mice exposed to 30 or 90 ppm of chloroform in air for two years (Nagano et al., 1998), male ICI mice treated with 60 mg/kg·day by gavage in toothpaste or arachis oil for 80 weeks (Roe et al., 1979), male Osborne–Mendel rats treated with 90 or 180 mg/kg·day by gavage in corn oil for 78 weeks (NCI, 1976), and male Osborne–Mendel rats exposed to 160 mg/kg·day in drinking water for two years (Jorgenson et al., 1982, 1985). These data were judged to provide sufficient evidence of the carcinogenicity of chloroform in experimental animals (IARC, 1979, 1999; U.S. EPA, 2001; California OEHHA, 2006).

NONCANCER TOXIC EFFECTS

In addition to cancer, toxic effects have been observed in the liver and kidney of rats, mice, guinea pigs, rabbits, and dogs (ATSDR, 1997). Characteristic pathological changes in the liver include congestion, enlargement, fatty infiltration, and centrilobular necrosis. Related observations were increased levels of serum biomarkers for hepatotoxicity, hepatic cell proliferation, and jaundice. In the kidney, chloroform produced a variety of degenerative, necrotic, and regenerative lesions in the proximal tubules.

A number of other effects have been observed in various tissues (in addition to hepatic and renal toxicity) following the administration of chloroform (Table 2). The observed multiplicity of effects is not unique to chloroform but typical of many toxicants. The EPA stated in 1993: "Exposure to a given chemical, depending on the dose employed, may result in a variety of toxic effects.

TABLE 1. Positive Findings of Carcinogenicity Studies of Chloroform Administered Orally to Rats or Mice

Sex	Applied Dose (mg/kg·day)	Tumor Site and Type	Incidence
NCI (1976), B6 C3F$_1$ Mice			
Male	0, 138, 277	Liver, hepatocellular carcinoma	1/18, 18/50, 44/45
Female	0, 238, 477	Liver, hepatocellular carcinoma	0/20, 36/45, 39/41
Roe et al. (1979), ICI Mice			
Male	0, 17, 60	Renal adenoma or carcinoma	0/72, 0/37, 8/38
Male	0, 60	Renal adenoma or carcinoma	1/50, 12/48
Eschenbrenner and Miller (1945), Strain A Mice			
Female	0, 150, 300, 600, 1200, 2400	Liver, hepatoma	0/5, 0/5, 0/5, 3/3, 4/4, 0/0
NCI (1976), Osborne–Mendel rats			
Male	0, 90, 180	Kidney, renal carcinoma and adenoma	0/19, 4/50, 12/50
Female	0, 100, 200	Thyroid, adenoma or carcinoma	1/19, 8/49, 10/46
Tumasonis et al. (1985), Wistar rats			
Male	0, 180	Liver, hepatic adenofibrosis	0/22, 17/28
Female	0, 240	Liver, hepatic adenofibrosis	0/18, 34/40
Jorgenson et al. (1985), Osborne–Mendel Rats			
Male	0, 19, 38, 81, 160	Kidney, adenoma or carcinoma	4/301, 4/313, 4/148, 3/48, 7/50

These may range from gross effects, such as death, to more subtle biochemical, physiologic, or pathologic changes. In assessments of the risk posed by a chemical, the toxic endpoints from all available studies are considered, although primary attention usually is given to the effect (the 'critical effect') exhibiting the lowest NOAEL" (U.S. EPA, 1993). Chloroform administration has also been associated with effects such as transient changes in body weight, organ weights, water consumption, and the serum enzymes alanine aminotransferase and sorbitol dehydrogenase, transient increases in labeling index in the liver or kidney; and increased levels of CYP2E1 in the liver. These effects may not themselves be considered to be toxicity, but may indicate that toxic effects are occurring.

Genotoxicity

Chloroform has been the subject of extensive genotoxicity testing (IARC, 1979, 1999; ATSDR 1997). A battery of in vitro genotoxicity tests included gene

TABLE 2. Toxic Effects Associated with the Oral Administration of Chloroform to Rats and Mice

Species/ Gender	Exposure	Endpoint	LOAEL (mg/kg·day)	Study
Rat	1 dose	Hematological	546	Chu et al., 1982a
Rat/male	4 doses	Decreased body weight	180	Larson et al., 1995a
Rat/female	10 doses	Hematological	100	Ruddick et al., 1983
	10 doses	Decreased body weight	100	Ruddick et al., 1983
	10 doses	Gastric erosions	516	Thompson et al., 1974
	10 doses	Alopecia	126	Thompson et al., 1974
Mouse	14 doses	Suppressed immunity	50	Munson et al., 1982
Rat/female	1 dose	Reduced lymphocytes	1071	Chu et al., 1982b
Mouse	1 dose	Ataxia, anesthesia, brain hemorrhage	500	Bowman et al., 1978
Mouse/male	10 doses	Taste aversion	30	Landauer et al., 1982
Rat/female	10 doses	Increased resorptions	316	Thompson et al., 1974
Rat	10 doses	Decreased fetal weight	400	Ruddick et al., 1983
Rat/male	28 days	Hematological	193	Chu et al., 1982b
	90 days	Decreased body weight	160	Jorgenson and Rushbrook, 1980
	21 days	Decreased body weight	106	Larson et al., 1995a
Rat/female	3 weeks	Respiratory	34	Larson et al., 1995b
Rat	91 days	Hematological	410	Palmer et al., 1979
Mouse	90 days	Depressed immunity	50	Munson et al., 1982
	60 days	Neurological	100	Balster and Borzelleca, 1982
Rat	91 days	Gonadal atrophy	410	Palmer et al., 1979
Mouse/male	105 days	Epididymal degradation	41	Gulati et al., 1988
Rat/male	1260 days	50% reduced weight gain	200	Tumasonis et al., 1985, 1987
Mouse/female	78 weeks	Pulmonary inflammation	238	NCI, 1976
	78 weeks	Cardiac thrombosis	238	NCI, 1976

Source: Adapted from ASTDR (1997), Table 2-2.

mutation studies with and without activation in bacteria, gene mutation studies in yeast and mammalian cells, sister chromatid exchange, and unscheduled DNA synthesis assays. In vivo genotoxicity studies include DNA binding, chromosomal aberrations, unscheduled DNA synthesis, micronuclei induction, and sperm abnormalities (reviews: by ATSDR, 1997; U.S. EPA, 2001). Because chloroform is volatile and activated by metabolism to a highly reactive intermediate phosgene, the most relevant studies have been conducted in a closed system to prevent loss of the chemical and would include activation, preferably by an endogenous system where the metabolites are formed. With a few exceptions, most of the genotoxicity assays were negative. The occasional positive finding for genotoxicity is characterized by weak genotoxicity activity. Overall, the weight of evidence indicates that chloroform is not a potent genotoxic agent.

MECHANISMS OF CARCINOGENICITY

Given the weak evidence of genotoxicity for chloroform, considerable effort has been focused on elucidating the mechanism by which chloroform caused tumors in rodent liver and kidney. Investigators have studied a role of peroxisome proliferation, apoptosis, cytotoxicity combined with tissue regeneration, and DNA methylation in the tumorigenesis of chloroform. Despite these efforts, the precise mechanism by which chloroform administration causes tumors is unclear. However, based on extensive studies in rats and mice, (U.S. EPA, 2001; Health Canada, 2006; ILSI, 1997) the appearance of tumors in the rodent liver and kidney was considered to be secondary to cytotoxicity and tissue regeneration:

Mechanistic Model of Tumor Development

Chloroform → Cytotoxicity → Tissue regeneration → Tumor development

The EPA (U.S. EPA, 2001) summarized the evidence for the aforementioned model and concluded:

> Available data also indicate that cytotoxicity and regenerative hyperplasia are required for liver cancer, although the strength of this conclusion is somewhat limited because most of the observations are based on short-term rather than long-term histological or labeling index measurements. For example, in the B6 C3F$_1$ mouse, corn oil gavage (bolus dosing) at the same doses that resulted in liver tumors in the study by the NCI 1976 also caused hepatic cytolethality and a cell proliferative response at 4 days and 3 weeks (Larson et al., 1994b,c). Similarly, exposure of female B6 C3F$_1$ mice to chloroform in drinking water at levels that did not induce liver tumors (Jorgenson et al., 1985) also did not induce hepatic cytolethality or cell proliferation at 4 days or 3 weeks (Larson et al., 1994c). This consistency of the data (i.e., evidence of cytolethality and/or regenerative hyperplasia is always observed in cases of increased liver tumors) supports the conclusion that this liver cancer also occurs via a mode of action involving regenerative hyperplasia.
>
> There are numerous cases where exposure to chloroform causes an increase in the labeling index without any observable increase in cancer incidence. These data indicate that chloroform exposures that are adequate to cause cytotoxicity and regenerative cell proliferation do not always lead to cancer.
>
> There are no cases in which a tumorigenic response has been observed where evidence of cell regeneration is not also observed at the same or lower dose as that which caused an increase in tumors. This consistency of evidence (i.e., cell regeneration is detected in all cases of tumorigenicity) is strong evidence supporting the conclusion that cell regeneration is a mandatory precursor for tumorigenicity.

The EPA (U.S. EPA, 1998a, 2001) and the ILSI expert panel (1997) concluded that the occurrence of tumors in the kidney and liver in animals exposed to chloroform was due to (or is secondary to) cytotoxicity and tissue regeneration. Therefore, based on the model, no tumors would occur at exposure levels where no cytotoxicity and tissue regeneration occur. Accordingly, dose–response assessment of the risk assessment should focus on determining the highest dose where cytotoxicity and tissue regeneration did not occur and therefore tumor development will not occur (the NOAEL). Alternatively, a benchmark dose associated with cytotoxicity could be determined.

Relationship Between Cytotoxicity, Cellular Regeneration, and Tumors

While evidence of cytotoxicity and tissue regeneration has been observed at the high doses of chloroform associated with tumors in rats in mice, other effects (Table 1) also occur at high dose levels of chloroform. Thus, observed cytotoxicity and tissue regeneration at the dose(s) of chloroform associated with tumors are themselves not compelling evidence that the effects are related. Much more persuasive evidence of a relationship between the effects would include consistency in the appearance of cytotoxicity, tissue regeneration, and then tumors; consistency in the dose–response relationship that characterizes the appearance of these effects; and a consistent temporal relationship in the appearance of these effects. Cytotoxicity and tissue regeneration were not investigated in the available cancer bioassays. However, there are a number of studies that have investigated the relationship between cytotoxicity and tissue regeneration in the liver and kidney (Tables 3 and 4). Are the findings of these studies consistent with the conclusion that tumor formation is a consequence of tissue regeneration, which is a consequence of or secondary to cytotoxicity (per the model)?

Increased organ weights, histopathology, and in the case of the liver, increased serum enzyme levels have been employed to identify renal and hepatotoxicity. Although toxicity in the liver and kidney appeared to increase with dose, in many studies toxicity appears to diminish with continued exposure to chloroform. Histopathology observed initially is absent or much less evident with continual exposure, suggesting the ability to better tolerate exposure to chloroform with time. Also, animals were able to tolerate much higher doses of chloroform if a smaller dose was administered initially (Nagano et al., 1998; Yamamoto et al., 1999).

The occurrence of tissue regeneration in the liver and kidney is demonstrated by an increase in DNA synthesis as measured by a change in the incorporation of radiolabeled DNA precursors. The *labeling index* is the proportion of labeled cells in the S phase. Akin to toxicity, tissue regeneration, typically peaked early and was markedly reduced or largely absent with continued exposure to chloroform (Pereira, 1994; Templin, 1996a). Unlike toxicity, increase in dose may decrease the level of tissue regeneration, perhaps because the increased insult to the tissue prevented regeneration. With both toxicity and tissue regeneration, the response is time dependent. Given the proposed model of carcinogenicity, toxicity should appear prior to tissue regeneration.

Consistency/Inconsistency

If tumors are secondary to cytotoxicity and tissue regeneration, neither tumors nor regeneration should occur without the occurrence of cytotoxicity. Furthermore, cytotoxicity should be followed by tissue regeneration and then tumors. However,

TABLE 3. Consistency or Inconsistencies of Changes in Cytotoxicity, Labeling Index, and Occurrence of Tumors in the Liver of Rats and Mice

Study	Species	Vehicle	Sex	Relationship Between Cytotoxicity, Labeling Index, and Occurrence of Tumors	Comment
Larson et al., 1993	Rat	Corn oil	M	Differences in dose–response	Cytotoxicity at lower doses without comparable increase in labeling index
	Mouse	Corn oil	F	Inadequate data	
Templin et al., 1996c	Rat, F344	Corn oil	M	Consistent changes in cytotoxicity and labeling index	
	Rat, Osborne–Mendel	Corn oil	M	Consistent changes in cytotoxicity and labeling index	
Larson et al., 1995b	Rat, F344	Corn oil	F	Differences in dose–response	Little cytotoxicity but increased labeling index at middle doses
				Divergent responses	Increased labeling index but no tumors at site
Larson et al., 1994b	Mouse	Corn oil	M	Consistent changes in cytotoxicity and labeling index	
Larson et al., 1994c	Mouse	Corn oil	F	Divergent responses	Cytotoxicity with no increase in labeling index at middle doses
				Temporal differences when response occurred	Labeling index peaked at 3 days, cytotoxicity more severe at 3 weeks
	Mouse	Drinking water	F	Divergent responses	Mild cytotoxicity with no increase in labeling index
Larson et al., 1995a	Rat	Corn oil	M	Differences in dose–response	Cytotoxicity with no increase in labeling index at middle dose
				Divergent responses	Increased labeling index but no tumors at site

(continued overleaf)

TABLE 3. ***(continued)***

Study	Species	Vehicle	Sex	Relationship Between Cytotoxicity, Labeling Index, and Occurrence of Tumors	Comment
	Rat	Drinking water	M	Divergent responses	Cytotoxicity with no increase in labeling index
				Divergent responses	Cytotoxicity with no tumors at this site
Larson et al., 1996	Mouse	Air	M	Differences in dose–response	Cytotoxicity with no increase in labeling index at middle dose (30 ppm)
				Temporal differences when responses occurred	Labeling index peaked at 3 weeks, cytotoxicity remained elevated at 13 weeks in high-dose group
	Mouse	Air	F	Differences in dose–response	Cytotoxicity but no increase in labeling index at middle dose (10 ppm)
				Temporal differences when responses occurred	Cytotoxicity more severe at 4 days while labeling index higher at middle dose (30 ppm) after 3 weeks
Templin et al., 1996a	Rat	Air	M	Differences in dose–response	Cytotoxicity but no increase in labeling index at middle dose
				Temporal differences when responses occurred	Labeling index peaked at 6 weeks, cytotoxicity more severe at 13 weeks
				Divergent responses	Increased labeling index but no tumors at site
	Rat	Air	F	Differences in dose–response	Cytotoxicity but no increase in labeling index at middle doses
				Divergent responses	Increased labeling index but no tumors at site
Templin et al., 1996b	Mouse	Air	M	Differences in dose–response	Little cytotoxicity but increased labeling index at middle dose
	Mouse	Air	F	Consistent changes in cytotoxicity and labeling index	
Constan et al., 2002	Mouse	Air	F	Differences in dose–response	No cytotoxicity but increased labeling index in one dose group

TABLE 4. Consistency or Inconsistencies of Changes in Cytotoxicity, Labeling Index, and Occurrence of Tumors in the Kidney of Rats and Mice

Study	Species	Vehicle	Sex	Relationship Between Cytotoxicity, Labeling Index, and Occurrence of Tumors	Comment
Larson et al., 1993	Rat	Corn oil	M	Differences in dose–response	Cytotoxicity increased with dose, labeling index decreased at higher dose
	Mouse	Corn oil	F	Divergent responses	No cytotoxicity but increased labeling index
				Divergent responses	Increased labeling index but no tumors at site
Templin et al., 1996c	Rat, F344	Corn oil	M	Differences in dose–response	Dose-dependent increase in labeling index at all doses, cytotoxicity only at high dose
	Rat, Osborne–Mendel	Corn oil	M	Divergent responses	No cytotoxicity but dose-dependent increase in labeling index
Larson et al., 1995b	Rat, F344	Corn oil	F	Differences in dose–response	No cytotoxicity but increased labeling index at middle dose
				Divergent responses	Cytotoxicity, increased labeling index, but no tumors at this site
				Temporal differences when responses occurred	Cytotoxicity but labeling index not diminished at 3 weeks
Larson et al., 1994b	Mouse	Corn oil	M	Consistent changes in cytotoxicity and labeling index	At 4 days
				Differences in dose–response	Cytotoxicity but no increased labeling index in low-dose groups at 3 weeks
Larson, et al., 1994c	Mouse	Corn oil	F	Divergent responses	No cytotoxicity but increased labeling index at high dose
				Divergent responses	Increased labeling index but no tumors at site

(*continued overleaf*)

TABLE 4. ***(continued)***

Study	Species	Vehicle	Sex	Relationship Between Cytotoxicity, Labeling Index, and Occurrence of Tumors	Comment
	Mouse	Drinking water	F	Divergent responses	No cytotoxicity but increased labeling index at higher doses
				Divergent responses	Increased labeling index but no tumors at site
Larson et al., 1995a	Rat	Corn oil	M	Differences in dose–response	Dose-dependent cytotoxicity, increased labeling index only at high dose
	Rat	Drinking water	M	Divergent responses	Mild cytotoxicity, no increase in labeling index
Larson et al., 1996	Mouse	Air	M	Divergent response (perhaps)	Marked increased labeling index at 3 weeks but minimal cytotoxicity at 3 or 13 weeks in middle (30 ppm)-dose group
	Mouse	Air	F	No cytotoxicity or changes in labeling index	
Templin et al., 1996a	Rat	Air	M	Differences in dose–response	Cytotoxicity but no increase in labeling index at low dose
	Rat	Air	F	Differences in dose–response	Cytotoxicity but no increase in labeling index at low dose
				Divergent responses	Cytotoxicity, increased labeling index, but no tumors at site
Templin et al., 1996b	Mouse	Air	M	Consistent changes in cytotoxicity and labeling index	
	Mouse	Air	F	Consistent changes in cytotoxicity and labeling index	

increased labeling index (a measure of tissue regeneration) without cytotoxicity was observed in the kidney of animals administered chloroform (Larson et al., 1993; Templin et al., 1996c). Conversely, following chloroform administration, cytotoxicity was observed in the kidney (Larson et al., 1995a) or liver (Larson et al., 1994c, 1995a) without an increase in labeling index.

Following chloroform administration, increased cytotoxicity or labeling index have been observed in the rat liver (Larson et al., 1993, 1995a,b; Templin et al., 1996a) and female mouse kidney (Larson et al., 1993, 1994c) or female rat kidney (Larson et al., 1995b; Templin et al., 1996a), tissues where no statistically significant increases in tumors were observed in cancer bioassays.

Dose–Response Relationships

Comparable (parallel) dose–response relationships for cytotoxicity, tissue regeneration, and tumor formation would provide compelling evidence that the effects are related. Conversely, evidence that these effects occur at or peak at substantially different doses suggests that the effects may be unrelated. Because tumors were observed with at most two doses, dose–response information that characterizes this effect is very limited. However, as discussed above, there is considerable information regarding the appearance of cytotoxicity and tissue regeneration in the liver and kidney.

Following chloroform administration, an increased labeling index was observed at dose levels where cytotoxicity was not observed in the liver (Larson et al., 1995b; Templin et al., 1996b; Constan et al., 2002) or kidney (Larson et al., 1995b; Templin et al., 1996c). Conversely, cytotoxicity was observed at dose levels where increased labeling index or comparable increased labeling index was not observed in the liver (Larson et al., 1993, 1995a, 1996; Templin et al., 1996a) or kidney (Larson et al., 1993, 1994b, 1995a; Templin et al., 1996a). The dose–response relationship for the labeling index for various trihalomethanes did not appear to parallel the dose–response relationship for hepatotoxicity for these related compounds (Melnick et al., 1998). Suggestions that the labeling index is a more sensitive measure of cytotoxicity is not supported by studies where toxicity was observed at doses without an increase in the labeling index in the liver (Larson et al., 1995a; Templin et al., 1996) or kidney (Larson et al., 1993, 1994c, 1995a, 1996; Templin et al., 1996a).

Temporal Relationship

Given the hypotheses that cytotoxicity and tissue regeneration are antecedent to tumor formation, the predicted temporal pattern of toxicity, tissue regeneration, and then tumors should be observed. Although toxicity did precede or accompany an increase in the labeling index in certain studies (Larson et al., 1993, 1995b), increases in the labeling index preceded toxicity or peaked and then was reduced or returned to near-background levels while toxicity continued or became more severe in the liver (Larson et al., 1994c, 1996; Templin et al., 1996a). Serum alanine aminotransferase and sorbitol dehydrogenase level peaked at 3 weeks of treatment at 238 and 477 mg/kg·day, while the hepatic labeling index peaked at 34 days and was markedly reduced at these dose levels at 3 weeks. Suggestions that the labeling index decreased because increased toxicity prevents regeneration were not supported by the findings that the labeling index was not increased at lower doses, where toxicity was not as severe (Larson et al., 1994a).

As mentioned earlier, toxicity and tissue regeneration occurs rapidly after chloroform administration and appears to peak quickly when the chemical is administered repeatedly. Toxicity and a marked increase in labeling index were observed within 1 to 2 days after only one dose of chloroform (Larson et al., 1993). Compared to 3 weeks, a much higher level of cellular regeneration as indicated by the labeling index occurs at 4 days post-chloroform administration (Larson et al., 1994a). Given the proposed model, no additional exposure to chloroform would appear to be necessary, as toxicity and cellular regeneration is most notable after 4 days. It is highly unlikely that the increase in tumors associated with chloroform administration would be detected if exposure were discontinued after 1, 4, or 21 days.

Corn Oil Vehicle

Because of volatility, palatability, or water solubility concerns, many bioassays have employed corn oil as a vehicle for administering test articles. Potential confounding effects from using corn oil as a vehicle are unclear. When administered alone, corn oil did not result in significant increases in tumors in the rat kidney or liver (NTP, 1994). Chloroform's toxicity and carcinogenicity have been linked to the use of a corn oil vehicle (Bull et al., 1986; NRC, 1987; Golden et al., 1997). The increase in tumors has been proposed to be secondary to cytotoxicity and tissue regeneration, a response that did not appear to occur when chloroform was administered in drinking water (ILSI, 1997; Bruckner and Warren, 2001; U.S. EPA, 2001). Although data on the effect of corn oil administration on chloroform-induced tumors is limited (tumors were observed only at a few dose levels), a number of studies have investigated the effect of vehicle or route of administration on the tissue regeneration in the rodent liver and kidney that followed chloroform administration.

The labeling index 4 days and 3 weeks following chloroform administration in corn oil, drinking water, or by inhalation were compared based on measured dose or an estimate of the dose the animals received (Figures 1 and 2). The dose of chloroform administered in corn oil and water are from the investigators, while the dose received by the inhalation route are estimated based on body weight, inhalation rate, and 50% absorption by the inhalation route. Contrary to the conclusions of the U.S. EPA (2001), the ILSI (1997), and Bruckner and Warren (2001), no significant differences in labeling index was observed, whether corn oil or water was employed as the vehicle, particularly at 4 days, when the increase in the labeling index was most notable. Interestingly, equivalent doses administered by the inhalation route appeared to yield a greater increase in the labeling index (the dose–response curve appears to have shifted to the left). Further study of this effect is warranted.

The difference in cytotoxicity and regenerative response when chloroform was administered in corn oil has been attributed to higher tissue levels of chloroform in the kidney and liver, which are responsible for the increased toxicity (Bruckner and Warren, 2001). However, more rapid chloroform uptake into the liver and

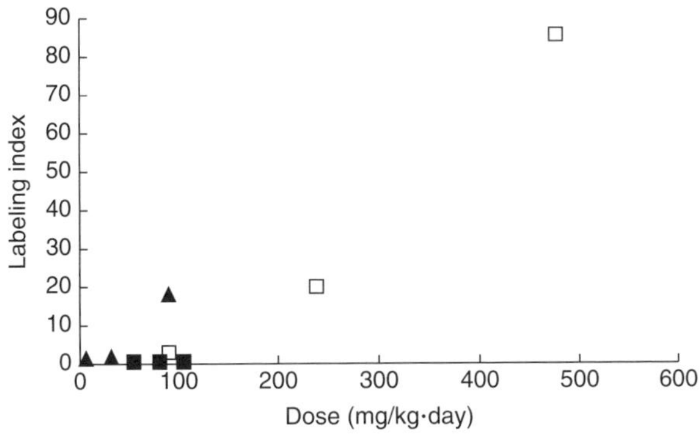

Figure 1. Labeling index in liver of female mice exposed to chloroform for 4 days. Open square, labeling index observed in mice administered chloroform in corn oil; black triangle, labeling index observed in mice exposed to chloroform in the air; black square, labeling index observed in mice administered chloroform in drinking water. (Adapted from Larson et al., 1994c, 1996.)

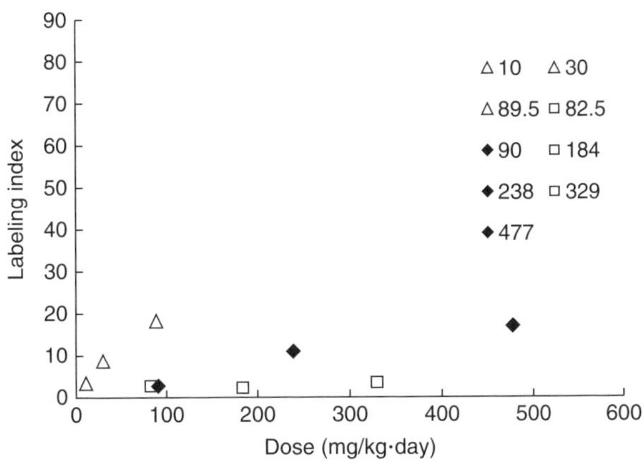

Figure 2. Labeling index in female mice exposed to chloroform for 3 weeks. Black diamond, labeling index observed in mice administered chloroform in corn oil; open triangle, labeling index observed in mice exposed to chloroform in the air; open square, labeling index observed in mice administered chloroform in drinking water. (Adapted from Larson et al., 1994c, 1996.)

kidney and more binding to macromolecules were observed when chloroform was administered in water compared to when corn oil was employed as the vehicle (Pereira, 1994). The findings of Pereira (1994) are not unexpected, given that lipids are known to delay gastric emptying. The administration of chloroform by inhalation over 6 hours resulted in an increase in labeling index in the liver at 4 days and perhaps at 3 weeks (Larson et al., 1996). The protocol used ensured that the agent was administered slowly (and therefore slowly absorbed) over a 6-hour time frame each day. Thus, it appears that a slower delivery of chloroform may result in a higher labeling index.

Other Chemicals

The possibility that tumors are secondary to cytotoxicity and tissue regeneration has also been investigated for other carcinogenic substances. The administration of 1,2,3-trichloropropane in corn oil to B6 C3F_1 mice resulted in increased labeling index in the kidney, liver, forestomach, and glandular stomach (La et al., 1996). The increase in cell proliferation occurred in two tissues with a tumorigenic response in a NTP bioassay (liver and forestomach) and in two tissues where increases in tumors were not detected. For trichloropropane, there did not appear to be a relationship between and increase in labeling index and the occurrence of tumors.

B6 C3F_1 male mice were administered the carcinogen dichloroacetate (0, 95, or 44 mg/kg·day) for either 5 to 15 days or 20 to 30 days. In male mice administered dichloroacetate, decreased labeling index was observed in the liver of the high-dose group treated for 5 or 10 days and the low-and high-dose group of animals treated for 20 or 25 days (Carter et al., 1995). Thus, contrary to the model, cell proliferation decreased following treatment with dichloroacetate. As discussed earlier, the dose–response relationship for labeling index for various THMs did not appear to parallel the dose–response relationship for hepatotoxicity for these related compounds (Melnick et al., 1998).

REGULATION OF CANCER RISK

Recent guidance that is being utilized to develop drinking water criteria have emphasized the use of all relevant information when considering the most appropriate approach for conducting a risk assessment. The U.S. EPA (1996) recognized an important change in its recent cancer guidelines: "The use of mode of action in the assessment of potential carcinogens is the main thrust of these guidelines." With respect to chloroform, many studies have recently been conducted to gain a better understanding of its metabolism, genotoxicity, and mechanism(s) of carcinogenicity. These studies have been very important in the risk assessment used to promulgate a new MCL for the THMs (U.S. EPA, 1998a) and a proposed MCLG for chloroform (U.S. EPA, 2003a).

The MCLG proposed for chloroform reflects mechanistic considerations as specified by recent guidance.

> In accord with EPA guidelines for cancer risk assessment (U.S. EPA, 1996), the method used to characterize and quantify cancer risk from a chemical depends on what is known about the mode of action of carcinogenicity and the shape of the cancer dose–response curve for that chemical. A default assumption of linearity is appropriate when evidence supports a mode of action of gene mutation due to DNA reactivity or supports another mode of action that is anticipated to be linear. The linear approach is used as a matter of policy if the mode of action of carcinogenicity is not understood. A default assumption of nonlinearity is appropriate when there is no evidence for linearity and sufficient evidence to support an assumption of nonlinearity. Alternatively, the mode of action may theoretically have a threshold, e.g., the carcinogenicity may be a secondary effect of toxicity that is itself a threshold phenomenon. (U.S. EPA, 1996)

Regarding chloroform's mechanism of action, U.S. EPA summarized the findings of available studies:

> Available data also indicate that cytotoxicity and regenerative hyperplasia are required for liver cancer, although the strength of this conclusion is somewhat limited because most of the observations are based on short-term rather than long-term histological or labeling index measurements. For example, in the B6C3F1 mouse, corn oil gavage (bolus dosing) at the same doses that resulted in liver tumors in the study by NCI (1976) also caused hepatic cytolethality and a cell proliferative response at 4 days and 3 weeks (Larson et al., 1994b,c). Similarly, exposure of female B6C3F1 mice to chloroform in drinking water at levels that did not induce liver tumors (Jorgenson et al., 1985) also did not induce hepatic cytolethality or cell proliferation at 4 days or 3 weeks (Larson et al., 1994c). This consistency of the data (i.e., evidence of cytolethality and/or regenerative hyperplasia is always observed in cases of increased liver tumors) supports the conclusion that this liver cancer also occurs via a mode of action involving regenerative hyperplasia. (U.S. EPA, 2001)

> EPA is proposing an MCLG for chloroform of 0.07 mg/L based on a cancer reference dose (RfD), an assumption that a person drinks 2 liters of water per day (the 90th percentile of intake rate for the U.S. population) and a relative source contribution (RSC) of 20 percent. The MCLG is proposed at a level at which no adverse effects on the health of persons is anticipated with an adequate margin of safety. This conclusion is based on toxicological evidence that the carcinogenic effects of chloroform are an ultimate consequence of sustained tissue toxicity. The MCLG is set at a daily dose for a lifetime at which no adverse effects will occur because the sustained tissue toxicity, which is a key event in the cancer mode of action of chloroform, will not occur. (U.S. EPA, 2003a)

DISCUSSION

In the case of chloroform, how well is the carcinogenic mechanism understood? Numerous effects, some clearly representative of toxicity, some not, have been associated with the administration of high doses of chloroform to rodents. Because an effect occurs at a relatively high dose does not necessarily establish that the effect is a precursor to, causative of, caused by the same mechanism, or related to the mechanism that caused tumors in the liver or kidney. That chloroform causes many effects in many tissues does not establish that the tumors are secondary to the other effects.

The relationship between cytotoxicity and tissue regeneration at the dose(s) of chloroform that yielded tumors in experimental animals has been studied extensively. While increases in labeling index tended to occur at high doses of chloroform, the increases often did not reflect the pattern of toxicity observed in the animals. Toxicity has been observed without an apparent increase in labeling index. Toxicity may increase with continual exposure while the labeling index diminishes. Increased labeling index has been observed without toxicity being evident. Finally, cytotoxicity or increases in the labeling index have been observed without an observed tumorigenic response.

The findings of tissue regeneration studies were not particularly supportive of the belief that the increase in tumors associated with chloroform is in essence an artifact of use of a corn oil vehicle. Differences in labeling index following administration of chloroform in corn oil or water appeared to be accounted for by differences in dose alone. However, the labeling index associated with inhalation exposure may be elevated compared to when chloroform was administered in corn oil or water. Mechanistic studies of other carcinogenic substances (trichloropropane, dichloroacetate, and other THMs) did not support the cytotoxicity-cell proliferation model of cancer development for these substances.

There have been extensive studies regarding the role of cytotoxicity and tissue regeneration in the mechanism of chloroform carcinogenicity. The findings of these studies do suggest that high doses of chloroform may result in toxicity and transient increases in labeling index. However, a number of other effects have also been associated with high doses of chloroform. Often, the pattern of increase in toxicity and cell proliferation (dose–response, temporal) is not particularly consistent with the predictions of the model, that tumor formation is secondary to tissue regeneration caused by cytotoxicity. It appears that cytotoxicity and tissue regeneration are one of many effects that have been observed following the administration of high doses of chloroform. Given the inconsistencies of the findings of numerous studies, cytotoxicity and tissue regeneration appear to be unrelated to the occurrence of the tumors in the rodent liver and kidneys. Therefore, the conclusion of the EPA that risk assessment for chloroform should be conducted with an assumption of a nonlinear (threshold) mechanism does not appear to be supported.

Choice of a linear or nonlinear mechanism, which depends on whether chloroform-related tumors are or are not secondary to cytotoxicity and regenerative hyperplasia in the liver and kidney, has had a considerable impact on the resulting health-based criterion developed by the EPA. In 1980, using a standard assumption that no exposure to chloroform is without risk of developing cancer (a nonthreshold assumption), the EPA established an Ambient Water Criterion for chloroform of 0.19 μg/L for ingestion of water and organisms, based on a 10^{-6} increased lifetime cancer risk (U.S. EPA, 1980). This was revised in 1992 based on a revised cancer slope factor to 5.7 μg/L. However, using the threshold assumption of the more recent assessment (U.S. EPA, 2001), a revised ambient water criterion for chloroform of 68 μg/L has now been proposed for ingestion of drinking water and organisms (U.S. EPA, 2003b). [This is essentially the same as the MCLG (U.S. EPA, 2003a, 2006) because the contribution from ingestion of organisms is extremely small.] In other words, the revised calculations and assumptions over this period of time have resulted in increasing the presumed safe exposure level for chloroform by about 360-fold since 1980 and 12-fold since 1992.

Chloroform was considered the prototype in using mechanistic information as a critical component of its risk assessment. Consequently, the amount of research focused on understanding the mechanism by which chloroform causes liver and kidney tumors is quite impressive and laudable. Unfortunately, the putative mechanism, that the tumors are secondary to the cytotoxicity and regenerative hyperplasia and that the aforementioned toxic response only occurs above a given threshold, is not particularly compelling. The much less stringent drinking water standard for chloroform that has resulted from the mechanistic information in the revised risk assessment may therefore be inappropriate.

Disclaimer

The opinions expressed in this chapter are those of the author and not necessarily those of the Office of Environmental Health Hazard Assessment or the California Environmenal Protection Agency.

REFERENCES

ATSDR (Agency for Toxic Substances and Disease Registry). 1997. *Toxicological Profile for Chloroform (Update)*. Public Health Service, U.S. Department of Health and Human Services, Atlanta, GA, September.

Balster RL, Borzelleca JF. 1982. Behavioral toxicity of trihalomethane contaminants of drinking water in mice. *Environ Health Perspect* 46: 127–136.

Bowman FJ, Borzelleca JF, Munson AE. 1978. The toxicity of some halomethanes in mice. *Toxicol Appl Pharmacol* 44: 213–215.

Bruckner JV, Warren DA. 2001. Toxic effects of solvent and vapors. In: *Casarett and Doull's Toxicology: The Basic Science of Poisons*. 6th ed. Klaassen CD, ed. McGraw-Hill, New York.

Bull RJ, Brown JM, Meierhenry EA, Jorgenson TA, Robinson M, Stober JA. 1986. Enhancement of the hepatotoxicity of chloroform in B6 $C3F_1$ mice by corn oil: implications for chloroform carcinogenesis. *Environ Health Perspect* 69: 49–58.

California OEHHA (Office of Environmental Health Hazard Assessment) 2006. Chemicals known to the state to cause cancer or reproductive toxicity. Safe Drinking Water and Toxic Enforcement Act of 1986 (Proposition 65). Accessed at: http://www.oehha.ca.gov/prop65/.

Carter JH, Carter HW, DeAngelo AB. 1995. Biochemical, pathologic and morphometric alterations induce in male B6 $C3F_1$ mouse live by short-term exposure to dichloroacetic acid. *Toxicol Lett* 92: 55–71.

Chu I, Villeneuve DC, Secours VE, et al. 1982a. Trihalomethanes: II. Reversibility of toxicological changes produced by chloroform, bromodichloromethane, chlorodibromomethane and bromoform in rats. *J Environ Sci Health B* 17: 225–240.

Chu I, Villeneuve DC, Secours VE, et al. 1982b. Toxicity of trihalomethanes: I. The acute and subacute toxicity of chloroform, bromodichloromethane, chlorodibromomethane and bromoform in rats. *J Environ Sci Health B* 17: 205–224.

Constan AA, Wong BA, Everitt JI, Butterworth BE. 2002. Chloroform inhalation exposure conditions necessary to initiate liver toxicity in female B6 $C3F_1$ mice. *Toxicol Sci* 66(2):201–208.

Environment Canada and Health Canada. 2001. *Canadian Environmental Protection Act, 1999. Priority Substances List Assessment Report: Chloroform*. Ottawa, Ontario, Canada.

Eschenbrenner AB, Miller E. 1945. Induction of hepatomas in mice by repeated oral administration of chloroform, with observations on sex differences. *J Natl Cancer Inst* 5: 251–255.

Golden RJ, Holm SE, Robinson DE, Julkunen PH, Reese EA. 1997. Chloroform mode of action: implications for cancer risk assessment. *Regul Toxicol Pharmacol* 26: 142–155.

Gulati DK, Hope E, Mounce RC, et al. 1988. *Chloroform: Reproduction and Fertility Assessment in CD-l Mice when Administered by Gavage*. Report by Environmental

Health Research and Testing, Inc., Lexington KY to National Toxicology Program, National Institute of Environmental Health Sciences, Research Triangle Park, NC.

Health Canada. 1996. *Guidelines for Drinking Water Quality: Canada Water Act*, 6th ed. Federal–Provincial Advisory Committee on Environmental and Occupational Health, Minister of Supply and Services, Health Canada, Ottawa, Ontario, Canada.

———. 2005. *Trihalomethanes in Drinking Water*. Document for Public Comment. Prepared by the Federal–Provincial–Territorial Committee on Drinking Water, Health Canada, Ottawa, Ontario, Canada.

———. 2006. *Guidelines for Canadian Drinking Water Quality*. Guideline Technical Document. Prepared by the Federal–Provincial–Territorial Committee on Drinking Water, Health Canada, Ottawa, Ontario, http://www.hc-sc.gc.ca/ewh-semt/alt_formats/hecs-sesc/pdf/pubs/water-eau/doc-sup-appui/trihalomethanes/trihalomethanes_e.pdf.

IARC (International Agency for Research on Cancer). 1979. *Some Halogenated Hydrocarbons*. IARC Monographs on the Evaluation of Carcinogenic Risks of Chemicals to Humans, Vol 20. World Health Organization, Lyon, France, 609 pp.

———. 1999. Chloroform. In: *Some Chemicals That Cause Tumors of the Kidney or Urinary Bladder in Rodents and Some Other Substances*. IARC Monographs Programme on the Evaluation of Carcinogenic Risks of Chemicals to Humans, Vol 73. World Health Organization, Lyon, France. pp 131–182.

ILSI (International Life Sciences Institute). 1997. *An Evaluation of EPA's Proposed Guidelines for Carcinogen Risk Assessment Using Chloroform and Dichloroacetate as Case Studies: Report of an Expert Panel*. ILSI Press, Washington, DC.

Jorgenson TA, Rushbrook CJ. 1980. *Effects of Chloroform in the Drinking Water of Rats and Mice: Ninety-Day Subacute Toxicity Study*. Report by SRI International, Menlo Park, CA to Health Effects Research Laboratory, Office of Research and Development, U.S. Environmental Protection Agency, Cincinnati, OH.

Jorgenson TA, Rushbrook CJ, Jones DCL. 1982. Dose–response study of chloroform carcinogenesis in the mouse and rat: status report. *Environ Health Perspect* 46: 141–149.

Jorgenson TA, Meierhenry EF, Rushbrook CJ, Bull RJ, Robinson M. 1985. Carcinogenicity of chloroform in drinking water to male Osborne–Mendel rats and female $B6C3F_1$ mice. *Fundam Appl Toxicol* 5: 760–769.

La DK, Schoonhoven R, Ito N, Swenberg JA. 1996. The effects of exposure route on DNA adduct formation and cellular proliferation by 1,2,3-trichloropropane. *Toxicol Appl Pharmacol* 140(1): 108–114.

Landauer MR, Lynch MR, Balster RL, et al. 1982. Trichloromethane-induced taste aversions in mice. *Neurobehav Toxicol Teratol* 4: 305–309.

Larson JL, Wolf DC, Butterworth BE. 1993. The acute hepatotoxicity and nephrotoxic effects of chloroform in male F-344 rats and female $B6C3F_1$ mice. *Fundam Appl Toxicol* 20: 302–315.

Larson JL, Wolf DC, Morgan KT, Mery S, Butterworth BE. 1994a. The toxicity of l-week exposures to inhaled chloroform in female $B6C3F_1$ mice and male F-344 rats. *Fundam Appl Toxicol* 22: 431–446.

Larson JL, Wolf DC, Butterworth BE. 1994b. Induced cytolethality and regenerative cell proliferation in the livers and kidneys of male $B6C3F_1$ mice given chloroform by gavage. *Fundam Appl Toxicol* 23: 537–543.

______. 1994c. Induced cytotoxicity and cell proliferation in the hepatocarcinogenicity of chloroform in female B6C3F$_1$ mice: comparison of administration by gavage in corn oil vs. ad libitum in drinking water. *Fundam Appl Toxicol* 22: 90–102.

Larson JL, Wolf DC, Butterworth BE. 1995a. Induced regenerative cell proliferation in liver and kidneys of male F-344 rats given chloroform in corn oil by gavage or ad libitum in drinking water. *Toxicology* 95: 73–86.

Larson JL, Wolf DC, Mery S, Morgan KT, Butterworth BE. 1995b. Toxicity and cell proliferation in the liver, kidney and nasal passages of female F344 rats induced by chloroform administered by gavage. *Food Chem Toxicol* 33: 443–456.

Larson JL, Templin MV, Wolf DC, Jamison KC, Leininger JR, Mery S, Morgan KT, Wong BA, Conolly RB, Butterworth BE. 1996. A 90-day chloroform inhalation study in female and male B6C3F$_1$ mice: implications for cancer risk assessment. *Fundam Appl Toxicol* 30(1):118–137.

Melnick RL, Kohn MC, Dunnick JK, Leininger JR. 1998. Regenerative hyperplasia is not required for liver tumor induction in female B6C3F$_1$ mice exposed to trihalomethanes. *Toxicol Appl Pharmacol* 148(1):137–147.

Munson AE, Sain LE, Sanders VM, et al. 1982. Toxicology of organic drinking water contaminants: Trichloromethane, bromodichloromethane, dibromochloromethane and tribromomethane. *Environ Health Perspect* 46: 117–126.

Nagano K, Nishizawa T, Yamamoto S, Matsushima T. 1998. Inhalation carcinogenesis studies of six halogenated hydrocarbons in rats and mice. In: *Advances in the Prevention of Occupational Respiratory Diseases. Proceedings of the 9th International Conference on Occupational Respiratory Diseases*, Kyoto, Japan, October 13–16, 1997. Chiyotani K, Hosoda Y Aizawa Y eds. Elsevier Science, New York, pp. 741–746.

NCI (National Cancer Institute). 1976. *Report on Carcinogenesis Bioassay of Chloroform*. NTIS PB-264018. National Institute of Health, Springfield, VA.

NRC (National Research Council). 1987. Disinfectants and disinfection by-products. In: *Drinking Water and Health*, vol 7. National Academies Press, Washington, DC.

NTP (National Toxicology Program). 1994. *Comparative Toxicology Studies of Corn Oil, Safflower Oil, and Tricaprylin (CAS Nos. 8001-30-7, 8001-23-8, and 538-23-8) in Male F344/N Rats as Vehicles for Gavage*. NTP Technical Report Series 426, National Institute of Environmental Health Sciences, Research Triangle Park, NC.

Palmer AK, Street AE, Roe JC, et al. 1979. Safety evaluation of toothpaste containing chloroform. II. Long term studies in rats. *J Environ Pathol Toxicol* 2: 821–833.

Pereira MA. 1994. Route of administration determines whether chloroform enhances or inhibits cell proliferation in the liver of B6C3F$_1$ mice. *Fundam Appl Toxicol* 23(1): 87–92.

Roe FJC, Palmer AK, Worden AN, Van Abbe NJ. 1979. Safety evaluation of toothpaste containing chloroform, I: Long-term studies in mice. *J Environ Pathol Toxicol* 2: 799–819.

Ruddick JA, Villeneuve DC, Chu I. 1983. A teratological assessment of four trihalomethanes in the rat. *J Environ Sci Health* B18: 333–349.

Templin MV, Larson JL, Butterworth BE, Jamison KC, Leininger JR, Mery S, Morgan KT, Wong BA, Wolf DC. 1996a. A 90-day chloroform inhalation study in F-344 rats: profile of toxicity and relevance to cancer studies. *Fundam Appl Toxicol* 32(1):109–125.

Templin MV, Jamison KC, Sprankle CS, Wolf DC, Wong BA, Butterworth BE. 1996b. Chloroform-induced cytotoxicity and regenerative cell proliferation in the kidneys and liver of BDF1 mice. *Cancer Lett* 108(2):225–231.

Templin MV, Jamison KC, Wolf DC, Morgan KT, Butterworth BE. 1996c. Comparison of chloroform-induced toxicity in the kidneys, liver, and nasal passages of male Osborne–Mendel and F-344 rats. *Cancer Lett* 104(1): 71–78.

Thompson DJ, Warner SD, Robinson VB. 1974. Teratology studies on orally administered chloroform in the rat and rabbit. *Toxicol Appl Pharmacol* 29: 348–357.

Tumasonis CF, McMartin DN, Bush B. 1985. Lifetime toxicity of chloroform and bromodichloromethane when administered over a lifetime in rats. *Ecotoxicol Environ Saf* 9: 233–240.

———. 1987. Toxicity of chloroform and bromodichloromethane when administered over a lifetime in rats. *J Environ Pathol Toxicol Oncol* 7: 55–64.

U.S. EPA (Environmental Protection Agency). 1980. *Ambient Water Criteria for Chloroform*. Office of Water Regulations and Standards, U.S. EPA, Washington, DC.

———. 1993. *Reference Dose (RfD): Description and Use in Health Risk Assessments*. Background Document 1A, March 15. Accessed at: http://www.epa.gov/iris/rfd.htm.

———. 1996. Proposed guidelines for carcinogen risk assessment. *Fed Reg* 61: 17960–18011.

———. 1998a. National primary drinking water regulations: disinfectants and disinfection byproducts; final rule. 40 CFR Parts 9, 141, and 142, December 16.

———. 1998b. *Occurrence Assessment for Disinfectants and Disinfection Byproducts in Public Drinking Water Supplies*. EPA/815/B/98/004. Office of Groundwater and Drinking Water, U.S. EPA, Washington, DC, November 13. PB99–111320.

———. 2001. *Toxicological Review of Chloroform* (*CAS No. 67-66-3*). In Support of Summary Information on the Integrated Risk Information System (IRIS). EPA/635/R-01/001. Office of Research and Development, National Center for Environmental Assessment, U.S. EPA, Washington, DC.

———. 2003a. National primary drinking water regulations: stage 2 disinfectants and disinfection byproducts rule; national primary and secondary drinking water regulations: approval of analytical methods for chemical contaminants; proposed rule. 40 CFR Parts 9, 141 and 142. *Fed Reg*, 49547–49681, August 18.

———. 2003b. *Water Quality Criterion for the Protection of Human Health: Chloroform*. Revised Draft. EPA/823/D-03/001. Office of Water, U.S. EPA, Washington, DC.

———. 2006. National primary drinking water regulations: stage 2 disinfectants and disinfection byproducts rule; national primary and secondary drinking water: final rule. 40 CFR Parts 9, 141 and 142. *Fed Reg* 387–493, January 4.

WHO (World Health Organization). 1994. *Chloroform*. Environmental Health Criteria Series, No. 163. International Programme on Chemical Safety, United Nations Environmental Program, International Labor Organization, World Health Organization, Geneva, Switzerland. Available online at: http://www.who.int/pcs/ehc/summaries/ehc_163.html and http://www.inchem.org/documents.ehc/ehc/ehc163.htm.

Yamamoto S, Nishizawa T, Nagano K, Aiso S, Kasai T, Takeuchi T, Matsushima T. 1999. Development of resistance to chloroform toxicity in male BDF1 mice exposed to a stepwise increase in chloroform concentration. *J Toxicol Sci* 24(5): 421–424.

11

RISK ASSESSMENT OF A THYROID HORMONE DISRUPTOR: PERCHLORATE

DAVID TING
California Environmental Protection Agency, Oakland, California

Some waterborne contaminants, such as perchlorate and nitrate, can reduce the uptake of iodide into the thyroid. When exposed to high levels and for a long time, these contaminants can deplete the iodide stored in the thyroid and cause thyroid enlargement and thyroid hormone imbalance. If this situation is not remedied in time, it can lead to low serum thyroid hormone levels that are associated with a wide range of adverse health effects, such as abnormalities in metabolism, growth, and development. Although many of these effects are reversible, some (e.g., abnormal neurological development in fetuses and infants) are not. These contaminants should be recognized as thyroid hormone disruptors, and it is important to consider their effect on the hormonal system in a risk assessment.

In this chapter we share our experience in evaluating the antithyroid properties of perchlorate and developing a human health risk assessment for the chemical. It is hoped that the approaches and methods described can be used in evaluating other environmental goitrogens.

BACKGROUND

Perchlorate is used primarily in the manufacture of solid propellants for rockets and missiles; it is also used in fireworks, road flares, blasting agents, and

Risk Assessment for Chemicals in Drinking Water, Edited by Robert A. Howd and Anna M. Fan

automobile air bags. Perchlorate is highly water soluble and stable at ambient temperature and pressure. It is mobile in an aqueous medium, as it is not adsorbed by organic or inorganic materials. Due to past disposal practices, extensive contamination of water sources in many states has been reported (U.S. EPA, 2002). Perchlorate is bioaccumulated to various degrees in a wide variety of edible plants, including lettuce, strawberry, and cantaloupe (Environmental Working Group, 2002; U.S. FDA, 2004); (Applied Biosystems, 2005). Low levels have also been detected in dairy milk and human breast milk (Kirk et al., 2003, 2005; U.S. FDA, 2004; Téllez et al., 2005). Although it is clear that most environmental perchlorate contaminations are attributable to human activities, there is evidence that there could be a natural perchlorate background of atmospheric origin (Dasgupta et al., 2005).

As of April 2006, there are no established federal or state regulatory standards for perchlorate. Before 1997, the detection limit of perchlorate in water was relatively high (400 ppb) and it was not recognized as an important environmental contaminant. Since then, the detection limit has been lowered by at least two orders of magnitude, and perchlorate has been detected in many drinking water systems. For instance, perchlorate at levels between 4 and 13 ppb had been detected in water samples taken from Lake Mead (U.S. EPA, 2006). The Colorado River is an important source of drinking water of 15 to 20 million people in communities in Arizona, southern California, and southern Nevada. There is a need to evaluate the oral toxicity of perchlorate and develop a drinking water level for public health protection.

Toxicological information on perchlorate has been reviewed (U.S. EPA, 2002; California OEHHA, 2004; Massachusetts DEP, 2004, 2006; NAS, 2005; New Jersey Drinking Water Institute, 2005). Perchlorate is a potent inhibitor of a transmembrane protein called sodium–iodide *symporter* (NIS) that is responsible for the active transport of iodide into thyroid cells. As iodide is a key ingredient of thyroid hormones, if there is a sustained decrease of iodine supply to the thyroid, synthesis and secretion of thyroid hormones can be impaired. Under normal circumstances, the thyroid hormones triiodothyronine (T_3) and thyroxine (T_4) are maintained within a narrow range by a negative feedback system. A lowered level of thyroid hormone in the circulation would stimulate the hypothalamus to secrete thyrotropin-releasing hormone (TRH), which in turn stimulates the pituitary to produce more thyroid-stimulating hormone (TSH). An increased TSH level would stimulate the thyroid and cause it to produce more thyroid hormones. However, prolonged and excessive stimulation can cause the thyroid cells to increase in size (hypertrophy) and number (hyperplasia) and lead to thyroid enlargement or goiter. Figure 1 shows the current understanding of how perchlorate affects the thyroid. A more detailed discussion of absorption, toxicokinetics, and health effects of perchlorate are provided in the following sections.

Absorption and Toxicokinetics of Perchlorate

Data from human and animal studies indicate that perchlorate is readily absorbed from the gastrointestinal tract and distributed systemically with total body water;

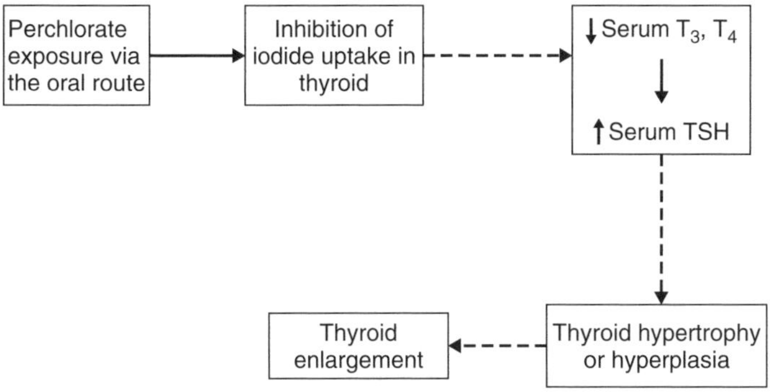

Figure 1. Suggested mode-of-action model of perchlorate toxicity. Solid arrow represents an outcome that has been observed in humans during perchlorate exposure. Dashed arrows represent outcomes observed in laboratory animals; they have not been clearly demonstrated in humans exposed to perchlorate but are biologically possible in the absence of adequate compensation. (Modified from NAS, 2005.)

it is essentially unmetabolized in vivo (Wolff, 1998). A higher concentration of perchlorate is associated with the thyroid than with other tissues. According to a human study reported by Greer et al. (2002) and an occupational study reported by Lamm et al. (1999), the biological half-life of perchlorate in humans is approximately 8 hours.

Animal Toxicity Data

In acute and subchronic animal studies, perchlorate administered through the oral route (0.01 to 30 mg/kg·day) reduced uptake of iodide into the thyroid, perturbed thyroid hormone regulation, induced hypertrophy and hyperplasia in the thyroid, and caused an increase in thyroid weight (Caldwell, et al., 1995; Argus Research Laboratories, 1998, 2001; Springborn Laboratories, 1998; Yu et al., 2000). In developmental and reproductive studies in rats, perchlorate at doses up to 30 mg/kg·day did not affect fertility and pregnancy outcome measures. There were changes in thyroid weight, thyroid histopathology, and thyroid hormone levels in the dams or the offspring. Some changes in the fetal brain development were noted, but due to methodological limitations, the interpretation of the data is unclear (Argus Research Laboratories, 1998, 1999, 2001; TERA, 2001; U.S. EPA, 2002). In a two-generation rat study, two male pups in the 30-mg/kg·day dose group were found to have thyroid follicular-cell adenomas (Argus Research Laboratories, 1999). Although the result was not statistically significant, it is noteworthy because of the low historical background incidence rate of the tumor and the relative short duration of exposure.

Ammonium perchlorate has been tested in a battery of in vitro and in vivo genotoxicity tests; perchlorate does not appear to be mutagenic or clastogenic (California OEHHA, 2004). A number of animal studies have shown that perchlorate at high doses (over 1300 mg/kg·day) causes thyroid tumors in rodents

(Kessler and Kruskemper, 1966; Gauss, 1972; Pajer and Kalisnik, 1991). As the occurrence of the tumors were mostly preceded by signs of thyroid hormone disruption and thyroid enlargement, they are generally interpreted as being secondary to the antithyroid effects of perchlorate (U.S. EPA, 2002; California OEHHA, 2004).

Human Toxicity Data

In the 1960s, potassium perchlorate was used to treat patients with Graves' disease. The therapeutic dose ranged from 500 to 2000 mg/day; most treatments lasted several weeks, but in a few cases treatment as long as a year was reported (Godley and Stanbury, 1954; Crooks and Wayne, 1960; Morgans and Trotter, 1960). In some patients getting high doses, side effects such as skin rashes, nausea, gastrointestinal problems, and a serious blood disorder were noted (Fawcett and Clarke, 1961; Hobson, 1961; Johnson and Moore, 1961; Barzilai and Sheinfeld, 1966). Due to these side effects, perchlorate is no longer used to treat thyroid disorder. Crooks and Wayne (1960) administered potassium perchlorate at 600 to 1000 mg/day to a group of pregnant women who were suffering from hyperthyroidism and observed a slightly enlarged thyroid in 1 of the 12 infants born to the mothers. The enlarged thyroid returned to normal size in 6 weeks, and no other abnormalities were reported. The use of the clinical data in a risk assessment is hampered by the fact that all the subjects had abnormal thyroid functions going into the studies, and it is difficult to extrapolate the observations to the general population.

In addition to the information gleaned from the therapeutic use of perchlorate, a variety of ecological, occupational, and human exposure studies are available and are discussed next. Two ecological studies reported by Brechner et al. (2000) and Schwartz (2001) that investigated the association between perchlorate in drinking water and abnormal serum thyroid hormone or serum TSH levels in neonates showed a positive correlation. Confidence in these studies is not high because of the small sample sizes, limited exposure data, and other methodological issues. The negative results of similar ecological studies in infants and adults (Lamm and Doemland 1999; Crump et al., 2000; Li et al., 2000a,b, 2001; Morgan and Cassady, 2002; Kelsh et al., 2003;) have also been questioned. Since the publication of these studies, perchlorate has been detected in many food products, including fresh vegetables, fruits, and cow's milk. It shows that perchlorate contamination is more widespread than previously thought and that it is not limited to drinking water. Thus the "unexposed subjects" in the ecological studies might have been exposed to perchlorate. The potential misclassification of perchlorate exposure reduces the statistical power of, as well as the confidence in, these study results.

The three occupational studies in the literature (Gibbs et al., 1998; Lamm et al., 1999; Braverman et al., 2005). indicated that exposure to perchlorate via the inhalation route at estimated daily doses ranged from approximately 0.01 0.5 mg/kg·day did not impair thyroid function, as indicated by the serum levels of TSH, T_3, and T_4. In the study reported by Gibbs et al. (1998), serum hormone levels were measured before and after a shift. Given the short exposure duration, it is not surprising that there was no change in the parameters monitored. There are difficulties in interpreting the results reported by Braverman et al. (2005) and Lamm et al. (1999) because of the small sample sizes, relatively high urinary iodine levels, and intermittent nature of exposure. Workers in these studies worked for three consecutive days, followed by three consecutive days off, as the recovery period is much longer than the biological half-life of perchlorate (8 hours), the workers might have replenished the stored iodide in the thyroid during the off days.

Many human exposure studies have been conducted. Stanbury and Wyngaarden (1952) showed that a single oral perchlorate dose as low as 2.2 mg caused detectable release of iodide from the thyroid, and reported a positive correlation between perchlorate dose and the fraction of stored iodide discharged from the thyroid. Brabant et al. (1992) administered an oral daily dose of 900 mg to five subjects for 4 weeks and found no effect on serum thyroid hormone levels. They reported a decrease in serum TSH level and an increase in serum thyroglobulin level, indicating the stress on the thyroid hormone balance. Although these three studies were not selected for dose–response evaluation, they do provide supporting information regarding the potency of perchlorate.

The two 14-day oral studies reported by Lawrence et al. (2000, 2001) investigated the effect of perchlorate on thyroidal iodide uptake and serum thyroid hormone levels and serum TSH levels. At a dose of approximately 0.14 mg/kg·day, they observed a significant decrease in iodide uptake but no effects on serum hormone levels. At a lower dose of approximately 0.04 mg/kg·day, they found a small but statistically insignificant decrease in iodide uptake. The usefulness of these studies is limited by the small number of subjects (less than 10 in both studies) and the lack of control regarding the dosing schedule.

A similar but better designed human exposure study was reported by Greer et al. (2002). In this study, groups of euthyroid male and female subjects were asked to ingest water dosed with perchlorate at 0.007, 0.02, 0.1, or 0.5 mg/kg for 14 days. They were asked to drink one-fourth of the perchlorate dose at four times spaced throughout each day. Thyroid iodide uptake was measured before (baseline), during, following a 2-day exposure period, following a 14-day exposure period (E14), and 15 days after the termination of exposure (P15). They found a statistically significant decrease in iodide uptake after 14 days of exposure in all except the lowest-dose group (Table 1). Greer et al. found that perchlorate had no effect on the serum thyroid hormone levels and serum TSH levels measured throughout the study period. There was no sex difference. Uptakes measured on post-exposure day 15 were not significantly different from the baseline, indicating that the inhibitory effect of perchlorate is reversible.

The data from Greer et al. (2002) show that even in the highest-dose group (0.5 mg/kg·day), with 67% reduction in thyroidal iodide uptake, there was no change in serum thyroid hormones and serum TSH. This is not unexpected, if

TABLE 1. Reduction of Thyroid Radioiodine Uptake Following a 14-Day Exposure to Perchlorate

Average Dose (mg/kg·day)	Change in Radioactive Iodine Uptake by the Thyroid (%) after 24 Hours of Dosing		Number of Subjects in Each Dose Group
	Average	Standard Deviation	
0.007	−1.8	22.0	7
0.02	−16.4[a]	12.8	10
0.1	−44.7[a]	12.3	10
0.5	−67.1[a]	12.1	10

Source: Derived from Greer et al. (2002) and G. Goodman (personal communication).

[a]Statistically significant, $p < 0.005$ (pairwise comparison to baseline) (Greer et al., 2002).

one realized that in an iodide sufficient adult, the amount of iodide stored in the thyroid is estimated to be able to sustain thyroid hormone production for several months (Greer et al., 2002), even with no iodide uptake.

HUMAN HEALTH RISK ASSESSMENT

Following the conventional practice of human health risk assessment, we organized our evaluation into three parts: hazard identification, dose–response evaluation, and risk characterization.

Hazard Identification

Hazard identification is the process of evaluating human, animal, and other relevant data to determine whether exposure to a chemical can cause an increase in the incidence of a health condition. Based on the toxicity information reviewed, it is clear that perchlorate can reduce thyroidal iodide uptake. With prolonged exposure and at sufficiently high doses, perchlorate can impair thyroid function and disrupt thyroid hormone balance. The body would try to counteract this disruption through a negative-feedback system that stimulates thyroid cells and increases the production of thyroid hormones. If the situation does not improve in time, it can cause hypertrophy and hyperplasia of the thyroid; in some cases, it can also cause thyroid enlargement or goiter.

Although perchlorate has been shown to induce thyroid tumors in rodents, it is not believed that it poses a significant cancer risk to humans. Perchlorate has not been shown to be genotoxic. There is evidence that humans may not be as sensitive quantitatively to thyroid–pituitary disruption as rodents (U.S. EPA, 1998). Thyroid hormones in rodents are not bound to thyroxine-binding globulin as in humans; they have a higher rate of destruction and thus have to be replenished at a higher rate. Rodent thyroid is chronically under stress and is more sensitive to chemicals that disrupt thyroid hormone balance (Hill et al., 1998).

Health information related to iodine deficiency indicates that pregnant women and their fetuses are likely to be the most sensitive to the antithyroid effects of perchlorate. Glinoer (2001) suggested that pregnancy itself represents a stress on the thyroid hormonal system and that iodine deficiency can compound the problem. Results of a prospective study reported by Kung et al. (2000) showed that in a borderline iodine-sufficient area (median urinary iodine level = 9.8 μg/dL), pregnancy resulted in higher rates of maternal goitrogenesis as well as neonatal hypothyroxinemia and hyperthyrotrophinemia. It is important to note that thyroid enlargement in these women persisted and failed to revert completely even three months after delivery.

Several epidemiological studies indicate that iodine deficiency during pregnancy may affect brain development and cause neurointellectual deficits in the offspring. The severity of effects depends on the timing as well as the severity of iodine deficiency and thyroid disorder (Morreale de Escobar et al., 2000). Evidence suggests that normal fetal brain development requires an adequate supply of maternal thyroid hormone throughout the first trimester, before the fetal thyroid begins to function (Hollowell and Hannon, 1997; Haddow et al., 1999; Pop et al., 1999, 2003). Even borderline maternal iodine deficiency, as observed in some European countries, may be accompanied by impaired school achievement in apparently normal children (Glinoer, 2001).

Dose–Response Evaluation

In dose–response evaluation, we identify the most appropriate biological endpoint as the basis for risk assessment. A suitable study is also identified and used to establish the dose-response relationship and the no-observed-effect level. As mentioned earlier, there is reason to believe that thyroid of rodents is more sensitive to thyroid hormone–disrupting chemicals than that of humans (Hill et al., 1998). Further evidence can be obtained by comparing toxicity data obtained from rodents and humans. Several 14-day drinking water studies showed significant depression in serum T_3, T_4, and elevation in serum TSH levels in rodents exposed to perchlorate doses as low as 0.01 or 0.1 mg/kg·day (Caldwell et al., 1995; Keil et al., 1998; Springborn Laboratories, 1998; Yu et al., 2000). Similar human studies of the same exposure duration showed no changes in serum T_3, T_4, and TSH levels in volunteers exposed to doses up to 0.5 mg/kg·day (Lawrence et al., 2000; Greer et al., 2002).

Given the potentially large uncertainty in extrapolating rodent toxicity to humans, we decided not to use rodent data in dose–response evaluation. Of the many biological changes related to perchlorate exposure, we identified reduction of thyroidal iodide uptake as the most appropriate endpoint for quantitative dose–response characterization. First it is the initial step in a chain of events that if unchecked may lead to serious health effects. By setting an exposure level that would ensure normal iodide uptake, all adverse health effects related to thyroid hormone disruption could be avoided. Second, in contrast to changes in serum thyroid hormones and TSH levels, change in iodide uptake is fast and reversible. Greer et al. (2002) data indicated that a steady state was reached in only 2 days and complete recovery was achieved 15 days following the termination of perchlorate exposure. Complete recovery following exposure cessation was also reported by Lawrence et al. (2000). As discussed earlier, while the short exposure duration of the Greer et al. study made observation of changes in serum TSH, T_3, or T_4 levels unlikely, it should not affect the validity of the iodide uptake results. Another advantage of choosing reduction of thyroidal iodide uptake as the critical endpoint is that it can minimize the effect of perchlorate on NIS in nonthyroidal tissues. Besides the thyroid, NIS has been found in stomach, lactating mammary gland, placenta, and to a lower extent in small intestine, skin, and brain (Perron et al., 2001). Breast milk is the sole source of nutrient for many infants, so the potential of perchlorate to reduce the secretion of iodide into the breast milk needs to be considered.

Of the three human exposure studies in the literature, we found the one reported by Greer et al. (2002) to be the most suitable for quantitative dose–response evaluation. A benchmark dose approach was used for this purpose; it is preferred over the conventional NOAEL/LOAEL approach because it uses all the data in the study, it is less affected by the spacing of the doses, and it accounts for the variability of the data as well as the slope of the dose–response relationship. We employed the Benchmark Dose Software version 1.3.1 provided by the U.S. EPA (2000) to evaluate the data (see Table 1) reported by Greer et al. (2002) and found the Hill model adequately describes the Greer et al. data (goodness-of-fit test, $p = 0.46$), shown plotted in Figure 2. The model was run with intercept set to zero, power parameter restricted to be greater than 1, and the assumption of a constant variance. The fit is generally considered adequate when the p-value is

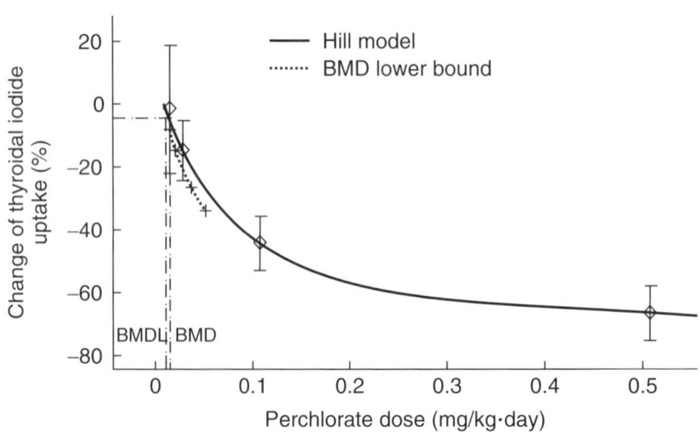

Figure 2. Analysis of the Greer et al. (2002) data by the benchmark dose approach.

greater than 0.05. The form of the response function estimated by the model is as follows:

$$\text{response} = \text{intercept} + \frac{\nu \times \text{dose}^n}{k^n + \text{dose}^n} \qquad (1)$$

where

$$\begin{aligned} \text{intercept} &= 0 \\ \nu &= -73.4469 \\ n &= 1.15067 \\ k &= 0.0663651 \end{aligned}$$

As the thyroidal iodide uptake data are continuous, not quantal, we need to identify a response level that is equivalent to a no-observed-effect level (NOEL). We experimented by setting the reference level to either minus 14% of the baseline (one mean standard deviation of all the dosed groups in the study is approximately 14%), minus 10% of the baseline, or minus 5% of the baseline, and found that they correspond to a calculated dose of 0.0188, 0.013, or 0.0068 mg/kg·day, respectively. As the calculated dose of 0.0188 mg/kg·day is close to the second-lowest dose (0.02 mg/kg·day) of the study (Table 1), it can be considered to give a positive response. The dose calculated 0.0068 mg/kg·day is close to the lowest dose (0.007 mg/kg·day) of the study (Table 1); it can be considered to give a negative response. Without additional data, it is difficult to determine if the calculated dose of 0.013 mg/kg·day (reference level = – 10%) represents a positive or a negative response. Given the options, we decided to use minus 5% of the baseline as the critical response level for our modeling. Using the parameters described above, we found that the benchmark dose (BMD) corresponds to 0.0068 mg/kg·day and the lower limit of a one-sided 95% confidence interval on the BMD or the BMDL corresponds to 0.0037 mg/kg·day. In the risk assessment, we consider the BMDL to be equivalent to a NOEL and use it as the basis of the quantitative evaluation, as it takes into consideration the limited sample size of the study and the variability exhibited in the data.

Calculation of a Health-Protective Drinking Water Level As pregnant women and their fetuses are identified as the sensitive subgroups, we used the following equation to estimate a health-protective concentration for drinking water (C, in mg/L):

$$\begin{aligned} C &= \frac{\text{BMDL} \times \text{RSC} \times (\text{BW/WC})}{\text{UF}} \qquad (2) \\ &= \frac{0.0037 \text{ mg/kg} \cdot \text{day} \times 0.6 \times (25.2 \text{ kg} \cdot \text{day/L})}{10} \\ &= 0.0056 \text{ mg/L (rounded to 0.006 ppm, or 6 ppb)} \end{aligned}$$

where

BMDL = lower limit of a one-sided 95% confidence interval of a perchlorate dose that reduces mean thyroidal iodide uptake by 5%

RSC = relative source contribution; a value of 60% is used for pregnant women because of the detection of perchlorate in farm produce and cows' milk

BW/WC = ratio of body weight (kg) and tap water consumption rate (L/day); the ratio for 95th percentile of the pregnant woman population is estimated to be 25.2 kg·day/L (California OEHHA, 2000)

UF = uncertainty factor of 10 to account for interindividual variability

The water concentration resulting from this calculation, 6 ppb, is judged adequate to protect all individuals, including potential sensitive subpopulations, from adverse health effects of perchlorate, from short-term to chronic exposures. An uncertainty factor of 10 for interindividual variability is used because the subject population in the Greer et al. (2002) study was small (37 subjects) and did not include pregnant women, infants, and individuals with thyroid problems. Dietary iodine intake and thyroidal iodide uptake are known to vary among individuals; they are affected by the type of food one eats (some food is rich in iodine, while other foods contain goitrogens), smoking habits (tobacco smoke contains goitrogens), medication (e.g., lithium), and exposure to environmental contaminants (such as nitrate, polychlorinated biphenyls, and dioxins). The interindividual variability in the general population is likely to be greater than that shown by the study population.

Preliminary survey results indicate that perchlorate is detected in lettuce, wheat, tomato, cantaloupe, cucumber, and cows' milk (Kirk et al., 2003; Smith and Jackson, 2003; Sharp, 2004; U.S. FDA, 2004; Jackson et al., 2005). Perchlorate has also been detected in human breast milk samples (Kirk et al., 2005; Téllez et al., 2005), thus confirming the viability of the breast-milk exposure pathway as well as indicating significant neonatal exposures to perchlorate from sources other than drinking water. However, the data on food sources were not sufficient for a precise calculation of the contribution of perchlorate from drinking water versus food. Based on the data available, we assumed that the majority of the perchlorate exposure would come from drinking water and determined that the most appropriate relative source contribution is 60%.

RISK CHARACTERIZATION AND CONCLUSIONS

In our human health risk assessment of perchlorate, we identified the reduction of thyroidal iodide uptake as the critical endpoint. It is reasoned that if this undesirable effect can be avoided, all the subsequent health effects related to thyroid hormone disruption can be prevented. We applied benchmark-dose modeling techniques to a set of human data reported by Greer et al. (2002) and determined a benchmark dose (equivalent to a NOEL) of 0.0037 mg/kg·day. Dividing by an uncertainty factor of 10 for human variability, a health-protective daily dose of 0.00037 mg/kg·day would be derived from this exercise. A drinking water level of 6 ppb was calculated by using the health-protective daily dose, a relative source contribution of 60%, and exposure assumptions specific to pregnant women.

As the endpoint selected is related to a physical property of a membrane protein, the NIS, the interindividual variability is likely to be less than that of changes in serum thyroid hormones or TSH level. The threshold of the endpoint chosen is also less likely to be affected by exposure duration, iodide status, and the physiological condition of the subject. Given these considerations and the fact that the critical endpoint was based on an early effect in the chain of possible perchlorate effects, derived from human studies, we concluded that the uncertainty factor of 10 would be adequate.

In our evaluation, pregnant women with marginal or inadequate iodine intake and their fetuses were identified as the sensitive subgroups. Our concern is supported by a National Health and Nutrition Examination Surveys (NHANES III) taken between 1988 and 1994 which indicates that some women of childbearing age in the United States may have less than the sufficient level of iodide intake (Hollowell et al., 1998). Low urinary iodine concentrations (< 5 μg/dL) were found in 6.7% of the pregnant women and 14.9% in women of childbearing age surveyed. Although the data do not indicate that there is an iodine deficiency, it does raise the possibility that a percentage of the sensitive subgroup may be getting less than an adequate amount of iodine.

An 18-month evaluation of perchlorate toxicity by a panel of the National Academy of Sciences (NAS, 2005) supported the approach and methods used in our assessment. The NAS also identified reduction of thyroidal iodide uptake as the critical biological effect and selected the data reported by Greer et al. (2002) as the basis for its dose–response evaluation. The NAS identified the lowest dose (0.007 mg/kg·day) as the NOEL and developed a reference dose (RfD) by dividing this NOEL by an uncertainty factor of 10, corresponding to an equivalent dose of 0.0007 mg/kg·day.

This risk assessment is particularly notable because of its use of human data on the effect of perchlorate on a physiological process rather than on a frank adverse effect. The basic consideration in this evaluation is that the public should not have to be concerned about the effects of drinking water contaminants on their bodily processes, and prevention of significant inhibition of iodide uptake by perchlorate should satisfy this goal. This approach might also be taken for other chemicals, such as nitrate, which also competes with iodine for uptake into the thyroid. The use of such endpoints, the appropriate statistical measures of effect levels, and uncertainty factors to be applied for such data to minimize health risks, particularly for sensitive populations, deserves a full consideration and extensive debate among risk assessors.

Disclaimer

The opinions expressed in this chapter are those of the author and not necessarily of the Office of Environmental Health Hazard Assessment or the California Environmental Protection Agency.

REFERENCES

Applied Biosystems. 2005. Analysis of perchlorate in foods and beverages by ion chromatography coupled with the API 2000 IC/MS/MS system. Accessed February 2006 at: http://marketing.appliedbiosystems.com/mk/get/APMMS_APP_PERCHLORATE?_A=6 1997&_D=41100&_V=0#.

Argus Research Laboratories. 1998. Oral (drinking water) two-generation (one litter per generation) reproduction study of ammonium perchlorate in rats. Protocol 1416-001. ARL, Horsham, PA.

______. 1999. Oral (drinking water) two-generation (one litter per generation) reproduction study of ammonium perchlorate in rats. Protocol 1416-001. ARL, Horsham, PA.

______. 2001. Hormone, thyroid and neurohistological effects of oral (drinking water) exposure to ammonium perchlorate in pregnant and lactating rats and in fetuses and nursing pups exposed to ammonium perchlorate during gestation or via maternal milk. Protocol 1416-003. ARL, Horsham, PA.

Barzilai D, Sheinfeld M. 1966. Fatal complications following use of potassium perchlorate in thyrotoxicosis: report of two cases and a review of the literature. *Israel J Med* 2: 453–456.

Brabant G, Bergman P, Kirsch CM, Kohrle J, Hesch RD, Von Zur Muhlen A. 1992. Early adaptation of thyrotropin and thyroglobulin secretion to experimentally decreased iodine supply in man. *Metabolism* 41: 1093–1096.

Braverman LE, He XM, Pino S, Cross M, Magnani B, Lamm SH, et al. 2005. The effect of perchlorate, thiocyanate, and nitrate on thyroid function in workers exposed to perchlorate long-term. *J Clin Endocrinol Metab* 90: 700–706.

Brechner RJ, Parkhurst GD, Humble WO, Brown MB, Herman WH. 2000. Ammonium perchlorate contamination of Colorado River drinking water is associated with abnormal thyroid function in newborns in Arizona. *J Occup Environ Med* 42: 777–782.

Caldwell DJ, King JH, Kinkead ER, Wolfe RE, Narayanan L, Mattie DR. 1995. Results of a fourteen day oral-dosing toxicity study of ammonium perchlorate. Tri-Service Toxicology Consortium, Armstrong Laboratory. Wright-Patterson Air Force Base, OH.

California OEHHA Office of Environmental Health Hazard Assessment. 2000. *Air Toxics Hot Spots Program Risk Assessment Guidelines*, Part IV. Exposure Assessment and Stochastic Analysis Technical Support Document. California OEHHA, Oakland, CA, September.

______. 2004. *Public Health Goal for Perchlorate in Drinking Water*. California Environmental Protection Agency, Sacramento, CA.

Crooks J, Wayne EJ. 1960. A comparison of potassium perchlorate, methylthiouracil, and carbimazole in the treatment of thyrotoxicosis. *Lancet* 1: 401–404.

Crump C, Michaud P, Tellez R, Reyes C, Gonzalez G, Montgomery EL, et al. 2000. Does perchlorate in drinking water affect thyroid function in newborns or school-age children? *J Occup Environ Med* 42: 603–612.

Dasgupta PK, Martinelango PK, Jackson WA, Anderson TA, Tian K, Tock RW, Rajagopalan S. 2005. The origin of naturally occurring perchlorate: the role of atmospheric processes. *Environ Sci Technol* 39(6): 1569–1575.

Environmental Working Group. 2002. High levels of toxic rocket fuel found in lettuce. (published online). Washington, DC and Oakland, CA., Accessed December 12, 2005 at: http://www.ewg.org/reports/rocketlettuce.

Fawcett JW, Clarke CWF. 1961. Aplastic anaemia due to potassium perchlorate. Br Med J, May 27, p 1537.

Gauss W. 1972. Das Verhalten einiger physiologischer und histologischer Kriterien der Schilddrüsenfunktion bei einmaliger oder längerer Verabreichung von Kaliumperchlorat an adulte Mäeuse (*Mus musculus L.*), I: Langzeitversuche. *Z Mikrosanat Forsch* 85: 469–500.

Glinoer D. 2001. Pregnancy and iodine. *Thyroid* 11(5): 471–481.

Godley AF, Stanbury JB. 1954. Preliminary experience in the treatment of hyperthyroidism with potassium perchlorate. *J Clin Endocrinol* 14: 70–78.

Greer MA, Goodman G, Pleus RC, Greer SE. 2002. Health effects assessment for environmental perchlorate contamination: the dose–response for inhibition of thyroidal radioiodine uptake in humans. *Environ Health Perspect* 110(9): 927–937.

Haddow JE, Palomaki GE, Allan WC, Williams JR, Knight GJ, Gagnon J, et al. 1999. Maternal thyroid deficiency during pregnancy and subsequent neuropsychological development of the child. *N Engl J Med* 341: 549–555.

Hill RN, Crisp TM, Hurley PM, Rosenthal SL, Singh DV. 1998. Risk assessment of thyroid follicular cell tumors. *Environ Health Perspect* 106: 447–457.

Hobson QJG. 1961. Aplastic anaemia due to treatment with potassium perchlorate. *Br Med J*, May 13, pp. 1368–1369.

Hollowell JG, Hannon WH. 1997. Teratogen update: iodine deficiency, a community teratogen. *Teratology* 55: 389–405.

Hollowell JG, Staehling NW, Hannon WH, Flanders DW, Gunter EW, Maberly GF, Braverman LE, Pino S, Miller DT, Garbe PL, DeLozier DM, Jackson RJ. 1998. Iodine nutrition in the United States. Trends and public health implications: iodine excretion data from National Health and Nutrition Examination Surveys I and III (1971–1974 and 1988–1994). *J Clin Endocrinol Metab* 83: 3401–3408.

Jackson WA, Joseph P, Laxman P, Tan K, Smith PN, Yu L, et al. 2005. Perchlorate accumulation in forage and edible vegetation. *J Agric Food Chem* 53: 369–373.

Johnson RS, Moore WG. 1961. Fatal aplastic anaemia after treatment of thyrotoxicosis with potassium perchlorate. *Br Med J* 5236: 1369–1371.

Keil D, Warren A, Jenny M, EuDaly J, Dillard R. 1998. Effects of ammonium perchlorate on immunotoxicological, hematological, and thyroid parameters in B6C3F1 female mice. Funded by Defense Special Weapons Agency, DSWA01-97-0008. Department of Medical Laboratory Sciences, Medical University of South Carolina, Charleston, SC, September 30.

Kelsh MA, Buffler PA, Daaboul JJ, Rutherford GW, Lau EC, Barnard JC, et al. 2003. Primary congenital hy pothyroidism, newborn thyroid function, and environmental perchlorate exposure among residents of a southern California community. *J Occup Environ Med* 45: 1116–1127.

Kessler, FJ, Kruskemper, HJ. 1966. Experimentelle Schilddrusentumoren durch mehrjahrige Zufuhr von Kaliumperchlorat. [Experimental thyroid tumors caused by long-term administration of potassium perchlorate]. *Klin Wochenschr* 44: 1154–1156.

Kirk AB, Smith EE, Tian K, Anderson TA, Dasgupta PK. 2003. Perchlorate in milk. *Environ Sci Technol* 37: 4979–4981.

Kirk AB, Martinelango PK, Tian K, Dutta A, Smith EE, Dasgupta PK. 2005. Perchlorate and iodide in dairy and breast milk. *Environ Sci Technol* 39: 2011–2017.

Kung AWC, Lao TT, Chau MT, Tam SCF, Low LCK. 2000. Goitrogenesis during pregnancy and neonatal hypothyroxinaemia in a borderline iodine sufficient area. *Clin Endocrinol* 53: 725–731.

Lamm SH, Doemland M. 1999. Has perchlorate in drinking water increased the rate of congenital hypothyroidism? *J Occup Environ Med* 41: 409–413.

Lamm SH, Braverman LE, Li FX, Richman K, Pino S, Howearth G. 1999. Thyroid health status of ammonium perchlorate workers: a crosssectional occupational health study. *J Occup Environ Med* 41: 248–260.

Lawrence JE, Lamm SH, Pino K, Richman K, Braverman LE. 2000. The effect of short-term low-dose perchlorate on various aspects of thyroid function. *Thyroid* 10: 659–663.

Lawrence JE, Lamm SH, Braverman LE. 2001. Low dose perchlorate (3 mg daily) and thyroid function. *Thyroid* 11: 295.

Li Z, Li FX, Byrd D, Deyhle GM, Sesser DE, Skeels MR, et al. 2000a. Neonatal thyroxine level and perchlorate in drinking water. *J Occup Environ Med* 42: 200–205.

Li FX, Byrd DM, Deyhle GM, Sesser DE, Skeels MR, Katkowsky SR, et al. 2000b. Neonatal thyroid-stimulating hormone level and perchlorate in drinking water. *Teratology* 62: 429–431.

Li FX, Squartsoff L, Lamm SH. 2001. Prevalence of thyroid diseases in Nevada counties with respect to perchlorate in drinking water. *J Occup Environ Med* 43: 630–634.

Massachusetts DEP. 2004. *Perchlorate Toxicological Profile and Health Assessment*. Final draft. Office of Research and Standards, Massachusetts Department of Environmental Protection, Boston, MA, May.

______. 2006. *Update to "Perchlorate Toxicological Profile and Health Assessment."* Office of Research and Standards, Massachusetts DEP, Boston, MA, May.

Morgan JW, Cassady RE. 2002. Community cancer assessment in response to long-time exposure to perchlorate and trichloroethylene in drinking water. *J Occup Environ Med* 44(7): 616–621.

Morgans ME, Trotter WR. 1960. Potassium perchlorate in thyrotoxicosis [letter]. *Br Med J*, October 8, pp. 1086–1087.

Morreale de Escobar G, Obregon MJ, Escobar de Rey F. 2000. Is neuropsychological development related to maternal hypothyroidism or to maternal hypothyroxinemia? *J Clin Endocrinol Metab* 85: 3975–3987.

NAS (National Academy of Sciences) 2005. *Health Implications of Perchlorate Ingestion*. Committee to Assess the Health Implications of Perchlorate Ingestion. National Academies Press, Washington, DC.

New Jersey Drinking Water Quality Institute. 2005. Maximum contaminant level recommendation for perchlorate. October. Accessed December 12 2005 at: http://www.state.nj.us/dep/watersupply/perchlorate_mcl_10_7_05.pdf.

Pajer Z, Kalisnik M. 1991. The effect of sodium perchlorate and ionizing radiation on the thyroid parenchymal and pituitary thyrotropic cells. *Oncology* 48: 317–320.

Perron B, Rodriguez AM, Leblanc G, Pourcher T. 2001. Cloning of the mouse sodium iodide symporter and its expression in the mammary gland and other tissues. *J Endocrinol* 170(1): 185–196.

Pop VJ, Kuijpens JL, van Baar AL, Verkerk G, van Son MM, de Vijlder JJ, et al. 1999. Low maternal free thyroxine concentrations during early pregnancy are associated with impaired psychomotor development in infancy. *Clin Endocrinol* 50: 149–155.

Pop VJ, Brouwers EP, Vadert HL, Vulsma T, van Baar AL, de Vijlder JJ. 2003. Maternal hypothyroxinaemia during early pregnancy and subsequent child development: a 3-year follow-up study. *Clin Endocrinol* 59: 282–288.

Schwartz J. 2001. Gestational exposure to perchlorate is associated with measures of decreased thyroid function in a population of California neonates. Thesis, University of California, Berkeley, CA.

Sharp R. 2004. *Rocket Fuel Contamination in California Milk*. Environmental Working Group, Washington, DC and Oakland, CA. Accessed at: http://www.ewg.org/reports/rocketmilk.

Smith PN, Jackson WA. 2003. Perchlorate in the environment: ecological considerations. Presentation material available at: www.tribalwater.net/perchlorate/Martinez.pdf. Accessed Dec 12, 2005.

Springborn Laboratories. 1998. A 90-day drinking water toxicity study in rats with ammonium perchlorate. Study 3455.1, June 3. *Health and Environmental Sciences*, Springborn Laboratories, Spencerville, OH.

Stanbury JB, Wyngaarden JB. 1952. Effect of perchlorate on the human thyroid gland. *Metabolism* 1: 533–539.

Téllez RT, Chacón, PM, Abarca CR, Blount BC, Van Landingham CB, Crump KS, Gibbs JP. 2005. Long-term environmental exposure to perchlorate through drinking water and thyroid function during pregnancy and the neonatal period. *Thyroid* 15(9): 963–975.

TERA (Toxicology Excellence for Risk Assessment). 2001. Report on five expert reviews of the Primedica 2001 Study report. Submitted to California OEHHA, December 12.

U.S. EPA (Environmental Protection Agency), 1998. *Assessment of Thyroid Follicular Cell Tumors*. EPA/630/R-97/002. U.S. EPA, Washington, DC, March. Accessed at: http://cfpub.epa.gov/ncea/cfm/recordisplay.cfm?deid=13102.

——. 2000. Benchmark dose technical guidance document, External review draft. Accessed December 12, 2005 at http://cfpub2.epa.gov/ncea/cfm/recordisplay.cfm?deid=20167.

——. 2002. *Perchlorate environmental contamination: toxicological review and risk characterization*. External review draft. NCEA-1-0503. U.S. Environmental Protection Agency, Office of Research and Development, U.S. EPA, Washington, DC.

——. 2006. Perchlorate in the Pacific Southwest. Region 9, U.S. EPA. Accessed March 15, 2006 at: http://www.epa.gov/region9/toxic/perchlorate/per_nv.html.

U.S. FDA (Food and Drug Administration). 2004. Exploratory data on perchlorate in food. Accessed at: http://www.cfsan.fda.gov/dms/clo4data.html.

Wolff J. 1998. Perchlorate and the thyroid gland. *Pharmacol Rev* 50(1): 89–106.

Yu KO, Todd PN, Young SM, Mattie DR, Fisher JW, Narayanan L, et al. 2000. *Effect of Perchlorate on Thyroidal Uptake of Iodide with Corresponding Hormonal Changes*. AFRL-HE-WP-TR-2000-0076. Air Force Research Laboratory, Wright-Patterson Air Force Base, OH, July.

12

EMERGING CONTAMINANTS IN DRINKING WATER: A CALIFORNIA PERSPECTIVE

STEVEN A. BOOK AND DAVID P. SPATH
California Department of Health Services, Sacramento, California

The California Department of Health Services' (DHS's) Drinking Water Program has been addressing emerging contaminants in drinking water supplies for several decades. *Emerging contaminants*, in this case, are chemicals that are unregulated as contaminants in drinking water, or those that have been found in drinking water, but are not widely recognized as contaminants, given their localized or only recently documented occurrence. In the 1980s, concerns about a number of industrial and agricultural chemical contaminants prompted the DHS to establish 60 advisory *action levels*. For many chemicals, action levels were established because chemicals were actually found in drinking water. For others, there was the potential for their occurrence, based on their presence nearby, as in a pesticide-handling facility or a hazardous waste site, as examples, and concerns about the chemicals' migration to drinking water wells.

These action levels were estimates of chemical concentrations in drinking water that would not pose an unacceptable health risk, calculated by standard risk assessment methods, based on an intake of 2 L/day. They were generally established at *de minimis* cancer risk levels (up to one excess case of cancer per million people per 70-year lifetime), or for noncarcinogens, at an approximation of the no-observable-adverse-effect level (NOAEL) divided by an uncertainty factor that reflects confidence in the available data. The relative source contribution (the fraction contributed by drinking water) has been added over the past decade (California DHS, 2005a). These estimates are similar to the approaches used

Risk Assessment for Chemicals in Drinking Water, Edited by Robert A. Howd and Anna M. Fan

elsewhere, most notably by the California Environmental Protection Agency's Office of Environmental Health Hazard Assessment, in its drinking water public health goals (PHGs) (California OEHHA, 2007). PHGs are risk assessments that contribute to the DHS's development of drinking water standards [maximum contaminant levels (MCLs)]. Prior to 1991, the California OEHHA was part of the DHS. The DHS used its action levels to provide advice to water systems and others who were faced with decisions about what to do when these unregulated chemicals appeared in water supplies. Eventually, MCLs were adopted for many of these contaminants.

Since then, the DHS has continued to use its action levels to provide information, although in 2005 legislation (Health and Safety Code Section 116455), changed the name to *notification levels*, to clarify the action to be taken: that is, notification of local government agencies such as city councils and county boards of supervisors. Additionally, the clarification minimizes confusion with the regulatory action levels for lead and copper.

The DHS has established 89 notification levels altogether since the early 1980s. Of these, 38 now have MCLs, and 25 have been archived or placed into semiretirement (still able to be referenced, but not considered part of the list of chemicals with notification levels). There are currently 26 chemicals with notification levels, and 18 of those have been established since 2000 (California DHS, 2005a).

EMERGING CHEMICALS OF THE RECENT PAST

Several drinking water contaminants occurred in sufficient numbers of drinking water sources to warrant regulation. Methyl *tert*-butyl ether (MTBE), perchlorate, and chromium-6 represent last decade's "emerging contaminants," and either have an MCL or are on their way toward one (California DHS, 2005b).

MTBE The DHS adopted regulations that established an enforceable secondary MCL of 5 μg/L in 1999 (to address taste and odor), and a 13-μg/L primary MCL for the gasoline additive MTBE. The DHS also administers a program that provides funding for MTBE treatment and research (California DHS, 2005b). MTBE, a relatively new chemical used to improve air quality, was found to degrade water quality because of leaking storage tanks and its ability to move through groundwater easily and rapidly. It was also a contributor to surface water contamination, the product of incomplete combustion by gasoline-powered watercraft. MTBE was truly an "emerging" contaminant in the 1990s, since it was relative new in commerce and recent as a contaminant. The PHG for MTBE is 13 μg/L, based on the carcinogenicity of the chemical in animals (California OEHHA, 1999a).

Perchlorate In 2006, the DHS proposed a primary MCL for perchlorate of 6 ppb. In California, perchlorate has contaminated hundreds of drinking water wells. Salts such as ammonium perchlorate are used as solid rocket propellant, or in fireworks, flares, and explosives. Perchlorate has been in use for decades, but there was little appreciation for it as an environmental toxin with potential health effects until the 1990s, and sensitive methods for its measurement in drinking water were not available until 1997. Perchlorate was thought to be confined to a few hazardous waste or Superfund facilities when its detection limit was greater than 100 to 400 μg/L. However, once it was able to be measured at levels 100 times lower, it was found to be considerably more widespread. Perchlorate as an emerging contaminant in the 1990s is probably a misnomer; it is better described as an old contaminant with emerging analytical methods. OEHHA's PHG for perchlorate is 6 μg/L, based on its effects on the thyroid gland derived from its interference with iodide uptake (California OEHHA, 2004).

Chromium-6 Chromium-6 is chromium in the hexavalent form. It has long been recognized as a carcinogen when inhaled and has appropriately received regulatory attention among occupational health and public health agencies because of risks associated with airborne exposures. It recently came to the fore because of pop culture, politics, and the press. The 2000 film *Erin Brockovich* made chromium-6 a household word, and local politicians and reporters jumped on the bandwagon that the film's theme provided: As chromium-6 data became available, they declared large areas of southern California to be at risk from its presence.

Much of the attention that emerged in 2000 resulted from OEHHA's PHG document for total chromium (California OEHHA, 1999b). The technical support document included an evaluation of chromium-6's carcinogenicity when ingested, based on a mouse study that showed the production of forestomach tumors from chromium-6 ingestion. Subsequently, a blue-ribbon panel was convened to evaluate the carcinogenicity of ingested chromium-6; the panel found evidence of its carcinogenicity when ingested to be lacking and suggested additional studies (CTRC, 2001). OEHHA subsequently withdrew the PHG document in 2001. However, the panel's determination was disputed by legislators who questioned the independence of the review panel, and the then-head of the California EPA stated that the panel's recommendations would not be used in developing the chromium-6 PHG.

The results of DHS's required monitoring using sensitive methods suggest that the chromium-6 can be present at low levels naturally. Whether it poses a risk by ingestion remains to be determined by animal testing by the National Toxicology Program and by OEHHA's new PHG document, which is expected in the near future. Although regulated under the total chromium MCL (established to be protective with regard to chromium-6, since chromium-3 is a nutrient), DHS was directed by statute to adopt an MCL specific for this form of the element and will eventually be doing so, following completion of the PHG. Today, MTBE, perchlorate, and chromium-6 are chemicals with actual or pending regulatory status, and as such, are losing their status in California as "emerging" chemicals. Nevertheless, as they move into the role of "everyday" regulated drinking water contaminants, they will be replaced by newer emerging contaminants.

NEWER EMERGING CONTAMINANTS

Two newer emerging contaminants are 1,2,3-trichloropropane (TCP) and *N*-nitrosodimethylamine (NDMA). These chemicals have been found in drinking water at low concentrations. Each is considered carcinogenic (California OEHHA, 2005; NTP, 2005), and they have high cancer potencies. 1,2,3-TCP and NDMA have *de minimis* cancer risk at very low levels (0.005 and 0.002 μg/L, respectively) (California DHS, 2005a). Their potencies have resulted in a need to improve analytical methods and lower detection limits to reach these low risk levels.

1,2,3-Trichloropropane 1,2,3-TCP has had prior industrial use and it was also used in the manufacture of pesticides. The DHS established a notification level for 1,2,3-TCP in 1999 after discovering the chemical at the Burbank Operable Unit (OU)—a southern California Superfund site—because of concerns that the chemical might find its way into drinking water supplies. It has also been found in several wells elsewhere in the state. In 2001 the DHS added 1,2,3-TCP to its list of nine unregulated chemicals for which monitoring was required to be completed by December 2003, with recommendations to use a method that could measure the chemical at 0.005 μg/L. 1,2,3-TCP has been found in industrial and agricultural areas of the state, in over 200 wells. The DHS has requested that OEHHA develop a public health goal for 1,2,3-TCP. A public health goal is an early step in the MCL process.

N-Nitrosodimethylamine (NDMA) NDMA is another chemical that the DHS believes may be appropriate for regulation. NDMA was first found to be drinking water contaminant associated with an aerospace facility. Although NDMA was not found in many drinking water sources, it was found in high levels in wells around a groundwater recharge project, and it was also found to be produced in some drinking water treatment plants, where chloramination is used to treat surface water sources. As a by-product of drinking water treatment, NDMA was not unlike a disinfection byproduct. Because of the treatment-related production, the DHS set its notification level for NDMA at 0.01 μg/L, five times higher than the usual *de minimis* risk level, both to acknowledge the difficulty in measuring NDMA at 0.002 μg/L, and to account for the treatment benefit that offsets the accompanying risk, similar to the way regulatory agencies balance the risks and benefits of water disinfection. The DHS requested that OEHHA develop a public health goal for NDMA so that the MCL process can proceed for this chemical. A PHG of 0.003 μg/L was established by OEHHA in December 2006 (California OEHHA, 2006).

FUTURE EMERGING CHEMICALS

Although 1,2,3-TCP and NDMA represent current emerging chemicals, we will probably see additional chemicals that reflect new substances in widespread use (like our experience with MTBE) or improvements in analytical capabilities for already existing chemicals in the environment (like our experience with perchlorate and chromium-6, as well as 1,2,3-TCP and NDMA). We will probably encounter new emerging chemicals that are related to particular sources of drinking water that have not previously been used to supply drinking water to the public. Stressors on drinking water supplies, principally population-driven requirements for increased household, commercial, industrial, agriculture, and recreational use, are such that high-quality drinking water supplies may not be present in amounts adequate to satisfy demands. Therefore, sources of lower quality will have to be used to satisfy demands for water. Contaminants in these lower-quality sources will require additional monitoring and care in the preparation for commercial and household use.

Examples of nontraditional sources of drinking water (or sources that historically would have been avoided altogether) include:

- Contaminated sources such as Superfund sites (which the DHS refers to as *extremely impaired sources*)
- Treated wastewater from sewage treatment plants
- Saline sources, including seawater

These sources, unlike traditional surface and groundwater sources permitted as high-quality sources of drinking water, by their very nature are known to be of lesser quality. Because of the presence of microbiological or chemical contaminants, they can be visualized as a chemical soup. Drinking water derived from these sources can be monitored to make sure that it complies with drinking water standards for regulated contaminants. But while regulated contaminants can be measured, how can unregulated and often unrecognized contaminants be addressed? The latter contaminants contain the future generations of emerging contaminants. Each of these new alternative sources of drinking water therefore requires special treatment and considerable regulatory oversight for the protection of public health.

Superfund and Other Hazardous Waste Sites

In California, several Superfund sites are providing treated water to public water systems, which is used to supplement drinking water supplies. These are sites where groundwater used previously as a drinking water source has become heavily contaminated as a result of years of land disposal of industrial chemicals. Although these operations have to meet drinking water standards, of course, the DHS has also had to address concerns about unregulated chemicals that are or may be present, applying additional constraints on projects that propose to use such alternative water supplies. For example, projects proposing to use water extracted from Burbank and Glendale OUs in southern California needed to address a number of concerns related to their operational permits. Regulators cleaning up those sites were looking for alternatives to surface water discharge of treated groundwater, and water systems were considering alternative water supplies to supplement drinking water supplies.

First and foremost, they needed to find a public water system that will oversee the delivery of such water and be subject to drinking water regulation and permitting rather than hazardous waste regulation and permitting. (Drinking water regulation and permitting is more stringent.) Project proponents must also fully characterize the source of water in terms of actual and potential contaminants and must evaluate the health risks associated with exposures that would be anticipated to occur from failures of the treatment system. DHS utilizes a procedure called the *97-005 Process*, named for the memorandum (California DHS, 1997), which outlines the steps, including public hearings that a water system must go through to have such a project permitted.

For the Glendale OU, the DHS established notification levels for approximately 10 chemicals that had been found in monitoring wells near the extraction wells at the site, to provide guidance in case they appeared in the water being produced. The use of water produced from hazardous waste sites requires protective barriers and redundancies in protection to ensure that contaminants do not reach the public. In areas in which water is at a premium, with the proper controls, it appears to be a preferred use of adequately treated water, particularly if the alternative fate of highly treated water is surface water discharge and ultimately flows into the ocean, as in southern California.

The identification of possible emerging contaminants from hazardous waste facilities, particularly Superfund sites, is enabled by the characterization of the site, which is required by the various laws that guide those cleanups. Thus, lists of chemicals are developed, soil and groundwater are monitored, and information is collected and evaluated. These lists of chemicals are also very useful when evaluating new sites where similar industrial activities were present.

Recycled Wastewater

Water recycling provides an opportunity to reuse wastewater. The use of recycled water for landscaping is widely accepted. Its use as a source of drinking water is developing. However, there are hurdles associated with the marketing of the so-called "toilet to tap" pathway. Nevertheless, with proper health protective barriers between toilet and tap, water supplies can be replenished. It is important to remember that the very nature of the water cycle implies that all water is recycled and reused on a long-term global scale. Water recycling occurs already when one community's treated wastewater is released into a surface water body such as a river, which also serves as another downstream community's source of drinking water. Effective controls that minimize industrial and commercial releases of chemicals and effective treatment of wastewater can protect the receiving water body. Those controls, and treatment of the water supply at the drinking water intake, protect the public health and assure consumers that their water supply is safe. In addition, should there be an accidental or otherwise unanticipated release of toxic materials into the surface water source, for example, from an industrial source that escapes treatment in the wastewater plant, drinking water intake can be avoided by ceasing intake until the contaminant bolus passes by. Thus, such an incident, although unfortunate and potentially risky, is of limited duration and of limited risk to consumers.

Treated wastewater can be introduced into groundwater by a variety of means. Some may be indirect and unintentional, that is, not associated with recharging projects. Examples include recharge from surface water (e.g., rivers and streams or from wastewater ponds). Other recharge may be direct and intentional, either by spreading, or application onto ground and seeping into aquifers, or by injection of treated wastewater. Treated wastewater that recharges groundwater is not as easily followed as when it is released into surface water, because the behavior of aquifers is not as predictable as that of a river or stream. Hence, where the water goes and at what speed is not as easily predicted. Also, should an accident or unintentional release of a contaminant occur into groundwater, retrieval is difficult. Depending on the nature of the contaminant, the aquifer may be rendered unfit for consumption. Thus, special attention is given to recharge projects, and even more special attention is given to projects that are specifically designed to provide drinking water.

In California, recycled water projects must be permitted by the applicable regional water quality control board. However, for projects involved in the recharge of groundwater for indirect potable reuse, the DHS evaluates the proposal and provides comments to the RWQCB. The DHS approach to the evaluation of these projects is guided by regulations in Title 17 of the California Code of Regulations and by draft regulations that have not yet completed the regulations process (California DHS, 2007).

In general, the DHS seeks to maximize the control of regulated contaminants and minimize the introduction of nonregulated contaminants for the protection of public health. It does so by encouraging the development of stringent industrial source control measures to keep unwanted materials from the wastewater treatment stream and to ensure that as new industrial practices are introduced, new contaminants are not released into the waste stream without properly informing the wastewater regulators and treatment plant operators.

The DHS also seeks to reduce the presence of nonregulated substances by recommending that standards be met for regulated substances and that total organic carbon and nitrogen compounds be held to higher, more restrictive limits to account for uncertainties in the presence of, or production of, chemicals in recycled water. It also recommends monitoring for a number of nonregulated chemicals for which the DHS has established advisory levels (called *notification levels*) and that the project proponents have a program of monitoring for nonindustrial chemicals, including household products, personal care products, and pharmaceuticals. The presence of these substances in water supplies has received considerable attention (U.S. EPA, 2004), as have hormones and endocrine-disrupting chemicals (U.S. EPA, 2003).

The specific chemicals listed in the endnotes of the draft regulations (California DHS, 2007) are those that a working group developed after discussions about chemicals that have been found in wastewater and groundwater. They include the following endocrine-disrupting chemicals, pharmaceuticals, and other chemicals for monitoring:

- *Hormones:* 17β-estradiol, estrone, ethinyl estradiol
- *"Industrial" endocrine disruptors:* bisphenol A, nonylphenol and nonylphenol polyethoxylate, octylphenol and octylphenol polyethoxylate, polybrominated diphenyl ethers
- *Pharmaceuticals and others substances:* acetaminophen, amoxicillin, azithromycin, caffeine, carbamazepine, ciprofloxacin, ethylenediaminetetraacetic acid (EDTA), gemfibrozil, ibuprofen, iodinated contrast media, lipitor, methadone, morphine, salicylic acid, triclosan

The DHS currently does not envision the regulation of these substances but seeks the collection of information, to enable more understanding about the presence of these substances in treated wastewater and in recycled water. Because in many cases the treated wastewater is only part of the water being recycled, the DHS also encourages the characterization of the other part of the recycled water, which is used to dilute or otherwise supplement the treated wastewater. For example, in southern California, stormwater runoff is captured and becomes part of the recycled water mix. This diluent water may make up a large portion of the water that is introduced into the underground aquifers, and as such, needs to be evaluated for its regulated and nonregulated constituents. Finally, information on the chemicals of interest mentioned above is also desirable for wastewater released into surface water sources, not only because those sources may be downstream users' drinking water, but because the surface source may also be indirectly recharging groundwater.

In addition to groundwater recharge, consideration is being given to augmenting surface water reservoirs with treated wastewater. In the mid-1990s the city of San Diego proposed to introduce highly treated wastewater into one of the City's drinking water reservoirs. The water from the reservoir undergoes standard drinking water treatment. The proposal would eventually have resulted in treated wastewater making up 50% of the water in the reservoir. Although the project was set aside as a result of public opposition, the city believes that the project is ultimately necessary to meet future water demands.

As is the case with groundwater recharge with recycled water, control of unregulated contaminants is of paramount concern for reservoir augmentation. Because this type of project is the closest to direct potable reuse, and as a result, has the least margin for uncertainty, understanding and monitoring the quality of the treated wastewater is essential.

Seawater and Other Saline Sources

Treating brackish water and seawater offers a great opportunity for extension of drinking water supplies (CCC, 1993; California DWR, 2003a). Although desalination plants are in use in some parts of the world (California DWR, 2003b), there are issues that may hinder the development of desalination plants in California and presumably elsewhere. These include costs of reverse osmosis membranes, costs associated with energy needs, the size of coastal facility footprints along with intake and discharge issues (e.g., environmental impact), and whether the use of desalination will supplement water needs or be used to promote population growth and a continued spiral of more people needing still more water. Besides of the economic and social aspects of desalination plants, there are concerns related to disposal of brine, to the release of metals from the reaction of materials with the salty water, and to other concerns related to protection of the marine environment.

With regard to the drinking water produced, there are also microbiological and chemical concerns that are common to freshwater sources, reflecting wastewater discharges and surface water runoff into coastal waters. Regulated contaminants would be covered by MCLs, but the presence of unregulated ones argues once again for attention to industrial source control and minimization of the presence of household hazardous wastes and other products. Agricultural runoff, including pesticides and fertilizers, may also eventually find its way to the sea and be taken up into the desalination plants.

There are also unique issues related to the presence of contaminants in drinking water that is produced from seawater. For example, boron is an unregulated element that is present in relatively high concentrations in seawater. It is unregulated in that it has no MCL, although it does have a California notification level of 1 mg/L (1000 μg/L). Because of uncertainty in the regulated community about what an MCL might be and whether they might invest in desalination plants only to find that they cannot meet a boron MCL in the future, there may be a push to develop such a standard.

Neurological and other toxins associated with dinoflagellates that are present in seawater also require some attention. Their concentrations are likely not high enough in seawater to bring about toxicities associated with paralytic shellfish poisoning (since they are more concentrated in shellfish, say, during a red tide episode), or with neurological, diarrheal, or amnestic shellfish poisoning. Nonetheless, there may be concerns about chronic exposures to low levels of these types of toxins. Therefore, it is reasonable to conclude that the potential for these toxins to penetrate the barriers of desalination membranes will need to be addressed to assure the public that drinking water from the ocean is safe and that desalination processes remove these toxins.

CONCLUSIONS

In California, for the past several decades, we have been able to maintain vigilance over drinking water supplies to protect public health. Our experience should inform regulators, health advocates, policy makers, and the public about the importance of recognizing that there are contaminants and potential contaminants beyond those that have undergone the formal regulatory process and are subject to regulation. Those that we regulate are the contaminants we know enough about to regulate. Those that we don't know enough about—or perhaps don't even know about *yet*—also require attention. Thus, the collection of information about contaminants in alternative sources such as those that are extremely impaired, or in recycled water, or in seawater or brackish water is very important, for that information can help identify contaminants and indicate whether they are present in concentrations high enough to warrant concern. The development of new or more sensitive analytical methods is also important. In many cases, contaminants may be present already but at levels undetectable by the usual analytical approaches. Finally, protection of drinking water supplies is important to ensure safe surface and groundwater sources. A dedicated focus on the handling of hazardous wastes and on limiting discharges to the environment from industry, agriculture, household, and defense-related activities can also protect current and future drinking water supplies from emerging contaminants.

Disclaimer

The opinions expressed in this chapter are those of the authors and not necessarily those of the California DHS, now the California Department of Public Health.

REFERENCES

California DHS (Department of Health Services). 1997. Guidance for direct domestic use of extremely impaired sources, Policy Memo 95-005, November 5. California DHS, Sacramento, CA. Accessed at: http://www.dhs.ca.gov/ps/ddwem/publications/policy/memo97-005.htm.

———. 2005a. *DHS drinking water notification levels: An overview*, California DHS, Sacramento, CA, May 9. Accessed at: http://www.dhs.ca.gov/ps/ddwem/chemicals/AL/notificationoverview.pdf.

———. 2005b. *Chemical Contaminants in Drinking Water: MTBE, perchlorate, and Chromium-6*. California DHS, Sacramento, CA. Accessed at: http://www.dhs.ca.gov/ps/dddwem/chemicals/chemindex.htm.

———. 2007. Groundwater recharge reuse regulations. Draft, January 4. California DHS, Sacramento, CA. Accessed at: http://www.dhs.ca.gov/ps/ddwem/waterrecycling/PDFs/rechargeregulationsdraft-01-04-2007.pdf.

California DWR (Department of Water Resources). 2003a. *Water Desalination: Findings and Recommendations*. Water Desalination Task Force, California DWR, Sacramento, CA, October. Accessed at: http://www.owue.water.ca.gov/recycle/desal/desal.cfm.

———. 2003b. California Water Desalination Task Force: Desalination Technology Working Paper. California Department of Water Resources, Sacramento, CA, September. Accessed at: http://www.owue.water.ca.gov/recycle/desal/Docs/DesalTechnology.doc.

Californa OEHHA. (Office of Environmental Health Hazard Assessment). 1999a. *Public Health Goal for Methyl Tertiary Butyl Ether (MTBE) in Drinking Water*. OEHHA, Oakland, CA, March. Accessed at: http://www.oehha.ca.gov/water/phg/pdf/mtbe_f.pdf.

______. 1999b. *Public Health Goal for Chromium in Drinking Water*. OEHHA, Oakland, CA, February. The PHG was withdrawn November 2001. Accessed at: http://www.oehha.ca.gov/water/phg/pdf/chrom_f.pdf.

______. 2004. *Public Health Goal for Chemicals in Drinking Water: Perchlorate*. OEHHA, Oakland, CA, March. Accessed at: http://www.oehha.ca.gov/water/phg/pdf/finalperchlorate31204.pdf.

______. 2005. *Chemicals known to the state to cause cancer,* Title 22, *California Code of Regulations*, Section 12000, pursuant to The Safe Drinking Water and Toxic Enforcement Act of 1986 (Proposition 65). OEHHA, Sacramento, CA, Accessed at: http://www.oehha.ca.gov/prop65/prop65_list/Newlist.html.

______. 2006. *Public Health Goals for Chemicals in Drinking Water: N-Nitrosodimethylamine*. OEHHA, Sacramento, CA, December. Accessed at: http://www.oehha.ca.gov/water/phg/pdf/122206NDMAphg.pdf.

______. 2007. *Public Health Goals as of December 22, 2006*. California Environmental Protection Agency, Sacramento, CA. Accessed at: http://www.oehha.ca.gov/phg/allphgs.html.

CCC (California Coastal Commission). 1993. *Seawater Desalination in California*. CCC, Sacramento, CA, October. Accessed at: http://www.coastal.ca.gov/desalrpt/dtitle.html.

CTRC (Chromate Toxicity Review Committee). 2001. Scientific review of toxicological and human health issues related to the development of a public health goal for chromium (VI), August 31. Accessed at: http://www.dhs.ca.gov/ps/ddwem/chemicals/Chromium6/reviewpanelreport.pdf.

NTP (National Toxicology Program). 2005. *Report on Carcinogens*, 11th ed. Public Health Service, U.S. Department of Health and Human Services, Washington, DC. Accessed at: http://ntp.niehs.nih.gov/index.cfm?objectid=32BA9724-F1F6-975E-7FCE50709-CB4C932.

U.S. EPA (Environmental Protection Agency). 2003. *Endocrine Disruptors Research Initiative* U.S. EPA, Washington, DC. Accessed at: http://www.epa.gov/endocrine/.

______. 2004. *Pharmaceuticals and Personal Care Products (PPCPs) as Environmental Pollutants: Pollution from Personal Actions, Activities, and Behaviors*. National Exposure Research Laboratory, U.S. EPA, Las Vegas, NV. Accessed at: http://www.epa.gov/nerlesd1/chemistry/pharma/index.htm.

______. 2005. *Endocrine Disruptors Screening Program*. U.S. EPA, Washington, DC. Accessed at: http://www.epa.gov/scipoly/oscpendo/.

13

U.S. EPA DRINKING WATER FIELD OFFICE PERSPECTIVES AND NEEDS FOR RISK ASSESSMENT

Bruce A. Macler
U.S. Environmental Protection Agency, San Francisco, California

First and foremost, risk assessments in the environmental and public health agencies are done almost exclusively to support management decisions involving some sort of hazardous or potentially hazardous situation. Regional (field-level) offices of the U.S. Environmental Protection Agency (EPA), as well as those for other federal, state, tribal, or local agencies, most often manage local or site-specific problems. Risk assessments generated in the field offices to address these problems can answer questions, inform decisions, comply with formal requirements, and provide a basis for communicating with the public. In addition, field offices frequently use risk information developed at the national level for broad applications, such as for drinking water regulations or guidance documents. This risk information is used to support regulatory implementation and enforcement, and to inform the public. In this chapter we discuss some of the applications where risk information is used by the EPA drinking water field offices and provide a few examples for consideration.

THE NATURE OF REGULATORY RISK ASSESSMENTS

It is important to understand at the onset the nature and limitations of risk assessments done for regulatory purposes. Since they exist to serve risk managers addressing specific environmental problems, these risk assessments are narrowly

Risk Assessment for Chemicals in Drinking Water, Edited by Robert A. Howd and Anna M. Fan

focused and generally idiosyncratic. They are generally not designed to examine the full range of risks, nor to quantify uncertainties and unknowables. They are often very difficult to use to compare different contaminants and different situations. Because of this, these risk assessments are also often very difficult to describe to the public.

Risk assessments done under environmental laws often vary markedly in both form and content, based on mandates in the applicable statutes (Clean Air Act vs. Toxic Substances Control Act vs. Clean Water Act) and on how these have been interpreted by the national EPA program offices that administer these statutes. Although efforts continue to standardize use of at least the toxicological components of risk assessments, most program offices, including those in the 10 EPA regional (field-level) offices, develop their assessments specific to their program and without general regard for other programs or applications. For example, the Safe Drinking Water Act amendments in 1986 and 1996 specified public health risk goals and procedures for developing national primary drinking water regulations (NPDWRs). The Act directs the EPA to establish maximum contaminant level goals (MCLGs) for contaminants of public health concern for drinking water: "Each maximum contaminant level goal established under this subsection shall be set at the level at which no known or anticipated adverse effects on the health of persons occur and which allows an adequate margin of safety" [Title XIV of the Public Health Service Act (the Safe Drinking Water Act), Section 1412(b)].

MCLGs are strictly health-based levels. They are developed and set at contaminant exposure levels believed to be without appreciable health risk to individuals, to be consistent with the provision "set at the level at which no known or anticipated adverse effects on the health of persons occur." Additionally, they must address the concern for the protection of sensitive subpopulations. A provision added to the act in 1996 specifies priorities for selecting contaminants for rule making to "take into consideration, among other factors of public health concern, the effect of such contaminants upon subgroups that comprise a meaningful portion of the general population (such as infants, children, pregnant women, the elderly, individuals with a history of serious illness, or other subpopulations) that are identifiable as being a greater risk of adverse health effects due to exposure to contaminants in drinking water than the general population" [Title XIV of the Public Health Service Act (the Safe Drinking Water Act), Section 1412(b)]. Therefore, risk assessments used to develop MCLGs must at a minimum provide an estimate for a *de minimis* risk exposure level for humans that may be more sensitive to the contaminant. These risk assessments are calculated from upper bounds for estimated risks, such that the true risks may be less or even zero; the assessments do not have to, nor are they designed to, estimate the full range of true risks to the average person.

Depending on the nature of the contaminant and the risk assessment approach used, MCLGs may be set at zero, at some nonzero nonquantifiable risk level, or at a nonzero quantifiable risk level (U.S. EPA, 1985, 1987, 1989, 1992, 1994a).

If a contaminant is considered to be an initiator carcinogen, EPA policy has been to consider no level of exposure as safe, so the MCLG is set at zero. For contaminants for which a reference dose approach to the risk assessment is used, the MCLG is a nonzero number but does not represent a particular quantitative risk level. For some contaminants, such as those that may be carcinogenic above a threshold, a probabilistic risk assessment can be used to set a MCLG at a *de minimis* risk level. As a consequence, it is often difficult to communicate the risks associated with a MCLG or with the associated NPDWR.

These MCLGs, which are not enforceable themselves, are to be used to set the enforceable NPDWRs: "Each national primary drinking water regulation for a contaminant for which a maximum contaminant level goal is established under this subsection shall specify a maximum contaminant level for such contaminant which is as close to the maximum contaminant level goal as is feasible. . . . For the purposes of this subsection, the term 'feasible' means feasible with the use of the best technology, treatment techniques and other means which the Administrator finds, after examination for efficacy under field conditions and not solely under laboratory conditions, are available (taking cost into consideration)" [Title XIV of the Public Health Service Act (the Safe Drinking Water Act), Section 1412(b)]. The NPDWR may include a maximum contaminant level (MCL) and/or require treatment and compliance with performance criteria. As part of this process, an economic impact analysis is developed that estimates potential regulatory costs and health benefits. The initial risk assessments are further refined to provide information on the number of people that might be affected by adverse consequences from different exposures and benefited from controlling those exposures. These consequences may be monetized to yield health cost and benefit values in dollars and cents. If the risk assessments do not allow adequate quantification, descriptive estimates may be used. For contaminants that have multiple adverse consequences, such as arsenic or some microbial pathogens, generally the better defined and more significant adverse endpoints are fully quantified for this purpose (U.S. EPA, 2001). These estimates appear in the NPDWR supporting documents as chapters of the economic analysis. After these considerations are addressed, the risks associated with any particular NPDWR may differ substantially from those of its MCLG(s).

The risks for different MCLs can span two orders of magnitude. For example, the MCL for dichloromethane is set at the 1 per million excess cancer risk level (i.e., one additional cancer in a population of a million similarly exposed persons over a lifetime) (U.S. EPA, 1992). The risk level for di(2-ethylhexyl)phthalate is set at 2 per million (U.S. EPA, 1992). On the other hand, the risk level for bromate is set at 2 per 10,000 (U.S. EPA, 1994a), while vinyl chloride, toxaphene, and ethylene dibromide are all set at 1 per 10,000 (U.S. EPA, 1989, 1992). To be sure, the uncertainties in the risk assessments allow that these differences may be unimportant or nonexistent. The EPA has long considered regulating carcinogenic contaminants within a "window" of risks between about 1 per 10,000 and 1 per million (U.S. EPA, 2001).

The Role of IRIS

The EPA maintains the integrated risk information system (IRIS) as a database containing information on human health effects that may result from exposure to various chemicals in the environment (U.S. EPA, 2007). The toxicity information in IRIS represents the consensus of opinions among scientists in the various EPA program offices and in the Office of Research and Development. This information goes through periodic updates and peer reviews. Depending on the nature of a given contaminant, sections for chronic health hazard assessment for noncarcinogenic effects, or carcinogenicity assessment for lifetime exposure, or both, may appear. The chemical files may contain information on oral reference doses (RfDs) and inhalation reference concentrations (RfCs) for the chronic noncarcinogenic health effects. They may contain hazard identification, oral slope factors, and oral and inhalation unit risks for carcinogenic effects. They also contain extensive references and documentation. Because EPA field offices almost always lack the resources to acquire hazard identification and dose–response data independently, IRIS serves as the primary resource for such information. However, it is recognized that risk assessments done by other entities may be substantially different in form and content than those in IRIS.

USE OF DRINKING WATER RISK INFORMATION IN EPA FIELD OFFICES

Pro Forma Use of Existing Risk Information

Field-level environmental protection programs around the country, be they federal, state, or local, do not employ many risk assessors or routinely prepare estimates of risks. By and large, these programs use what risk information is given to them by their headquarters units or is available through external sources in terms of regulations, guidance, and support materials to adequately inform the regulated community and the public. This is generally true for all media and environmental situations that occur, directed predominantly by the specific laws that apply to a given program.

The risk assessments that EPA regional (field) office drinking water programs mostly see are those for the NPDWRs under the SDWA. These assessments are done in one or another of EPA's national offices, usually in the Office of Science and Technology. The form and content of these assessments are dictated by the SDWA and the interpretations of this Act, other laws, precedents, and the views of the larger scientific community, as memorialized in agency risk policies (U.S. EPA, 1985, 1987, 1989, 1992, 1994a, 2001). In fact, few in the field offices actually access and review the full risk assessments that accompany a NPDWR. As noted before, these assessments are components of the economic analysis used for cost–benefit considerations. Most field office staff reference the risk assessment summaries provided in the regulatory preambles.

EPA field offices work primarily with state and tribal agencies that have delegated oversight ("primacy") for implementation and enforcement of the federal regulations. These agencies need to understand the regulations well enough that they can manage them effectively. Most field offices also provide direct regulatory oversight for some states, islands, and tribes. The field offices serve as a major point of contact for drinking water utilities, for the public, and for others in the larger drinking water community. These people often want to have the regulations and their underpinnings explained to them. Thus, the nominal use of risk information at the field office level is to explain and support NPDWRs in terms of health concerns and regulatory protection. Communicating to the public, either directly or via the media, is the dominant activity that uses risk information in the typical EPA field office.

Effective communication of this risk information becomes key. For most circumstances, this requires that the available materials for public use provide much more that merely the risk levels. It requires understandable descriptions of the assumptions, biases, and uncertainties that are dealt with and incorporated into full estimation of the given risk. As these are components of a risk characterization, this essentially means presenting the risk characterization in ways that a nontechnical audience can understand.

Field offices also benefit from having staff that understand the components of risk assessments and are comfortable considering other applicable risk assessments made by state drinking water programs, and those from related programs such as those under the Clean Water Act. Often, staff are called upon to talk to the public at large, who are unlikely to be knowledgeable in risk assessment, but want their questions answered in a way they can understand. These staff need more than a superficial understanding of risk assessment practices and the risk numbers associated with a particular situation. They also need skills in communications with the varied publics that seek them out for information.

Although the most common risk communication request is for an explanation that is understandable, two difficult communication situations frequently occur. These are when those that are upset or opposed to a situation involving environmental risk want to dispute EPA's version of the risk, and when there is a disagreement between agencies about risks for a given agent. The former requires that the risk assessor be sensitive to the anger and outrage factors that are affecting the communications. Seldom will externally developed risk information be organized in a way to help with this. The latter, when experts disagree, is frequently exposed by the media, especially if it is about a contentious or noteworthy contaminant. Field office risk assessors must be as versed as possible about how the information provided to them was generated, in order to communicate these factors most effectively to the public.

More Flexibility for Specific Situations

Besides using provided risk information to support regulatory actions or to communicate about regulations with the public, EPA field offices may prepare less-formal risk assessments for these and a number of other public health purposes associated with field office drinking water programs. For these purposes, risk assessments may not necessarily have to conform to the risk assessment approaches mandated for regulatory applications. A wider view of what risk means to the public can be brought forward and assessments done to answer particular kinds of questions. These may be quantitative estimates, but frequently semiquantitative or qualitative assessments are both useful and practical.

Regional EPA offices often have to address contaminant risks or situations that have not reached national significance or rule making. Some contaminants may occur or have been used only in limited geographical areas, so will never become national priorities for control. Yet they may pose unacceptable threats to those exposed. Some materials may be of emerging concern, such as endocrine disrupters or certain pharmaceuticals and personal care products, but currently lack sufficient information to allow regulation. Field offices also have to address situations, such as a bioterrorism attack on a water system, that can only be evaluated on a site-specific basis. Sometimes, enforcement actions under general water quality provisions in the SDWA trigger the need to find remedies that address the specific risk. More often, the realities of resource limitations for disadvantaged areas may require risk-based prioritization of compliance efforts.

Risk Analyses of Health Implications from Site-Specific Environmental Exposures Where Toxicological Data Are Available Examples of these include efforts to estimate health risks resulting from full body contact recreation at drinking water source reservoirs. In many areas, open water lakes and reservoirs used as sources for drinking water have bathing beaches and may have heavy seasonal use. EPA field offices, local health departments, and the drinking water agencies are often concerned for the fecal loadings that may result, especially for the protozoan parasite *Cryptosporidium* and for the pathogenic viruses. Most often these contaminants will ultimately be dealt with by treatment, but source water protection to control them under the SDWA or Clean Water Act may be possible. Because surface water treatment rule health goals are based on acceptable levels of risk, agencies may want estimates on what risks might remain following different treatment approaches. Sometimes public health agencies wish to know the risks to the bathers themselves from incidental ingestion of contaminated water. This information can be used to better manage reservoir activities or to suggest other protective mitigations for the source. While infectivity (dose–response) information is generally limited for most microbial pathogens, enough is known that probabilistic assessments can be done and be of value (U.S. EPA, 1994b, 2000, 2006).

Risk Analyses for Health Risk Situations with Limited or No Toxicological and/or Exposure Information (Systematic, Nonquantitative Analyses) Current concerns for possible bioterrorism attacks on drinking water systems have led EPA field offices, their state counterparts, and the drinking water utilities themselves to pursue risk assessments to assist in developing countermeasures and supporting emergency response planning. Although the likelihood of a terrorist attack on any given water system is remote, a successful attack might have significant consequences. Agencies want to identify possible vulnerabilities and potential consequences for different scenarios. Because information on the toxicological properties of weapons-grade chemical and biological agents is essentially nonexistent or unavailable, standard toxicity and exposure risk assessments for such contaminants are impossible. Even for better known and understood agents for which some toxicological information is available, exposure estimates are specific to a utility's circumstances and system design. Additionally, such assessments have to address the terrorist's perspective on the desirability of attack on the system (i.e., the probability that the system or a component is a target for attack), as well as the possible consequences of a successful attack. Risk assessors have turned to semiquantitative and qualitative approaches used in engineering domains to examine systems and system components for vulnerabilities, then classify their likelihoods and magnitudes of consequence on a relative basis (U.S. EPA, 2005). For example, the entire water system from its sources to treatment to storage to distribution is examined for single points of failure, or access points for introduction of a contaminant, or other vulnerabilities. Their physical availability and suitability for attack may then be estimated, generally using a low–medium–high structure. Similarly, the barriers to attack that might exist are evaluated. Direct consequences on public health, as well as longer-term impacts, are judged for nature and magnitude. These are combined to provide a relative listing of higher- and lower-priority situations.

A second example of this is being used to address water quality problems for aircraft drinking water. These tiny water systems are addressed and regulated under the SDWA, primarily for their microbiological safety. Water is stored in tanks and is available from taps in the galleys and restrooms. This water comes from local sources, so may vary substantially in quality. Although most airlines have maintenance procedures for these systems, there are few testing requirements, few inspections, and limited enforcement. The nature of the industry itself makes it difficult to monitor and track water quality in these systems. Because there is virtually no information on specific microbial pathogens in these systems, standard toxicity and exposure risk assessment procedures are unsuitable. Approaches used for food handling, based on the hazard assessment critical control point (HACCP) management system (U.S. FDA, 2006), have shown promise. Like the bioterrorism vulnerability assessments, they examine the water delivery system from sources to taps to identify vulnerabilities and consequences (hazard assessment). This allows for management by establishing protective criteria for the points or processes where contamination can occur (critical control points).

Prioritization of Risks and Risky Situations to Assist Effective Management
For some water systems, technical and financial limitations preclude full compliance with all NPDWRs. While each drinking water MCL is set to be protective of human health over a lifetime's exposure, as noted above, some MCLs are more protective than others. As well, some regulated contaminants are far more common than others. Some contaminants, such as the microbial pathogens and disinfection by-products, exist in a risk–risk trade-off situation where increasing

control for one class of contaminants can exacerbate risks from another. EPA field offices and their regulatory partners frequently must consider which problem contaminants to address first when a system has too many difficulties to address them all. They often conduct order-of-magnitude risk assessments for each contaminant at a specific system or site. These may be based on individual risk, or extended to estimate the numbers of people who might be affected by different contaminants. Managers will prioritize control activities based in part on the risk rankings, as well as other considerations.

A similar process may be done over an entire region or for specific groups of water systems, such as those on tribal lands or on the U.S.-controlled Pacific islands. Field offices often have to determine priorities for use of limited national funding for grants and loans. Risk assessments may be made to identify and rank problematic contaminants for priority attention. Individual systems may then consider their risks for these contaminants in their funding applications. These often are based on toxicity information in IRIS and use exposure profiles generated with site-specific data.

CONCLUSIONS

By and large, EPA drinking water field offices most often use risk information generated as a consequence of national SDWA regulatory development or residing in the IRIS database, rather than performing their own assessments. This is because most of the need for risk information at the field level is to communicate risks about regulated contaminants or those with a national or international interest where consistency within the EPA is desirable. However, risk assessments are frequently done in field settings for a wide variety of purposes. These are done to answer regional- or system-specific questions, especially to address unique exposure settings. These questions often require or benefit from the use of nonstandard risk assessment approaches. Some risk questions cannot be addressed using standard toxicity and exposure approaches used for regulations. Others require only approximate answers, or answers applicable to the average person, or answers that can be easily compared. These pose challenges for field office risk assessors, yet also make the tasks intellectually satisfying as well as useful.

Disclaimer

The opinions expressed in this chapter are those of the author and not necessarily those of the U.S. EPA.

REFERENCES

U.S. EPA (Environmental Protection Agency). 1985. NPDWRs; synthetic organic chemicals, inorganic chemicals and microorganisms; proposed rule. *Fed Reg* 50 (219): 46936–47022.

———. 1987. NPDWRs; filtration and disinfection; turbidity, *Giardia lamblia*, viruses, *Legionella* and heterotrophic bacteria; proposed rule. *Fed Reg* 52 (212): 42178–42222.

———. 1989. NPDWRs, proposed rule. *Fed Reg* 54(97): 22062–22160.

———. 1992. NPDWRs; synthetic organic chemicals and inorganic chemicals; final rule. *Fed Reg* 57 (138): 31776–31849.

———. 1994a. NPDWRs; disinfectants and disinfection byproducts; proposed rule. *Fed Reg* 59 (145): 38668–38829.

———. 1994b. NPDWRs; enhanced surface water treatment requirements; proposed rule. *Fed Reg* 59 (145): 38832–38858.

———. 2000. NPDWRs; ground water rule; proposed rules. *Fed Reg* 65 (91): 30194–30274.

———. 2001. NPDWRs; arsenic and clarifications to compliance and new source contaminants monitoring; final rule. *Fed Reg* 66 (14): 6976–7066.

———. 2005. Security and vulnerability assessment criteria. Accessed at: cfpub.epa.gov/safewater/watersecurity/index.cfm.

———. 2006. NPDWRs; ground water rule; notice of data availability. *Fed Reg* 71 (58): 15105–15109.

———. 2007. *Integrated Risk Information System: Database for Risk Assessment*. Office of Research and Development, U.S. EPA, Washington, DC. Accessed at: www.epa.gov/iris.

U.S. FDA (Food and Drug Administration). 2006. *Hazard Analysis and Critical Control Point Criteria*. Center for Food Safety and Applied Nutrition *U.S.* FDA, Washington, DC. Accessed at: www.cfsan.fda.gov/~lrd/haccp.html.

14

RISK ASSESSMENT: EMERGING ISSUES, RECENT ADVANCES, AND FUTURE CHALLENGES

ANNA M. FAN AND ROBERT A. HOWD
California Environmental Protection Agency, Oakland, California

Risk assessment for chemicals in the environment has been advanced extensively and the approach has continued to evolve since the definition of the field by the National Academy of Sciences (NAS, 1983). The process has been used by government agencies for setting priorities, policies, and regulations. The general principles and practices follow the protocols outlined in various U.S. Environmental Protection Agency (EPA) guidelines that have included public and peer reviews and are discussed throughout various chapters in this book. Certain situations require risk assessments tailored to meet more specific needs, such as addressing specific legislative mandates or sensitive subpopulations. In addition, specific guidance continues to develop for consideration of details of chemical properties (physicochemical, toxicologic, pharmacologic) and exposure media (air, water, food, and soil or dust). In this chapter we present an overview of some of the emerging issues, recent advances in biological and toxicological research, approaches and methodologies used in health risk assessment, and their implications for the future direction in risk assessment of chemicals. A focus and some examples are given in relation to chemicals in drinking water. Some aspects discussed here are also major topics of other chapters of this book (e.g., children's sensitivity, mixtures), to which the readers are directed for more detailed discussions. Here we highlight some of the major activities that present new issues with great implications for risk assessment.

Risk Assessment for Chemicals in Drinking Water, Edited by Robert A. Howd and Anna M. Fan

EMERGING ISSUES

Pharmaceuticals and Personal Care Products

Potential contamination of water resources by pharmaceuticals and personal care products (PPCPs) is emerging as an environmental and human health issue in the evaluation of water quality. The presence of pharmaceuticals in water is not new, but advancements in analytical technology have enabled the detection of trace amounts of these contaminants in water at extremely low concentrations not previously detected. Parallel to this is a developing concern regarding the potential implications of exposure to dozens of trace chemicals, representing untested combinations.

PPCPs are a diverse collection of chemical substances that are consumed or used by humans or animals for health or cosmetic reasons, including pharmaceuticals, fragrances, cosmetics, sunscreens, personal hygiene products, and nutritional supplements (U.S. EPA, 2000a, 2004a), and are disposed of or released incidentally into the environment. These are generally synthetic organic compounds, although there is currently no official list and no official terminology. Such contaminants originate from industrial, agricultural, medical, and common household uses (Kolpin et al., 2002). Substances resulting from human use include ingredients of cosmetics, detergents, and toiletries, and a variety of pharmaceuticals that include painkillers, anti-inflammatory agents, tranquilizers, antidepressants, antibiotics, birth control pills, estrogens for replacement therapy, blood pressure medications, chemotherapy agents, and antiseizure medications. Environmental contamination results from human excretion and flushing unused medication down the toilet, as well as from landfill leachate. Agricultural practices are a major source also, as antibiotics and hormones are fed to livestock as growth enhancers. Manure containing traces of pharmaceuticals is often spread on land as fertilizer, which could result in leaching into streams and rivers. In addition, animal wastes in holding ponds or heaps may run off into surface water during storms or leach into the groundwater.

In natural waters, PPCPs have been found in concentrations ranging from hundreds of micrograms per liter (μg/L) to nanograms per liter (ng/L) (U.S. EPA, 2002a). In a recent study by the Toxic Substances Hydrology Program of the U.S. Geological Survey (USGS) (Kolpin et al., 2002) water from 139 streams in 30 states was sampled during 1999 and 2000 and showed that a broad range of chemicals derived from residential, industrial, and agricultural uses commonly occurred at low levels. Concentrations were mostly less than 1 part per billion (ppb) downstream from areas of intense urbanization or animal production. The major chemicals found included human and veterinary drugs (including antibiotics), natural and synthetic hormones, detergent metabolites, plasticizers, insecticides, and fabric fire retardants. One or more chemicals were detected in 80% of the streams sampled, and 82 of the 95 chemicals surveyed were detected at least once. Mixtures of the chemicals were common; 75% of the streams had more than one, 50% had seven or more, and 34% had 10 or more. The most frequently detected chemicals (found in more than half of the streams) were coprostanol (fecal steroid), cholesterol (a major animal and minor plant steroid), *N*-*N*-diethyltoluamide (DEET, an insect repellent), caffeine (stimulant), triclosan (antimicrobial disinfectant), tri(2-chloroethyl)phosphate (fire retardant), and 4-nonylphenol (nonionic detergent metabolite). Steroids, nonprescription drugs, and insect repellent were the chemical groups detected most frequently. Detergent metabolites, steroids, and plasticizers generally were found at the highest concentrations.

The two issues that have received the most attention relating to PPCPs are endocrine disruption by naturally occurring and synthetic sex steroids and increased resistance of microorganisms to antibiotics (Daughton and Ternes, 1999; U.S. EPA, 2002a). There is presently no evidence of adverse human health effects due to traces of PPCPs in water, but studies in Britain (Sumpter, 1995; Kaiser, 1996) and the United States (Folmar et al., 1996) have linked ecosystem changes to certain PPCPs, such as feminization of fish that lived or were placed in cages downstream of sewage treatment plants. Sewage effluent is now known to contain ethinyl estradiol, the active ingredient in birth control pills, as well as natural female estrogens. In the British studies male fish were found to have testes laden with eggs (Kaiser, 1996) and male trout were found to have elevated levels of the protein vitellogenin in their blood (Sumpter, 1995). This protein is responsible for making egg yolks in female fish, and ordinarily, little, if any, vitellogenin is found in the blood of male fish (Sumpter and Jobling, 1995). Male fish have a gene which if triggered by estrogens can produce vitellogenin, but male fish ordinarily lack sufficient estrogen for this to occur. Sumpter and Jobling 1995 also tested a few common industrial chemicals to see if they could stimulate the production of vitellogenin in male trout under laboratory conditions. Chemicals found to induce vitellogenin in males included octylphenol and nonylphenol (alkyl phenols commonly used in detergents, toiletries, lubricants, and spermicides); bisphenol A (the building block of polycarbonate plastics); *o*, *p* '-DDT (one of the isomers of the common pesticide, banned in the United States in 1972 but still used against mosquitoes to control malaria in some industrializing parts of the world); and Aroclor 1221 [a commercial mixtures of polychlorinated biphenyls (PCBs) used as industrial chemicals, now banned in the United States but still found widely in the environment]. In the U.S. study (Folmar et al., 1996), carp sampled near the Minneapolis sewage treatment plant showed "a pronounced estrogenic effect," observed as the production of vitellogenin and reduced levels of testosterone (male sex hormone). Carp from the pesticide-contaminated Minnesota River had sharply reduced testosterone levels but did not show an effect on vitellogenin. Carp from the St. Croix River (Minnesota and Wisconsin), which is classified as a national wildlife and scenic river and is not heavily contaminated, were normal.

In response to the environmental concerns, the European Agency for the Evaluation of Medicinal Products (EMEA) developed guidelines to recommend long-term ecotoxicity testing for environmental risk assessment of pharmaceuticals, taking into account the possibility of effects of low concentrations of bioactive substances, such as endocrine disruptors (Hemminger, 2005). For a new drug the EMEA guidelines would initially calculate a rough aquatic predicted environmental concentration (PEC). The trigger concentration of pharmaceuticals that prompt risk assessment under this guidance is 0.01 μg/L, which is 10-fold lower than the U.S. Food and Drug Administration (FDA) trigger of 0.1 μg/L. The EMEA trigger of 0.01 μg/L is calculated from the maximum daily doses of the drug per patient and the assumption that the population is treated daily with the drug. This is divided by the amount of wastewater per person per day and a dilution factor of 10. The FDA trigger level of 0.1 μg/L corresponds to a PEC in surface water and assumes a dilution factor of 10, calculated from manufacturers' sales estimates. Details on this guidance for ecotoxicity assessment are not within the scope of this chapter, but the guidance development points out the need for concern posed by pharmaceuticals.

Liebig et al. (2006) noted that in the regulatory context, an environmental risk assessment has become essential for new PPCPs, and reliably predicted or measured environmental concentrations (PECs or MECs) of chemicals are essential for the exposure assessment. The authors reported measured data on four pharmaceuticals and one personal care product in surface waters compared with environmental concentrations predicted from the proposed EMEA exposure models and the European Technical Guidance Document on Risk Assessment for New Notified and Existing Chemical Substances (TGD). Calculations with exposure models provided by EMEA and the TGD resulted in PECs very close to the corresponding MECs in most cases, but under specific conditions the models applied in this study underestimated MECs. The report concluded that further development of the calculation models appears to be necessary.

The health implications of long-term exposure to low levels of PPCPs is not known, but since pharmaceuticals are designed to have a biological effect in humans, the potential exists for unexpected impacts at low dose levels, especially when exposure is to a mixture. The possibility of subtle effects over a lifetime, or over many generations, cannot be ruled out. Little is known about the occurrence, transport, fate, and synergistic or cumulative effects of many of these chemicals, and the chemicals may be mobile and/or persistent, or create other bioactive breakdown products.

PPCPs may be a particular concern in drinking water sources dominated or heavily supplemented by wastewaters. Added to the potential concern is an increasing practice of recharge of groundwater with treated sewage effluent. Some pharmaceuticals and other anthropogenic chemicals have been shown to persist in the waste stream and be mobile in the groundwater. The growing and aging population is likely to provide an increasing load of chemical products into municipal waste streams. Based on the low levels currently reported, these chemicals are not likely to be an acute or imminent health concern, yet they could be a concern for long-term, chronic, and combined exposures. This provides an obligation for attention to these products in drinking water risk assessment.

Currently, there is no national coordinated effort or requirement on the monitoring or focused treatment of waters and wastes for the presence of pharmaceuticals. There are presently no national primary drinking water regulations for PPCPs; EPA does not require routine monitoring at this time, but may add certain representative PPCPs to the drinking water contaminants candidate list (DWCCL) and unregulated contaminants monitoring rule (UCMR) in the future (U.S. EPA, 2002a). Of the 95 chemicals analyzed by the USGS (Kolpin et al., 2002), drinking water standards or other human or ecological health criteria have been established for 14, and measured concentrations rarely exceeded any of the standards or criteria. Thirty-three are known or suspected to be hormonally active, and 46 are pharmaceutically active. In California the Department of Health Services (DHS) recommends monitoring for a number of nonregulated chemicals for which the department has established notification levels (advisory levels) and recommends that proponents of projects which might result in significant releases have a program of monitoring for nonindustrial chemicals, including household products, personal care products, and pharmaceuticals. Some specific chemicals found in wastewater and groundwater are listed in the endnotes of the draft regulations for groundwater recharge reuse (California DHS, 2004) but the department does not currently envision regulation of these substances. The DHS notes that it is collecting information in order to gain a better understanding of the presence of these substances in treated wastewater and in recycled water.

Nanotechnology

The rapid proliferation of many different nanotechnology-based materials presents a challenge in the risk assessment of nanomaterials, with a focus at this time on the initial hazard identification process. Nanoscale science and nanotechnology are science and engineering that enable us to manipulate and characterize matter at the level of single atoms and small groups of atoms (NRC, 2002), permitting the study of structures and properties of materials and systems at a scale of 10^{-9}, 1/100,000 the width of a human hair. Nanomaterials are materials with at least one dimension of 100 nm or less (Oberdorster et al., 2005). Engineered nanomaterials are designed and produced from numerous bulk substances and encompass many forms. Nanoparticles are particles with at least one dimension smaller than 100 nm, including engineered nanoparticles, ambient ultrafine particles, and biological nanoparticles. Nanoparticles form the basis for many engineered nanomaterials and are currently being produced in a wide variety of types for a variety of applications. Among the most common ones are fullerenes (C60 or buckyballs), carbon nanotubes (CNTs), metal, and metal oxide particles. Synthetic polymer structures have the potential for new delivery systems, electronic circuits, catalysts, and light-harvesting materials, and many as yet unknown applications are expected in the future. For an additional perspective, a sheet of paper is about 100,000 nm thick, the diameter of DNA is in the 2.5-nm range, and red blood cells are approximately 2500 nm across (NRC, 2002).

Engineered nanomaterials typically possess unique nanostructure-dependent chemical, mechanical, electrical, optical, magnetic, or biological properties which make them desirable for commercial or medical applications. Potentially, the same properties may result in nanostructure-dependent biological activity that differs from and is not predicted directly by the bulk properties of the constituent compounds. Currently, few studies have addressed the health effects of nanomaterials, yet the existing research raises some concerns about the safety of nanomaterials. As part of the toxicological characterization of nanomaterials, it is important to differentiate between naturally occurring nanomaterials, nanoscale by-products of natural or chemical processes, and manufactured (engineered) nanomaterials, as they can have wide differences in properties.

That CNTs may have unusual toxic properties has been suggested in the available peer-reviewed literature: for example, in two studies on pulmonary toxicity of single-wall carbon nanotubes (SWCNTs) in mice and rats, employing an intratracheal instillation route of exposure (Dreher, 2004; Lam et al., 2004; Warheit et al., 2004; Donaldson et al., 2006). Using carbon black and quartz particles as low- and high-pulmonary-toxicity particle controls, carbon black and SWCNTs were shown to have different fates and reactions following pulmonary deposition in mice (Lam et al., 2004). Carbon black–laden macrophages were found to be scattered in the alveolar spaces, while SWCNT-laden macrophages moved to centrilobular locations, where they enter alveolar septa, resulting in epitheloid granulomas. Extrapolation from graphite-based occupational permissible limits may not be protective for exposure to SWCNTs, due to their unique physicochemical properties and pulmonary toxicity. Currently, SWCNTs are classified as a new form of graphite on material safety data sheets provided by manufacturers of these nanoparticles. Lam et al. (2004) concluded that SWCNTs could be more toxic than quartz, a particle recognized for its occupational health hazards after chronic inhalation exposures and commonly used as a benchmark for particle toxicity. In another study, pulmonary granuloma formation was also observed in rats following intratracheal instillation of SWCNTs without evidence of ongoing pulmonary inflammation or cellular proliferation (Warheit et al., 2004). This absence of pulmonary biomarkers of ongoing inflammation, cell proliferation, or cytotoxicity does not appear to follow the normal paradigm of toxic dust such as quartz and silica, suggesting a new mechanism of pulmonary toxicity and injury induced by SWCNTs.

Apart from the specific chemical surface reactivity of nanoparticles, a particle load or burden in the lung can induce toxicological response(s) that differ considerably from soluble or nonparticulate toxicants (Borm et al., 2006). Nanoscale particles having two or three dimensions in the range 1 to 100 nm may be considered to be a toxicological entity in their interactions with the environment, cells, and molecules within the cell. At these dimensions there is increased concern that nanoscale particles may translocate from the respiratory tract or other portals of exposure, with potential effects in other organ systems. Dose, deposition, dimension, durability, and defense are important determinants of response that should be noted for interpretation of inhaled particle effects. The dose deposited and its location depend on the inhaled concentration as well as the dimension of the particles. The dose metric for nanoscale particles provides an additional complexity, as particle number, surface area, and shape, among other factors may play a role in addition to the traditional mass-based metric. Solubility is a key factor for the clearance mechanisms involved in nanoparticle removal, as the kinetics of dissolution of inhaled particles determine whether a low-toxicity particle will dissolve in the epithelial lining fluid or whether such particles are engulfed by alveolar macrophages.

The National Nanotechnology Initiative (NNI) is a federal research and development program established to coordinate the multiagency efforts in nanoscale science, engineering, and technology (NAS, 2002). The goals of the NNI are to maintain a world-class research and development program aimed at realizing the full potential of nanotechnology; facilitate transfer of new technologies into products for economic growth, jobs, and other public benefit; develop educational resources, a skilled workforce, and the supporting infrastructure and tools to advance nanotechnology; and support responsible development of nanotechnology. Twenty-three federal agencies participate in the initiative, which is managed within the framework of the National Science and Technical Council (NSTC). The Nanoscale Science Engineering and Technology (NSET) Subcommittee of the NSTC coordinates planning, budgeting and program implementation and review to ensure a balanced and comprehensive initiative and is composed of representatives from each participating agency, the Office of Science and Technology Policy, and the Office of Management and Budget. A number of working groups have been formed under the NSET subcommittee to improve the efficiency of its operations and focus interagency attention and activity. Current working groups are focused on environmental and health implications of nanotechnology, liaison activities with industries, nanomanufacturing, and public engagement. Although a review by the NAS of the NNI has provided 10 recommendations to strengthen implementation of the initiative, they relate primarily to organizational structure and none was related specifically to toxicology and risk assessment.

To address the issues in toxicology of nanomaterials, an effort has been made via a toxicity-screening workgroup to identify the key elements of a toxicity screening strategy for engineered nanomaterials (Oberdorster et al., 2005). The strategy has three key elements: elucidation of relevant physicochemical characteristics, development or description of appropriate cellular and noncellular in vitro assays, and development or description of in vivo assays. Physicochemical characteristic considerations include particle size and size distribution, agglomeration state, shape, crystal structure, chemical composition, surface area, surface chemistry, surface charge, and porosity. In vitro tests suggested include portal-of-entry toxicity for lungs, skin, and mucosal membranes, and target organ toxicity for endothelium, blood, spleen, liver, nervous system, heart, and kidney. Noncellular assessment of nanoparticle durability, protein interactions, complement activation, and pro-oxidant activity are also considered. In vivo assays are proposed for two tiers. Tier 1 assays are proposed for pulmonary, oral, skin, and injection exposures, which include markers of inflammation, oxidant stress, and cellular proliferation in portal-of-entry and selected remote organs. Tier 2 assays are proposed for pulmonary exposures, which include deposition, translocation, and toxicokinetic and biodispersion studies; effects of multiple exposures; effects on reproductive systems, placenta and fetus; alternative models; and mechanistic studies.

Nanoparticles appear to present unique situations with major implications for toxicological evaluation, interpretation and future risk assessment, much of which is yet to be determined. Critical risk assessment issues for manufactured nanoparticles recently identified by the National Science Foundation (NSF) and the EPA (NSF and U.S. EPA, 2003) include exposure assessment; toxicology; ability to extrapolate using existing particle and fiber toxicological databases; environmental and biological fate, transport, persistence, and transformation; and recyclability and overall sustainability. Additional considerations have been presented regarding evaluation of consumer exposure to nanomaterials in a limited subset of products, which include evaluating the potential hazards associated with these products, the capacity of U.S. regulatory agencies to address their potential risks (Thomas et al., 2006a), and international efforts to develop risk-based safety evaluations for nanomaterials (Thomas et al., 2006b).

ADVANCES IN SCIENCE, APPROACHES, AND METHODS

Endocrine Effects

Risk assessment of endocrine effects of chemicals has gained increased interest since the mid-1990s. Efforts have been made to develop data and approaches for risk assessment of endocrine disruptors (Kavlock et al., 1996; U.S. EPA, 2006a). A health concern was first engendered by observations of the disruption of the endocrine systems of animals in laboratory studies. Subsequently compelling evidence was developed showing that endocrine systems of certain fish and wildlife have been affected by chemical contaminants in the environment, resulting in developmental and reproductive problems in the ecosystem (Colburn et al., 1993; NRC, 1999; U.S. EPA, 2006a). The 1996 Food Quality Protection Act, which amended the Federal Food, Drug, and Cosmetic Act, directed the EPA to develop a screening program, using appropriate validated test systems and other scientifically relevant information, to determine whether certain substances may have hormonal effects in humans. The 1996 amendments to the Safe Drinking Water Act authorize the EPA to screen substances that may be found in drinking water sources for endocrine disruption potential. The EPA has since initiated the Endocrine Disruptor Screening Program (EDSP) after numerous national and international public workshops on endocrine disruptors and meetings of the Endocrine Disruptors Screening and Testing Advisory Committee (EDSTAC). An overview of the EDSP with updates containing information on endocrine disruptors, important milestones of the program, and key program documents can be found on the agency Web site (U.S. EPA, 2006a).

One of the activities prior to the EDSP was an independent study by the National Research Council (NRC, 1999) which convened a multidisciplinary expert committee, the Committee on Hormonally Active Agents (HAAs) in the Environment, in response to a request from the EPA, the Department of the Interior, the Centers for Disease Control and Prevention, and the Congress. The committee was charged with the following tasks:

1. Review critically the literature on HAAs in the environment; identify the known and suspected toxicologic mechanisms and impacts on fish, wildlife, and humans; identify significant uncertainties, limitations of knowledge, and weaknesses in the available evidence; develop a science-based conceptual framework for assessing observed phenomena; and recommend research, monitoring, and testing priorities.
2. To the extent practicable with available information and study resources, identify particular chemical substances, geographic areas, contaminant sources, human subpopulations, and fish and wildlife populations of special concern with respect to HAAs.
3. If possible and warranted, suggest general approaches for identifying and mitigating toxicologic problems.

At the conclusion of the study, the committee developed recommendations for screening and monitoring of HAAs that are consistent, in principle, with those of the EPAs EDSTAC. Those recommendations include:

1. Development of a battery of short-term assays (for rapid and inexpensive screening of putative HAAs) that need to be validated, replicated, and deployed appropriately.
2. Development and validation of additional biomarkers to screen for embryonic and fetal events that predict long-term, delayed effects.
3. Monitoring of wildlife as environmental sentinels.
4. Further investigation of species- and tissue-specific effects resulting from exposure to HAAs.
5. Better characterization of dose–response relationships of various HAAs through in vitro and in vivo assays at concentrations encountered in the environment.

The Endocrine Disruptor Screening Program (EDSP) focuses on the estrogen, androgen, and thyroid hormones (U.S. EPA, 2006a). Estrogens are hormones responsible for female sexual development that are produced primarily by the ovaries and in small amounts by the adrenal glands. Androgens are hormones responsible for male sex characteristics, such as testosterone, produced by the testicles. Thyroid hormone is produced by the thyroid gland, and includes two main forms, thyroxine and triiodothyronine, also known as T_4 and T_3. These hormones stimulate all the cells in the body and modulate biological processes such as growth, reproduction, development, and metabolism.

The endocrine system as a whole is made up of many glands located throughout the body which release hormones into the bloodstream or the fluid surrounding cells, with corresponding receptors in various organs and tissues that recognize and respond to the hormones. The endocrine system influences or regulates all biological processes in the body from conception through adulthood and into old age, including development of the brain and nervous system, growth and function of the reproductive system, metabolism and energy consumption rate, including regulation of blood sugar levels. The female ovaries, male testes, and pituitary, thyroid, pancreas, and adrenal glands are major constituents of the endocrine system.

Disruption of the endocrine system can occur in various ways (U.S. EPA, 2006a), such as mimicking a natural hormone, which may induce an overresponse (e.g., increased muscle mass from growth hormone effects), responding at inappropriate times (e.g., producing insulin when it is not needed or causing early puberty from exposure to estrogens), or disruption of normal development (e.g., delaying neuronal maturation or delaying puberty). Some endocrine-disrupting chemicals block the effects of a hormone at its normal receptors. Others directly stimulate or inhibit the endocrine system and cause overproduction or underproduction of hormones (e.g., an over- or under-active thyroid). Certain drugs, such as birth control pills, are intended to produce such effects. However, unanticipated and undesirable endocrine disruption may occur as a side effect of beneficial effects of drugs. A classic case is that of the drug diethylstilbestrol (DES, a synthetic estrogen), which was given to mothers during pregnancy to block spontaneous abortion and promote fetal growth. Much later it was found to have affected the development of the reproductive system in the female offspring, and decades later resulted in an increased rate of vaginal or cervical cancer in the daughters.

The EPA has noted that the relationship of human diseases of the endocrine system to exposure to environmental contaminants is poorly understood and scientifically controversial (Kavlock et al., 1996; U.S. EPA, 1997a, 2006a). The science related to measuring and demonstrating endocrine disruption is relatively new, and appropriate testing methods are still being developed and validated. The U.S. EPA (1997a, 2006b) also stated that it "does not consider endocrine disruption to be an adverse effect per se, but rather to be a mode of action or mechanism of action potentially leading to other outcomes, for example carcinogenic, reproductive, or developmental effects, routinely considered in reaching regulatory decisions. Evidence of endocrine disruption alone can influence priority setting for further testing and the assessment of the results of the testing could lead to regulatory action if adverse effects are shown to occur." It stated further that "the current Agency position is consistent with a broad definition of endocrine disruption that must of necessity entail research questions, but also recognizes that regulatory decision-making is generally based on adverse effects using legislatively mandated risk-based criteria" (U.S. EPA, 2006b).

Issues surrounding the risk assessment and management of endocrine disruptors include difficulties in understanding ecological effects, deciding whether (or what type of) wildlife data can be used as a model for humans, the need for improvement in testing guidelines and screening for health effects, lack of a link of exposure data to effects for quantitative risk assessment, whether endocrine disruption is, in itself, an endpoint for adverse effect, and whether assessment of endocrine disruption should use a threshold approach. An additional point of contention is whether the precautionary principle should be applied to endocrine disruptors, considering that existing regulatory programs (i.e., EPA, FDA) are risk based and do not leave room for the precautionary principle.

Several of these issues were discussed in a workshop prior to establishment of the EDSP in an effort to identify research gaps related to adverse health effects following exposure to endocrine-disrupting chemicals and to establish priorities for future research activities (Kavlock et al., 1996); these issues were considered in developing the EDSP. In the workshop, an *environmental endocrine disruptor* was broadly defined as "an exogenous agent that interferes with the production, release, transport, metabolism, binding, action or elimination of natural hormones in the body responsible for the maintenance of homeostasis and the regulation of developmental processes." This definition reflects awareness that the issue of endocrine disruptors in the environment extends considerably beyond that of exogenous estrogens and includes antiandrogens and agents that act on other components of the endocrine system, such as the thyroid and pituitary glands. The majority of the workshop participants agreed that the endocrine disruptor hypothesis was of sufficient concern to warrant a concerted research effort, particularly study of potential effects on the development of reproductive capability at multiple phylogenetic levels. Suggestions for improving the testing guidelines were discussed. Further studies on the presumed receptor-based mechanisms, which are responsible for at least some adverse endocrine-modulated effects, were thought to present a unique opportunity to establish a common biologically relevant risk assessment process for all effects (i.e., developmental, immunologic, neurological, and carcinogenic effects). On the other hand, the group was not aware of a biological basis for selecting different models to quantitatively estimate cancer or noncancer effects for chemicals that act by endocrine-mediated mechanisms. Emphasis was placed on evaluating the developing embryo, fetus, and neonate, because the development processes are especially vulnerable to brief periods of endocrine disruption. It was noted that for many of the effects reported in both wildlife and humans that have been associated with endocrine disruption, exposure assessment has generally been inadequate for quantitative risk assessment. In this regard, linking specific exposures to specific effects in the general environment would often be difficult because of the complexities of exposure, the latency of the effects, and at times, the subtle nature of the outcomes. Therefore, the workgroup concluded that confirmation of the validity of the hypothesis will have to rely heavily on application of the Hill criteria (Hill, 1965; Fox, 1991) for causality (strength of the association, presence of a dose–response relationship, specificity of the association, consistency across studies, biological plausibility, and coherence of the evidence).

One drinking water contaminant that is receiving intense attention because of its widespread presence in water sources nationwide, that produces effects on the endocrine system, is perchlorate. Development of scientific consensus on a health-protective level of this contaminant in drinking water has been difficult, largely due to disagreements over what constitutes an adverse effect on a hormonal system, and what degree of protectiveness for potentially susceptible populations (pregnant women and their fetuses) is adequate. California has established a public health goal (PHG) of 6 ppb in drinking water (Ting et al., 2006) and is in the process of promulgating the drinking water standard or maximum contaminant level. The PHG is based on inhibition of iodine intake by the thyroid. A committee of the National Academy of Sciences (NAS, 2005) used the same data on iodine uptake inhibition as a basis for its risk assessment.

Other chemicals in the California PHG program being evaluated for their effects on the endocrine system include methoxychlor, atrazine, simazine, and PCBs (separate from the general approaches generally followed in EPA's risk assessment guidelines for reproductive and developmental toxicity).

Mode of Action/Mechanism of Action

Mode of Action Human Relevance Framework Increased attention is being given by the scientific and regulatory communities to mode-of-action (MOA) information from animal and human studies to evaluate the applicability of animal tumors to humans in risk assessment. Pertinent information in humans is less available, and the MOA of chemicals that produce neoplasia in animal bioassays is used to evaluate the relevance of these tumors to humans. Progress that has been made in understanding MOAs with a distinction between DNA reactivity and non-DNA reactivity, commonly referred to as genotoxic versus nongenotoxic MOAs, plays an important role in this regard. A mode-of-action human relevance Framework (HRF, or framework) developed by the International Programme on Chemical Safety (IPCS) (Sonich-Mullen et al., 2001) and the EPA (1999) provides guidance for animal tumor MOA analysis. The framework has since been expanded into a four-part analysis to include evaluation of human relevance of the MOA determined in animals and case studies methodology applying the framework to the MOAs postulated for animal carcinogens (Cohen et al., 2004; Meek et al., 2003).

The framework is intended as part of the hazard identification/characterization step of the risk assessment process and is not a complete risk assessment itself. It is based on three fundamental questions, plus a conclusion, which include the following (Meek et al., 2003):

- Is the weight of evidence sufficient to establish the MOA in animals?
- Are key events in the animal MOA plausible in humans?
- Taking into account kinetic and dynamic factors, is the animal MOA plausible in humans?
- *Conclusion:* statement of confidence, analysis, implications.

The framework itself is based on three observations: limited utility of animal MOA information, necessity for human information, and comparative analysis (i.e., MOA concordance, or looking at key events identified for the animal MOA and determining if there are or may be comparable events in humans). Each human relevance conclusion is specific as to the MOA and tissue or cell type or endpoint analyzed.

The case studies used in the framework include two groups (Meek et al., 2003):

1. Data findings probably irrelevant because the MOA is unlikely to have a human counterpart due to actions at a protein site specific to test animals (*d*-limonene), effects resulting from hormone suppression typical of laboratory animals but not humans (atrazine), or chemical-related enhanced hormone clearance rates in animals relative to humans (phenobarbital).
2. Relevant tumors but with insufficient data (acrylonitrile), the MOA involving cytotoxicity and cell proliferation in animals and humans (chloroform), and formation of urinary tract calculi (melamine).

Two of the case studies were on drinking water contaminants evaluated for PHG development in California, atrazine and chloroform.

Atrazine This herbicide produced mammary tumors in treated female Sprague–Dawley (SD) rats but not in F344 rats (Meek et al., 2003). Atrazine affects the hypothalamus, leading in female SD rats to the inhibition of the luteinizing hormone (LH) surge during the estrous cycle, which results in persistent secretion of estrogen and prolactin, leading to mammary tumors. Thus, the mammary tumor response in atrazine-treated female SD rats is mediated by a nongenotoxic, threshold-based mechanism leading to LH suppression, failed ovulation, and estrous-cycle disruption. The estrous-cycle disruption leads to an endocrine environment (i.e., prolonged exposure to endogenous estrogen associated with extended estrus) that favors mammary tumor formation. These hormonal changes do not occur in F344 female rats or CD-1 mice that are resistant to mammary tumor induction by atrazine. The framework analysis identified a lessened need to continue a full risk assessment for these tumors after identifying sufficient weight of evidence to establish the MOA in animals, and that the key events in the animal MOA are not plausible in humans (Meek et al., 2003).

Chloroform As discussed by Meek et al. (2003), this chemical induces liver tumors in mice and renal tumors in mice and rats. The MOA involved the induction of sustained cytotoxicity by metabolites (oxidation to the reactive intermediates phosgene and hydrochloric acid) and subsequent persistent regenerative cell proliferation. The weight of evidence is greatest for liver and kidney tumors in mice, and is limited for kidney tumors in rats. The MOA is considered qualitatively applicable to humans. The case analysis presents considerable evidence for an obligatory role for cytotoxicity in chloroform carcinogenicity, consistent with a nonlinear dose–response relationship for tumor induction in animals, and by inference, in humans. The mode of action (sequence of metabolism and cytotoxicity) is qualitatively and quantitatively possible in humans. Thus, a full assessment is necessary (Meek et al., 2003). The above represents the discussion by

Meek et al. OEHHA's PHG is still under development, but the evidence regarding chloroform carcinogenicity is discussed more extensively in Chapter 10.

The HRF was developed using MOA data for non-DNA reactive carcinogens, but it is applicable for endpoints other than cancer. As more information becomes available on the MOA for risk assessment purposes, further applicability of the HRF has been described for reproductive, developmental, neurologic, renal and cancer endpoints (Seed et al., 2005). The MOA/human relevance analysis is believed to enable identification of data needs and issues relating to dose–response and exposure assessment steps in the overall risk assessment. The case studies for noncancer endpoints included several chemicals in our program for PHG development: molinate, diethylhexyl phthalate (DEHP), and polychlorinated biphenyl mixtures (PCBs).

Molinate This rice herbicide is associated with decreased sperm production and reduced fertility observed in rats (Seed et al., 2005). The MOA appeared to involve several steps. First, molinate is metabolized to molinate sulfoxide, which inhibits high-density lipoprotein (HDL) deesterification to release cholesterol. This leads to decreased androgen availability in the testis, which results in a decreased sperm count and reduced fertility. However, analysis of the temporal and dose–response relationships revealed that while the sulfoxidation of molinate in animals occurs at dose levels associated with reduced fertility, inhibition of the esterase has not been shown at dose levels that reduce fertility. Lacking such evidence, the data are believed to be incomplete to establish the MOA for this effect for molinate.

Diethylhexyl Phthalate For DEHP the key events include effects on male reproduction in rats which involve hydrolysis of the chemical to the monoester, followed by a decrease in production of fetal testicular androgen, which leads to malformations of the prostate and other organs (Seed et al., 2005). The authors did not find human data relating exposure to phthalate esters to adverse reproductive outcomes or, therefore, on related key events in an MOA. On the other hand, data on similarities in rats and humans, metabolism of phthalates in general, plus similarities in the underlying developmental biology of androgen-dependent male reproductive organs all suggest that the animal findings may have human relevance. Therefore, it was concluded that animal–human comparability indicates the potential for human relevance of the male reproductive system effects for DEHP.

PCBs The case study for PCBs is based on the loss of low-frequency hearing following developmental exposures in rats (Seed et al., 2005). Key events considered in the animal MOA for this effect include upregulation of hepatic uridine diphosphoglucuronyltransferases, which leads to hypothyroxinemia. Subsequent reductions in tissue triiodothyronine (T_3) then inhibit development of hair cells in the upper turns of the cochlea. A comparative analysis showed species difference in the relative timing of cochlear development and in PCB kinetics which contributed to increased risk of adverse outcomes in rodents compared to humans, thus requiring a higher dose in the developing human than in the developing rat for the effect.

Role of Mode of Action/Mechanism in Risk Assessment Guidelines and Carcinogen Evaluation/Classification The significance of information on the MOA and on the mechanism can be seen in its role described in various aspects in this chapter. Two examples are the focus on the MOA in the 2005 EPA guidelines for carcinogen risk assessment (U.S. EPA, 2005a), and the focus on mechanism in the International Agency for Research on Cancer's (IARC) 2006 update of its carcinogenic classification system (IARC, 2006).

U.S. EPA 2005 Cancer Guidelines The use of mode of action in the assessment of potential carcinogens is a main focus of the 2005 EPA cancer guidelines (U.S. EPA, 2005a). The guidelines provide a framework for critical analysis of mode-of-action information to address the extent to which the available information supports the hypothesized mode of action, whether alternative modes of action are also plausible, and whether there is confidence that the same inferences can be extended to populations and life stages that are not represented among the experimental data.

In the guidelines (U.S. EPA, 2005a), the term *mode of action* is defined as a sequence of key events and processes, starting with interaction of an agent with a cell, proceeding through operational and anatomical changes, and resulting in cancer formation. A *key event* is an empirically observable precursor step that is itself a necessary element of the mode of action or is a biologically based marker for such an element. Mode of action is contrasted with *mechanism of action*, which implies a more detailed understanding and description of events, often at the molecular level, than is meant by *mode of action*. Examples of possible modes of carcinogenic action include mutagenicity, mitogenesis, inhibition of cell death, cytotoxicity with reparative cell proliferation, and immune suppression. Mechanism of action would include MOA, but not vice versa.

The risk assessment process emphasizes that significant information should be developed to ensure that a scientifically justifiable mode of action underlies the process leading to cancer at a given site (U.S. EPA, 2005a). On the other hand, in the absence of sufficient, scientifically justifiable mode-of-action information, the EPA generally takes public health-protective, default positions regarding the interpretation of toxicologic and epidemiologic data (i.e., animal tumor findings are judged to be relevant to humans, and cancer risks are assumed to conform with low dose linearity). Understanding mode of action is anticipated to help to identify processes that may cause chemical exposures to affect a particular population segment or life stage differentially. Some modes of action are anticipated to be mutagenic and are assessed with a linear approach. Other modes of action may be modeled with either linear or nonlinear approaches after a rigorous analysis of available data according to the guidance provided in the framework for mode-of-action analysis. The term *nonlinear* is used in the guidelines in a narrower sense than its usual meaning in the field of mathematical modeling. It refers to threshold models (which show no response over a range of low doses that include zero) and some nonthreshold models (e.g., a quadratic model, which shows some response at all doses above zero).

Criteria that are generally applicable for judging the adequacy of mechanistically based data include mechanistic relevance of the data to carcinogenicity, number of studies of each endpoint, consistency of results in different test systems and different species, similar dose–response relationships for tumor and mode of action-related effects, conduct of the tests in accordance with generally accepted protocols, and degree of consensus and general acceptance among scientists regarding interpretation of the significance and specificity of the tests (U.S. EPA, 2005a). Although important information can be gained from in vitro test systems, a higher level of confidence is generally given to data that are derived from in vivo systems, particularly results that show a site concordance with the tumor data.

The HRF provides guidance on how to assess the relevance of an animal carcinogen to humans, based on MOA data. The view is that once an MOA has been established and accepted for the tumorigenic effects of a chemical in experimental animals, a systematic analysis of the individual key events comprising that MOA, both qualitatively and quantitatively, should enable the relevance of this MOA to humans to be determined. Hence, the expanded framework is intended to provide a defined procedure whereby assessment is rigorously structured, with a clear and consistent documentation of the facts and reasoning that includes assessment of inconsistencies and uncertainties in the available data (Meek et al., 2003).

Chemical-specific data often, but not always, form the basis of the concordance analysis in applying the framework, as considerable emphasis is placed on the underlying biology of the processes involved. An example is the key event in the renal tumors induced by *d*-limonene in the male rat, which is irreversible binding to α 2u-globulin. Knowledge that humans do not synthesize this protein is sufficient to exclude this key event from the effects of the chemical in humans, without necessarily demonstrating that indeed *d*-limonene and its metabolites do not bind to a protein in human plasma. Therefore, even if the relevance of an MOA to humans cannot be dismissed, the structured application of the framework is believed to provide a robust basis for risk characterization, by establishing the biological basis of a threshold dose for carcinogenicity and by identifying key events. The framework is anticipated to require a detailed evaluation of the human relevance only once for a given MOA, and future assessments of chemicals that share the same MOA and do not demonstrate additional key events contributing toward other possible MOAs would require progressively fewer data. It is recognized that the continual clarification of modes of action by filling data gaps is important and that many chemicals will not be able to be identified clearly as demonstrating "hazard not relevant to humans."

International Agency for Research on Cancer Carcinogen Evaluation/Classification

In the evaluation of carcinogenic risk to humans by the International Agency for Research on Cancer (IARC), mechanistic and other evidence judged to be relevant to an evaluation of carcinogenicity and of sufficient importance to affect the overall evaluation is shown to play an important role in the final carcinogen classification as described in the recently updated preamble to the *IARC Monographs on the Evaluation of Carcinogenic Risks to Humans* (IARC, 2006). The IARC monographs are scientific evaluations of carcinogenic risks to humans of chemicals, complex mixtures, occupational exposures, lifestyle factors, physical and biological agents, and other potentially carcinogenic exposures. Each monograph includes a critical review of the pertinent peer-reviewed scientific literature as the basis for an evaluation of the weight of the evidence that an agent may be carcinogenic to humans. The preamble discusses the principles and procedures used in developing the monographs, including the scientific criteria that guide the working group's evaluations.

The IARC critical review of the pertinent scientific data for evaluating mechanistic data is intended to improve the analysis of studies in both humans and experimental animals, provide insight into the biology of cancer, and help identify stages where intervention may be possible (IARC, 2006). The nature of the mechanistic data depends on the biological activity of the agent being considered. In the evaluation process the possible mechanisms by which the agent may increase the risk of cancer are identified and a representative selection of key data from humans and experimental systems is summarized for each mechanism, noting data gaps and any data suggesting more than one operating mechanism. The relevance of the mechanism to humans is addressed, particularly when mechanistic data are derived from experimental model systems. Changes in the affected organs, tissues, or cells can be divided into three nonexclusive levels as changes in physiology, functional changes at the cellular level, and changes at the molecular level.

In the IARC evaluation process, the strength of the evidence that any carcinogenic effect observed is due to a particular mechanism is evaluated and designated as "weak," "moderate" or "strong" (IARC, 2006). Whether the particular mechanism is likely to be operative in humans is then assessed. In the classification process, mechanistic data are used to evaluate evidence of carcinogenicity, help in assessing the relevance and importance of findings of cancer in animals and in humans, and to further determine the final classification of a chemical under evaluation. The initial two partial evaluations are combined into a preliminary default evaluation and designated in one of the categories that classify whether the agent is "carcinogenic to humans" (group 1), "probably carcinogenic to humans" (group 2A), "possibly carcinogenic to humans" (group 2B), "not classifiable as to its carcinogenicity to humans" (group 3), or "probably not carcinogenic to humans" (group 4). Then the mechanistic and other relevant data are considered to determine whether there is a need for modification of the default evaluation. This determination considers the strength of the mechanistic evidence and whether the mechanism operates in humans, with a greater weight on data derived from humans or human (exposed) biological specimens, especially if they show that the agent under evaluation has caused changes in exposed humans that are on the causal pathway to carcinogenesis. The conclusion that a mechanism operates in experimental animals is strengthened by findings of consistent results in different experimental systems, by the demonstration of biological plausibility, and by coherence of the overall database. Strong support can be obtained from studies that challenge the hypothesized mechanism experimentally, by demonstrating that the suppression of key mechanistic processes leads to the suppression of tumor development. The evaluation also includes whether multiple mechanisms might contribute to tumor development, whether different mechanisms might operate in different dose ranges, whether separate mechanisms might operate in humans and experimental animals, and whether a unique mechanism might operate in a susceptible group. The possible contribution of alternative mechanisms must be considered before concluding that tumors observed in experimental animals are not relevant to humans. In the final evaluation, the body of evidence is considered as a whole in order to reach an overall evaluation of the carcinogenicity of the agent to humans. The agent is described according to the wording of one of the four categories, and the designated group is given. The final categorization of an agent is a matter of scientific judgment that reflects the strength of the evidence derived from studies in humans and in experimental animals and from mechanistic and other relevant data.

In considering all relevant scientific data, the working group may assign the agent to a higher or lower group than the default would indicate. A recent example of a drinking water contaminant included in our PHG development for which mechanistic data have played a role in determining (elevating) its carcinogen classification by the IARC is benzo[*a*]pyrene (BaP). Under IARC evaluation BaP is carcinogenic to humans (group 1). Without mechanistic evidence, it would have been classified as possibly carcinogenic (group 2B) (Straif et al., 2006). BaP is an indicator compound found in all polycyclic aromatic hydrocarbon (PAH) mixtures. The evidence for cancer in humans is inadequate as BaP occurs in mixtures and one cannot attribute the observed increased risk from exposure to those mixtures to BaP alone. The data on cancer in experimental animals provide sufficient evidence and BaP is used as a positive control. The mechanisms involve metabolic activation to a diol-epoxide for lung and skin tumors and to a radical cation for skin tumors. In the case of the diol-epoxide mechanism, the polycyclic aromatic hydrocarbons form oxides and dihydrodiols and then diol epoxides; these form stable or depurinating adducts with guanines and adenines, which can induce mutations (e.g., in ras proto-oncogenes) strongly associated with tumorigenesis. For the radical-cation mechanism, one-electron oxidation creates radical cations; these cations result in depurinating DNA adducts with guanines and adenines, which generate apurinic sites that can induce mutations in ras proto-oncogenes. Overall, BaP has demonstrated the complete sequence of steps in the metabolic activation to mutagenic diol epoxides in animals, human tissues, and humans. As evaluated by the IARC 2005:

> Following exposure, humans metabolically activate benzo[*a*]pyrene to benzo[*a*]pyrene diol epoxides that form DNA adducts: the *anti*-benzo[*a*]- pyrene-7,8-diol-9, 10-oxide–deoxyguanosine adduct has been measured in populations (e.g., coke-oven workers, chimney sweeps) exposed to PAH mixtures that contain benzo[*a*]pyrene.

> The reactive *anti*-benzo[*a*]pyrene-7,8-diol-9,10-oxide induces mutations in rodent and human cells. Mutations (G→T transversions) in the K-*Ras* proto-oncogene in lung tumours from benzo[*a*]pyrene-treated mice are associated with *anti*-benzo[*a*] pyrene-7,8-diol-9,10-oxide–deoxyguanosine adducts. Similar mutations in the K-*RAS* proto-oncogene and mutations in *TP53* were found in lung tumours from nonsmokers exposed to PAH-rich coal combustion products that are known to contain benzo[*a*]pyrene (as well as many other PAHs). In an in-vitro study, the codons in the tumour-suppressor gene *TP53* that are most frequently mutated in human lung cancer were shown to be targets for DNA adduct formation and mutations induced by benzo[*a*]pyrene.

The IARC 2006 did not define mechanism of action versus mode of action. Although the IARC evaluation and classification process is not a risk assessment, the classification often lends support to the hazard identification step in the risk assessment process. The evaluation of whether a particular mechanism is operative in humans not only affects the evaluation of the agent being considered but can also set a precedent for other agents that operate through similar mechanisms. Sufficient data available to identify a mechanism of carcinogenesis are also believed to be the key to identifying susceptible populations and life stages, including the prenatal and early postnatal periods. Another implication of using mechanistic data will be carcinogen identifications that are based on scientific inference in the absence of tumor studies in humans or experimental animals.

Sensitive Populations

The attention given to sensitive or age-related susceptible populations has been increasing for the past decade. The EPA has earlier developed guidance for developmental toxicity risk assessment (hazards to children that may result from exposures during preconception and prenatal or postnatal development to sexual maturity) (U.S. EPA, 1991), reproductive toxicant risk assessment (U.S. EPA, 1996), and neurotoxicity risk assessment (U.S. EPA, 1998). More recently it published the Supplemental Guidance for Assessing Susceptibility from Early-Life Exposure to Carcinogens (U.S. EPA, 2005c). The overall characterization of risk is conducted within the context of broader policies and guidance such as Executive Order 13045 (1997), "Protection of Children from Environmental Health Risks and Safety Risks," which is the primary directive to federal agencies and departments to identify and assess environmental health risks and safety risks that may disproportionately affect children (U.S. EPA, 2005b).

In California, consideration of sensitive populations is mandated in our risk assessment programs in drinking water, air toxics, school sites, and Proposition 65 (Fan et al., 2005; Fan and Howd, 2006). Much of the work in evaluating age differences in susceptibility to environmental chemical toxicity in risk assessment programs has been done on infants, children, and the pregnant women. More details on issues related to infants and children as sensitive subpopulations are provided in Chapter 7. California is the most populous state during the 1993–2020 period, projected to add over 16 million persons (Campbell,

2006) during this period. One effect of the increasing population of various ages is that the number of highly sensitive people and the consequent demand for very safe drinking water will increase. Certain members of the population may be especially susceptible to chemical contaminants in infancy, while undergoing childhood growth and development, during pregnancy, when physiologically impaired during illness, and when they become older adults. Consequently, at some point in life, many people are likely to be in a susceptibility group.

Another aspect relating to sensitive populations is evaluating the aging population, or older adults, as a sensitive subgroup (Geller and Zelnick, 2005; California OEHHA, 2006). Between 1993 and 2020, the proportion of youth is expected to decline as the elderly population increases in all the states (Campbell, 2006). Most of the projected growth of the older adult population (65 years old and over) is concentrated in the West and South. Eight states (Nevada, Arizona, Colorado, Washington, Georgia, Utah, Alaska, and California) are projected to see a doubling in their number of older adults. Florida would continue to have the highest proportion (19% in 1993, 26% in 2020; 2 million added). During the period from 2010 to 2020, the aging of the Baby Boom population (persons born between 1946 and 1964) should contribute to rapid increases of the elderly in all states. From a public health perspective, the implication is that people are living longer, and the U.S. population is not only growing, but also aging. Among older adults living longer are those with chronic diseases such as diabetes, asthma, or chronic heart conditions. An increasing population of frail elderly will provide a challenge to all aspects of public health protection. In the case of drinking water contaminants, elevated water arsenic has been linked to atherosclerosis and vascular diseases, lead has been associated with age-related decline in kidney function and bone loss in women after menopause, and lead, aluminum, iron, copper, manganese, and organochlorine compounds have been implicated in common neurodegenerative diseases in old age such as Alzheimer's disease, Parkinson's disease and amyotrophic lateral sclerosis (Adler, 2003; U.S. EPA, 2005d).

At the national level, in recognition that the United States is a rapidly aging nation, the EPA noted that in 2006, the first of the country's 76 million baby boomers began to turn 60, and by 2030, the number of older persons is expected to double to more than 70 million. The agency has initiated efforts to protect the environmental health of older persons under the aging initiative (U.S. EPA, 2006c). Included in the aging initiative is the National Agenda for the Environment and the Aging, which consists of three main components: to identify research gaps in environmental hazards on older persons; to prepare for an aging society in a smart growth context, and to encourage older persons to become involved in communities to reduce environmental hazards and protect the environment. One goal of this initiative is to provide risk assessors with the tools needed to consider special sensitivities in order to conduct a life stage–specific risk assessment. A report developed under the aging initiative on issues relevant to risk assessment (U.S. EPA, 2005d) has identified research needs, including better understanding of the pathophysiological mechanisms of aging; increased monitoring of the development and prevalence of chronic diseases, particularly through the use of longitudinal studies; more chemical-specific toxicity data in human epidemiological studies in the elderly population and in aged experimental animals; more dialogue between epidemiologists and mechanistic researchers of elderly people; focused research on the interactions between drugs and environmental contaminants; continued study of gene–environment interactions; more research on biomarkers as a reliable measure of biological age; and improved understanding of latent effects from early life stage exposures.

Presently there is no specific guidance for assessing health risk to the elderly or older adults. Older adults as a subpopulation will require careful consideration regarding exposure, dose, and health effects through targeted research to better characterize potential risks of exposures to environmental chemicals and their interactions and relationships with genetic traits, nutrition, changing physiological functions, pharmacologic parameters, and disease factors. Efforts are on the way to gain a better understanding of these issues in order to evaluate differential exposures to and responses of the elderly population to environmental agents, and risk assessors should keep abreast of the developments in this area.

In the process of risk assessment the variations in biological factors among different age groups need to be considered, and the commonly used exposure factors need to be reexamined (e.g., water consumption rate per kilogram of body weight vs. default drinking water consumption rates). Behavioral factors also differ with age, and genetic factors also need to be explored. Estimates of the relative source contribution for chemicals in drinking water versus other sources (including but not limited to food) may need to be expanded to consider age and lifestyle differences. Nutritional essentiality of chemicals and the age-related differences in use of supplements should be more fully considered. The availability of food and nutritional intake data from national surveys such as the National Health and Nutrition Examination Surveys (NHANES) and Continuing Survey of Food Intakes of Individuals (CSFII) database will be useful. All these factors will have an implication in risk assessments for sensitive subpopulations in terms of refining the parameters used in the approaches to characterize exposures rather than relying on default values.

Toxicogenomics

Recent advances in toxicogenomics are anticipated to have significant implications for risk assessment and regulatory guidelines. The goal is to understand responses at the gene and protein level and the scientific basis for the development and application of genomic methodologies to mechanism-based risk assessment. *Genomics* is the study of the genes of a cell or tissue at the DNA (genotype), mRNA (transcriptome), or protein (proteome) level (U.S. EPA, 2002b, 2004b). *Proteonomics* is the study of a cell's protein composition. *Metabonomics* is the biological analysis of biofluids and tissues to determine the profiles of endogenous metabolites present under normal conditions or following environmental chemical exposure, also referred to as *metabolic profiling*. *Bioinformatics* includes the technologies developed for data acquisition and processing designed to store, retrieve and analyze data generated from genomic analysis. The use of these "omic" technologies to study toxicological questions is called *toxicogenomics* (U.S. EPA, 2004b), a new paradigm in toxicology (Inoue and Pennie, 2003). The latest approach to data analysis of the huge whole-genome data sets led to the concept of *functional genomics*, defined as the development and application of global (genome-wide or system-wide) experimental approaches to assess gene function by making use of the information and reagents provided by structural genomics (Cunningham, 2006). It is characterized by high-throughput or large-scale experimental methodologies combined with statistical and computational analysis of the results. It expands the scope of biological investigation from studying single genes or proteins to studying many genes or proteins at once in a systematic fashion, to discover the biological function of particular genes, and to uncover how sets of genes and their products work together in health and disease. Thus, the science of toxicogenomics combines studies of genetics, genomic scale mRNA expression (transcriptomics), cell- and tissue-wide protein expression (proteomics), metabolic profiling (metabonomics), and bioinformatics along with conventional toxicology information in efforts to understand the modes of action of chemicals and the potential role of gene–environment interactions (ICPS, 2003).

In studying genetic disposition, it is worth noting that only about 1 to 2% of human DNA actually codes for RNA messages that can be translated into proteins; this portion is considered the theoretical functional genome (U.S. EPA, 2004b). The assessment of mRNA profiles (functional genomics) uses microarrays that contain all or a sampling of a cell's functional genome. Hybridization of the mRNA that is being actively produced by the cells in these microarrays demonstrates which genes are active in that cell. The noncoding DNA is also important, as it contains information that affects the activity of the functional genome by influencing where and when genes are active in an organism. DNA sequences are evaluated on separate microarray chips, which can identify structural variations that may be of relevance to toxicological responses. Small DNA alterations that might or might not affect gene function (and the corresponding response) are called *single nucleotide polymorphisms* (SNPs). As data are accumulated, genomics technologies and genomics-based data are anticipated to identify unique patterns of gene expression in organisms and cell-based models correlated with sensitivity to toxicants or induced by exposure to environmental stressors.

Among the challenges are linking genomics information to adverse outcome and interpreting genomics information for risk assessment, establishing a framework for analysis and acceptance criteria for genomics information, recruitment of people who possess genomics competencies in application of genomic data for risk assessment, and training risk assessors to interpret and understand genomics data in the context of risk assessment (U.S. EPA, 2004b). Genomics data are anticipated to lead to a better understanding of the mode of action of chemicals and improve the validity of data on adverse effects and exposure and enhance risk assessment (U.S. EPA, 2004b). Specifically in relation to risk assessment, genomics data are anticipated to improve hazard identification (evaluate MOS, define metabolic pathways, replace standard toxicity test with rapid response tests), improve dose–response assessment (possible nonlinear extrapolation for nongenotoxic carcinogens or alter points of departure for potency extrapolation based on biomarkers of effects), improve extrapolations (demonstrate relevance of animal data to humans with examination of key genes and expression pattern, pharmacodynamic responses, interspecies genomic link), and improve exposure assessment (with improved biomarkers of exposure). It is anticipated that genomics data will help in the hazard, dose–response, and exposure assessment components of the cumulative assessment process that also includes mixtures. Such data are also anticipated to help define and identify susceptible people, populations, and life stages.

Genomics technologies and genomics-based data are anticipated to provide more mechanistic and molecular-based data for risk-based prioritization of chemicals for toxicological testing. Such data will be useful in the voluntary high-production-volume screening process, to supplement computer model results or expert judgment for hazard estimation and prioritization, to advance screening of individual chemicals, and to help harmonize risk assessment approaches for different outcomes for which the development of a list of common MOAs is essential. The new approach will be used in groundwater-, drinking water–, and ambient water–related monitoring and regulations, and can be applied in many other programs and situations. Overall, genomics is not anticipated to alter the risk assessment process fundamentally, but is expected to serve as a more powerful tool for evaluating the exposure to and effects of environmental stressors (U.S. EPA, 2004b).

The completion of the Human Genome Project has provided the foundation to analyze the expression of all genes transcribed in a specific cell, as well as reference against which to assess genetic variability and its impact on susceptibility (Cunningham et al., 2003). Advances in genomics technologies will provide a better understanding and prediction of individual adaptive and toxicological responses after toxicant exposure. Simultaneous assessment of expression levels for thousands of different genes using DNA microarrays, and assessment of post-transcriptional and post-translational events using high-throughput proteonomics, are now possible. In contrast to previous estimation of risk of toxicant exposure–induced disease across populations with widely varying responses, new high-throughput genomic technologies are believed to have the potential to greatly improve the accuracy of risk assessment, allowing identification of sensitive populations at risk, and leading ultimately to personal profiles based on genetic composition. It is also recognized that there are ethical, legal, and social issues associated with the approaches. Achieving a better understanding of the linkages and correlations between toxicological endpoints, conventional toxicology, biochemical and cellular toxicology and toxicogenomic information is a first step before strategies and future applications for risk assessment purposes can be developed (ICPS, 2003).

Mixtures Toxicology and Cumulative Risk

Mixture Assessment is a complex process that is continually being advanced. *Mixtures* are any combination of two or more chemical substances with concurrent exposures, regardless of source or spatial or temporal proximity. Environmental exposures occur with mixtures, not with single or individual chemicals. Advances in chemical mixture toxicology made in the last two decades have been described (Yang et al., 2004), and guidelines for health risk assessment of chemical mixtures have been published (U.S. EPA, 1999, 2000a). The Agency for Toxic Substances and Disease Registry (ATSDR) has identified studies on chemical mixtures as one of six priority areas in its agenda for public health environmental research (De Rosa et al., 2004).

In the absence of toxicity data on the specific mixture of concern or a similar mixture, risk assessment has traditionally been based on the components of the mixture, with minimal considerations given to the potential interactions between the components. In the absence of sufficient data on interactions, which is often the case, the approach for systemic toxicity assessment has been based on dose additivity. However, human risk from exposure to multiple chemicals may not always obey the rule of additivity.

The methods used for combining exposures to estimate the risk of groups of chemicals having common mechanisms, with different potencies and exposure characteristics, include calculation of the hazard index (HI), toxicity equivalence factor (TEF), combined margin of exposure (MOE_T), point-of-departure index (PODI) (reciprocal of the HI), and cumulative risk index (CRI) (reciprocal of the MOE_T) (Wilkinson et al., 2000), summarized below.

The *hazard index* (HI) is the sum of the hazard quotients (HQs) of individual chemicals with a common mechanism; that is, it is the sum of potential hazards resulting from exposures to each of the chemicals in the group expressed as a fraction of its respective reference dose (RfD), which is defined as the amount of chemical to which a person can be exposed over a given period of time without incurring an adverse effect. The HI value is assumed to be health protective when it is less than 1. This provides the risk units or the fraction of a *risk cup*, which is a combined RfD for the chemical group having a common mechanism of action. Conceptually, the cup holds the total amount (100%) of a given chemical that a person could be exposed to daily, for 70 years, without additional health risks. Additional exposure would cause an excess risk and an overflow of the cup.

The *print-of-departure index* (PODI) is the sum of the exposures to each chemical expressed as a fraction of their respective PODs, or PODFs (POD fractions). The summation is done in terms of relative potencies and applies a single group UF as the last step (rather than in an earlier step, as in the case of establishing a RfD).

The *toxicity equivalence factor* (TEF) approach normalizes exposures to a group of chemicals with a common mechanism with different potencies to yield a total equivalent exposure to one of the chemicals, referred to as the *index compound* (IC) (e.g., for structurally related compounds such as polychlorinated dibenzo-*p*-dioxins, dibenzofurans, and biphenyls). TEFs are obtained as the ratio of the POD of the IC to that of each of the other members in the group. The exposure to each chemical is then multiplied by the appropriate TEF to express all exposures in terms of the IC. Summation of these values provides a total combined exposure to all chemicals expressed in terms of the IC. The combined risk from concurrent exposures to several products may be obtained by comparing the total exposure in product equivalents, or IC equivalents, against the RfD for the IC. The resulting value represents a combined quotient (fraction of a risk cup, which is obtained by multiplying the PODI by a group UF).

The *margin of exposure* MOE is the ratio of the POD (ED_{10}) to the level of exposure. The *combined MOE* (MOE_T) is the sum of the MOEs of each chemical being combined. The general practice is that the MOE or MOE_T should be equal to > 100, which is numerically the default UF applied to an NOAEL for deriving a RfD (this may not be true in all cases). The *cumulative risk index* (CRI) is an extension of the MOE approach as a way to combine MOEs for chemicals with different UFs. The risk index is the MOE divided by the UF for that chemical or simply the RfD divided by exposure. The risk index is the reciprocal of the HQ, and the CRI is the reciprocal of the HI. The risk increases as its value falls below 1.0.

A common mechanism might exist for two compounds if both cause the same critical effect, act on the same molecular target at the same target tissues, and act by the same pharmacological mechanism of action and may share a common toxic intermediate (Mileson et al., 1998), but it is not clear whether all three criteria must be met before a firm determination can be reached. One major assumption in these methodologies is that the chemicals are acting by a common mechanism, exposure to the various chemicals is concurrent, and the combined cumulative effect will reflect the sum of the POD-normalized exposures of the individual chemicals (i.e., neither synergistic nor antagonistic interactions occur). The assumption of additivity is discussed in more detail by the EPA (1999) in its guidance for conducting health risk assessment of chemical mixtures. Any procedure used is hypothetical given the current knowledge, data, and methodology available.

The hazard index (HI) approach for component-based assessment is most commonly used, assuming no interaction (other than additivity) between the components of the mixture. The approach is most appropriately applied to chemicals that cause the same effect by the same mechanism of action, assuming similar dose–response curve but varying potencies. It is intended to calculate the plausible toxicity index that would have been calculated had the mixture itself been tested. In practice, the guidance is being applied only to the critical effect of each component chemical, and thus the toxicity to other target organs is not being included in the overall toxicity assessment process.

Questions arise as to whether the HI approach is adequate to address multiple chemical exposures. Recent efforts have been made in cumulative risk assessment (CRA) to address the growing public concerns about multiple exposures and risks, and provides an approach to manage cumulative risk (U.S. EPA, 1997b, 2000a, 2003a, 2006d). The Food Quality Protection Act of 1996 (FQPA) requires the U.S. EPA to consider effects of exposure to all pesticides and other chemicals that act by a common mechanism of toxicity when they derive tolerances for pesticide use on crops. The focus on a common mechanism refers to chemicals that cause the same critical effect, act on the same molecular target tissue, and act by the same pharmacological mechanism of action and may share a common toxic intermediate. HI is one of the tools that can be used for cumulative risk characterization. *Cumulative risk* is the risk associated with exposure to multiple agents by multiple routes. Cumulative risk assessment combines the complexity of risk assessment of mixtures of chemicals with the added complexity of assessment of exposure by multiple routes. It also considers the aggregate ecological and human health risk caused by the accumulation of risk from multiple stressors. *Aggregate risk* is risk associated with exposure to a single agent by multiple routes, and it may be characterized by aggregate exposure and risk assessment.

The EPA (U.S. EPA 2006d) recently made available the completed cumulative risk assessment for organophosphate pesticides. The risk assessments were conducted to meet current standards under the Federal Insecticide, Fungicide, and Rodenticide Act (FIFRA) and the Federal Food, Drug, and Cosmetic Act (FFDCA), as amended by the FQPA. Section 408(b)(2)(D)(v) of the FFDCA directs the agency to consider available information on the cumulative risk from substances sharing a common mechanism of toxicity. The EPA has determined that the organophosphate pesticides share a common mechanism of toxicity, cholinesterase inhibition. The Organophosphate Cumulative Risk Assessment (2006 update) considered an addendum to the June 2002 assessment, includes improvements and refinements in assessing the cumulative risks of the organophosphate pesticides. The U.S. EPA concluded that the cumulative risks associated with the remaining uses of the organophosphate pesticides are below the agency's level of concern.

Progress has been made to bring descriptive work to mechanistic-based research. This physiologically based physicokinetic leads to the integration of (PBPK) or PBPK/physiodynamics (PD) modeling with chemical mixture toxicology and the toxicology of more complex mixtures (Yang et al., 2004). An example is the PBPK/PD modeling of toxicological interaction between kepone and carbon tetrachloride, based on mechanisms of interactive toxicology (impairment of the liver's regeneration process by kepone). This brings a higher computer-intensive technology to acute toxicity studies. Seven areas of scientific achievements in chemical mixture toxicology are noted (Yang et al., 2004): (1) application of better and more robust statistical methods, (2) exploration and incorporation of mechanistic bases for toxicological interactions, (3) application of PBPK modeling, (4) studies on more complex chemical mixtures, (5) use of science-based risk assessment approaches, (6) utilization of functional genomics, and (7) application of technology. These applications for mixture analysis led to elucidation of two concepts: dose-dependent toxicologic interactions and interaction thresholds. Understanding of these factors is critical in the prediction of effects of low-level environmental mixtures.

A future challenge in mixture analysis is how to address the infinite number of combinations of chemicals and other stressors. The authors believe that the initial focus should be on understanding the finite (biologic processes), as all stimuli and insults from external stressors are merely perturbations of the processes. This should be followed by capturing the biological process(es) by integrating recent advances in computational technology and modern biology into standard risk assessment, as discussed further in the next section.

Computational Toxicology/Physiologically Based Pharmacokinetic Modeling

Computational toxicology is the application of mathematical and computer models to predict adverse effects and to better understand the mechanism(s) through which a given chemical induces harm (U.S. EPA, 2003b, 2006e). It involves the following disciplines:

1. *Computational chemistry*, which refers to the mathematical modeling of physicochemical properties and processes at the molecular level and includes such topics as quantum chemistry, force fields, molecular mechanics, molecular simulations, molecular modeling, molecular design, and cheminformatics
2. *Computational biology* or *bioinformatics*, which involves the development of molecular biology databases and analysis of the data
3. *Systems biology*, which refers to the application of mathematical modeling and reasoning to the understanding of biological systems and the explanation of biological phenomena

Development of the field of computational biology offers the possibility that with advances in its subdisciplines (e.g., genomics, proteomics, metabonomics), scientists may gain the ability to develop a more detailed understanding of the risks posed by a much larger number of chemicals. Recently, there has been increased attention to utilization of these novel technologies in toxicological risk assessment. The computational approach has been dubbed *in silico toxicology* (Yang et al., 2004). In silico refers to integrating computer modeling with focused, mechanistic animal experimentation such that experiments that are impractical (e.g., too large, too expensive) or impossible to perform (e.g., human experiments with carcinogens) are simulated on a computer (referring to silicon-based microprocessors). Approaches derived from these technologies have been used to address the questions of screening and prioritizing chemicals for testing, as well as hazard identification and improvement of quantitative dose–response assessment (U.S. EPA, 2003b). Efforts are being made to apply computational toxicology to address issues in quantitative risk assessment such as defining the shape of the dose–response curve at low exposures, using molecular indicators of response, developing biomarkers for use in analysis of low dose responses, validating the interpretation of molecular indicators of response, defining the relevance of modes of action for risk assessment, constructing biologic based dose–response (BBDR) models of high-priority outcomes, assessing population-level effects in ecosystems, cross-species extrapolation, assessing toxicity of mixtures, and integrating human and ecological risk assessment. The overall goal of computational toxicology is to use emerging technologies to improve quantitative risk assessment by reducing uncertainties in the source-to-adverse outcome continuum. Specifically, three strategic objectives of the EPA (U.S. EPA, 2006e) computational toxicology initiative are to improve understanding of the linkages in the continuum between the source of a chemical in the environment and adverse outcomes, provide predictive models for screening and testing, and improve quantitative risk assessment.

The new in silico toxicology approach is demonstrated in the integration of biochemistry, physiology, and metabolism of three commonly used organic solvents with a computer simulation (Dobrev et al., 2002). For this modeling, an interactive PBPK model was developed to predict the individual kinetics of trichloroethylene (TCE), perchloroethylene (PERC), and methylchloroform (MC) in humans exposed to different mixtures of the three solvents. The mixture model was used to explore the general pharmacokinetic profile of two common biomarkers of exposure—peak TCE blood levels and total amount of TCE metabolites generated—in both rats and humans. Increases in the TCE blood levels led to higher availability of the parent compound for glutathione conjugation, a metabolic pathway associated with kidney toxicity and carcinogenicity. The simulated change in production rates of toxic conjugative metabolites exceeded 17% for a corresponding 10% increase in TCE blood concentration, indicating a nonlinear risk increase due to combined exposures to TCE. Evaluation of metabolic interactions and their thresholds illustrates a unique application of PBPK modeling in risk assessment of occupational exposures to chemical mixtures. PBPK models were used for each single chemical and the integrated mixture model to calculate interaction thresholds for both species for a variety of occupationally relevant exposure scenarios. Finally, the relative changes in each biomarker of exposure at the simulated interaction thresholds in humans were calculated. The approach illustrates a novel application of PBPK modeling to predict the occurrence of metabolic interactions during mixed exposures and quantitatively assess their impact on the kinetics of each mixture component.

Examples of in silico experimentation conducted in the same laboratories using PBPK/PD modeling and other computer modeling techniques, such as Monte Carlo simulation, include the following (Dobrev et al., 2002): (1) acute toxicity studies of toxicologic interactions between kepone and carbon tetrachloride, (2) human biological exposure index studies on six industrial solvents, (3) interaction threshold studies on binary and ternary chemical mixtures, (4) studies on age- and dosing-related pharmacokinetic changes in mice in a two-year chronic toxicity/carcinogenicity bioassay, and (5) clonal growth modeling of early stages of carcinogenesis. Application of computer technology as an alternative research method can help to minimize the use of animals in toxicity tests. Some of the chemicals involved in the studies conducted to date are primary drinking water contaminants such as chlorinated solvents; additional in-depth studies will undoubtedly be conducted on the complex disinfection by-product mixtures.

The genomic technologies available today allow for increasingly broader molecular profiling at multiple levels of biological organization (U.S. EPA, 2003b). Parallel to efforts in computational biology, there have been major advances in computational speed and access to data. In the field of bioinformatics, major advances are being made from the integration of computational and data acquisition technologies. It is now possible to begin to evaluate the vast amounts of information generated by "omic" technologies using data mining tools and high-speed desktop computers. Further advances can be expected as software is developed to utilize these technologies and more data are developed on the biological responses to toxic insults from a variety of chemicals, as observed using these new tools.

The qualitative and quantitative evaluations of the relationship between dose and response, or *dose metrics*, are key components of the quantitative risk assessment process that could be improved with the use of computational methods (U.S. EPA, 2003b). The choice of the chemical species and the actual dose metric for the risk assessment process depends on the particular mode or modes of action being assessed. As the biological steps between the external exposure and some internal toxicologically relevant dose are often nonlinear, pharmacokinetic (PK) models are often used to link an exposure of interest to an observed adverse outcome. PBPK models that depend on knowledge of anatomy, physiology, and biochemistry are the most common and typical examples of such models. For PBPK models to be used, several pieces of key information are needed. Some of this information, such as body size, organ volumes, and blood flows are known for several species, including the human. Other important pieces of information, such as metabolic transformation rates, are chemical specific and may vary from species to species. The expense and time required to gain all the required information in laboratory studies has limited the use of this modeling technology. Detailed knowledge of the molecular events that lead to toxicity is often lacking, so it can be unclear what chemical metric should be related to the ultimate toxic effect(s). Computational toxicology may enable broader use of PBPK by providing better indicators of the relevant doses and receptors within the target organism (U.S. EPA, 2003b) to overcome the limitations. Advancements made in genomics and proteomics are anticipated to better define the indicator of the most relevant doses for environmental chemicals entering the body. For example, specific binding to a particular part of the DNA, RNA, receptors, or enzymes might be a much more relevant dose metric than just the amount of chemical in a particular tissue. The identification of such interactions as biomarkers of toxic effects in easily obtained and measured biological fluids will be facilitated by the advances in computational toxicology, and provide more accurate indicators of a critical event inside key target cells.

Integration of studies of stressor dose, transformation (proteonomics), and metabolic sequelae (metabonomics) is seen as having the potential to reveal the clearest perspective yet on relevant dose and its variation across organisms and populations (U.S. EPA, 2003b). Advances in molecular biology allow for characterization of the genetic variation (polymorphisms) within populations. In turn, the genetic marker data are defining the contribution of genetic variation to the overall level of variation in dose–response in organisms. These data are providing correlations between genotypes and phenotypes in the population representative of sensitive or vulnerable subpopulations. Dosimetry models such as PK and PD models can then incorporate these data to reduce the uncertainties associated with assuming that populations are homogeneous or normally distributed regarding toxic response to stressors. Indicators in body fluids of persons in susceptible subpopulations may provide simple methods for the inclusion of relevant population data to exposure characterization within human risk assessments.

Computational toxicology techniques provide excellent promise to focus research on reducing uncertainties in both ecological and human health risk assessments (U.S. EPA, 2003b). However, the key predictive toxicology tools and approaches, including PBPK and QSAR (quantitative structure–activity relationship) models and/or alterations in gene (or protein) expression profiles, are useful only in the context of a thorough understanding of toxicity pathways of concern (i.e., the mechanism or mode of action of a chemical). Specifically, it is necessary to link adverse outcomes (e.g., reproductive or developmental changes, cancer) to initiating events, ideally through the cascade of biochemical and physiological changes that occur as a result of the initial interaction(s) of xenobiotics with biological molecules (e.g., receptor binding, enzyme inhibition).

In the context of applying computational toxicology in quantitative risk assessment, computational chemistry (QSAR) and mathematical biology [PBPK/biologically based dose–response (BBDR) modeling] are applicable to PK and BBDR models useful for risk assessment (U.S. EPA, 2003b). A systems biology approach to dose–response modeling would integrate PK and PD. QSARs can be used to estimate parameters of PBPK models and cellular response. PBPK data and models are being evaluated to adjust default uncertainty factors, used in inter- and intraspecies extrapolation and for developing an assessment of target organ dose for use in dose–response assessment of mixtures.

Benchmark Dose Approach

The *benchmark dose* (BMD) approach has been used increasingly as an alternative to the no-observed-adverse-effect level/lowest-observed-adverse-effect level (NOAEL/LOAEL) approach for deriving a point of departure for low-dose extrapolations. This approach does not require that a study identify a NOAEL but that at least one dose level needs to be near the range of the response level for the benchmark dose or benchmark concentration (BMD or BMC; i.e., a 5 or 10% response level). Usually, a statistical lower confidence limit on the dose producing a predetermined level of biological change in response, called the *benchmark response* (BMR), relative to controls, is used in the evaluation. The BMR is based on a biologically significant level of response or on the response level at the lower end of the observable range for a particular endpoint. The EPA is developing guidance for the use of the BMD approach (U.S. EPA, 2000b) and has provided an overview of the approach (U.S. EPA, 2000c), as summarized below.

The BMD approach is based primarily on modeling of dose and response limited to the experimental range, with no extrapolation to doses far below the experimental range (U.S. EPA, 2000c). As the models used are mainly statistical rather than biologically based models, they cannot be used reliably to extrapolate to low doses without incorporating detailed information on the mechanisms involved relating to the toxic agent and the particular effect being modeled. In the process, a mathematical dose–response curve and its corresponding confidence limits are first established, and a point on the lower confidence dose–response curve corresponding to the BMR chosen is selected. This point on the lower confidence curve is the lower confidence bound of the effective dose for that BMR, referred to as the BMDL. A BMDL may be calculated for each biological effect endpoint for which there are adequate data.

The determination of an RfD using the BMD approach involves four basic steps: (1) selection of the experiments and responses that will be used for modeling the BMD, (2) calculation of BMDs for the responses selected for all endpoints that have a potential for yielding the critical BMD, (3) selection of a single BMD from among those calculated, and (4) calculation of a RfD by dividing the chosen BMD or BMDL by appropriate uncertainty factors (UFs). Each step is associated with various decision points (Crump et al., 1995; U.S. EPA, 2000c). In selecting response data to model, ideally, BMD calculations would be performed for the complete set of relevant effects; however, that can be resource intensive, and it is difficult to interpret results from a large number of dose–response analyses (U.S. EPA, 2000c). An effect seen only at doses above the LOAEL but having a shallow dose–response could produce a lower BMD than an effect seen at the LOAEL, which has a steeper dose–response curve. One option is to focus on those responses for which there is evidence of a dose–response relationship (e.g., statistically increasing or decreasing trends in the response as the dose level increases while still considering biological significance). Another option is to focus on modeling the most critical effects as seen at the LOAEL. However, limiting the number of responses modeled may potentially misrepresent the minimum BMD. With regard to the form of the data used (U.S. EPA, 2000c), categorical data or quantal data are relatively straightforward to use in the BMD approach, since the data are expressed as the number (or percent) of subjects exhibiting a defined response at a given dose. Data in the continuous form, where results are expressed as measurements of a continuous biological endpoint, such as a change in organ weight or serum enzyme level, can also be modeled. This requires professional judgment to determine what degree of change in the measurement constitutes an abnormal (adverse) or nonadverse effect.

The EPA benchmark dose modeling program (version 1.2) includes dose–response models for quantal and continuous data and other options. Fitting the models to experimental data yields estimated parameters of goodness of fit of the model to the data (U.S. EPA, 2000c); usually, the model that best fits the data is selected for the risk assessment. For the measurement of altered response for quantal data, two measures of increased response, additional risk and extra risk, have been proposed (Crump, 1995; U.S. EPA, 2000c). For continuous data (Crump, 1995; U.S. EPA, 2000c), altered response can be expressed as the difference between the mean response to dose *d* minus the mean control response. More recent consideration of BMDs for continuous endpoints has suggested other alternatives. A critical decision for deriving the BMD is the selection of the benchmark response (BMR) (U.S. EPA, 2000c). The BMD is used like a NOAEL in derivation of the RfD, and the BMR should be selected near the low end of the range where effects were detected in a study. The lowest confidence limit on the dose predicted to cause a 10% increase in the incidence of the effect in the test population (ED_{10}) is frequently chosen as the BMR. For some data it may be possible to adequately estimate the ED_{05} or ED_{01}, which is closer to a true no-effect dose. However, in many cases the ED_{10} is the lowest level of risk that can reliably be estimated from standard toxicity studies (Crump et al., 1995; U.S. EPA, 2000c). The appropriate BMR has most commonly been accepted to be either 5% or 10%. BMDs defined in terms of 10% increases in probability of response have been shown to tend to be, on average, similar to corresponding NOAELs for quantal developmental toxicity studies (Allen et al., 1994a, 1994b; Faustman et al., 1994).

For derivation of water quality criteria, the EPA recommends use of the ED_{05} or ED_{10} when deriving a BMD (U.S. EPA, 2000c). The lower confidence bound on the dose corresponding to the BMD selected is used to account for population variability (U.S. EPA, 2000c). Use of a lower confidence limit rather than a maximum likelihood estimate (MLE) increases the confidence that the results from a study of a small group of animals can be extrapolated to (i.e., are protective of) the entire population. The agency recommends the use of one-sided 95th percentile confidence limits for BMD modeling, which is consistent with the size of the confidence limits used in cancer dose–response modeling (U.S. EPA, 2000c).

When multiple BMDs are calculated because several models fit a single data set, the smallest BMD or the geometric mean of combined BMDs may be selected. When multiple BMDs are calculated from different responses or different studies that examine the same endpoint, the choice among BMDs may also involve selection of the "critical effect" and the most appropriate species, sex, or other relevant feature of experimental design. Graphic representations of the model output and experimental data, as well as an understanding of the biological mode of action, would be helpful. Following the selection of a single or averaged BMD, the RfD can be calculated by dividing the BMD by appropriate uncertainty factors. As a default, all applicable uncertainty factors used in the traditional NOAEL-based RfD approach, except for the LOAEL–NOAEL extrapolation factor, should be considered. Other factors, such as the size of the BMR and confidence bounds, biological considerations (such as the possibility of a threshold), severity of the modeled effect, and the slope of the dose–response curve, may affect the choice and magnitude of uncertainty factors (Crump et al., 1995; U.S. EPA, 2000c).

The following major points can be derived from an overview of the BMD approach (U.S. EPA, 2000c):

1. The BMD approach offers a number of advantages over the traditional NOAEL/LOAEL approach for deriving the RfD, including considering the entire dose–response curve, including its shape, accounting better for statistical variability in the data; and not being overly sensitive to dose spacing, thus not limited to the experimental doses for determining the effect level.
2. The BMD approach allows for objective extrapolation of animal response data to human exposures across the different study designs encountered in noncancer risk assessment, but it does not reduce uncertainties inherent in extrapolating from animal data to humans (except for that in the LOAEL-to-NOAEL extrapolation).
3. The BMD approach is not recommended for routine use; it may be used when data are available and justify the extensive analyses required. The data requirements are more extensive than those for the NOAEL/LOAEL approach. Studies with small group sizes and evaluation of a limited number of endpoints will tend to yield lower BMD values because the confidence bands will be wider; more adequate studies give narrower confidence bands.
4. The BMD approach should not be applied to data sets with only two experimental groups (a control and one positive dose). In such cases, much of the advantage of the BMD approach with respect to consideration of the dose–response shape will be lost, as such data provide no information about the shape of the dose–response curve. When more doses are available, especially at lower response levels, the expected benefit of the BMD approach will be greater than that of the NOAEL-based approach.
5. Although the method does not require that a study identify a NOAEL, at least one dose level needs to be near the range of the response level for the BMD or BMC.
6. Overall, the BMD approach is proposed as an alternative procedure, where appropriate, that can be used until other more biologically motivated approaches are available.

The benchmark dose approach has been used in the OEHHA drinking water risk assessment program; a recent example of this is for the perchlorate risk assessment (Ting et al., 2006) discussed in Chapter 11.

CONCLUSIONS

This chapter constitutes an overview of some of the emerging issues and recent advances in biological and toxicological research, approaches, and methodologies used in health risk assessment and their implications for future directions in risk assessment of chemicals. Some examples are given in relation to chemicals in drinking water. Two emerging issues involve PPCPs (pharmaceuticals and personal care products) and engineered nanomaterials (physicochemical structures on a scale greater than typical atomic and molecular dimensions, but with at least one dimension of 100 nm or less, which exhibit physical, chemical, and/or biological characteristics associated with their nanostructure). For PPCPs the concerns relate to the finding of unregulated PPCPs in water sources and the uncertainty associated with potential human exposure to and health risk from these chemicals in drinking water sources. For engineered nanomaterials the concerns relate to the properties and characteristics of these materials that render them different from the traditional chemical contaminants, thus requiring different approaches in toxicity evaluations and risk assessment. Much is yet unknown about the hazards, dose–response characteristics, and exposure to such materials.

Recent advances are highlighted in areas relating to endocrine effects, mode of action/mechanism, sensitive populations, toxicogenomics, mixtures and cumulative risk, computational toxicology and PBPK modeling, and benchmark dose approach as they relate to risk assessment. The challenges of toxicological research and health risk assessment of chemicals that affect future directions include the following:

- Continuing research and evolution to reflect advances and changes made in an understanding of the underlying science (e.g., biological relationship between chemical toxicity and human response, toxicogenomics)
- Filling data gaps and reducing uncertainties (e.g., testing for endocrine effects, refining of exposure factors)
- Performing evaluations or reevaluations of an increasing number of environmental chemical agents (regulated and unregulated) because of new scientific data or understanding (e.g., mode of action), and the increasing numbers of chemicals (carcinogens and noncarcinogens) being identified that were not previously evaluated (e.g., PPCPs)
- Addressing the increasing need for refined methodologies, approaches (e.g., computer-assisted methods/computational toxicology, address age sensitivity, mixtures), resources, and skilled risk assessors (i.e., education and training)
- Ensuring quality peer reviews, considering all sources of input and applying scientifically based approaches to address issues raised while maintaining independence of scientific evaluations
- Providing transparency for underlying assumptions
- Ensuring commitment of the risk assessors to the highest standards of scientific and ethical conduct
- Generating and interpreting data for health effects evaluation and risk assessment relating to new technologies (e.g., nanotechnology, toxicogenomics technology, computational toxicology)
- Updating chemical evaluations and risk assessments as appropriate
- Developing guidance and addressing public health and regulatory impacts

Disclaimer

The opinions expressed in this chapter are those of the authors and not necessarily those of the Office of Environmental Health Hazard Assessment or the California Environmental Protection Agency.

REFERENCES

Adler T. 2003. Aging research: the future face of environmental health. *Environ Health Perspect* 111: A761–A755.

Allen BC, Kavlock RJ, Kimmel CA, Faustman EM. 1994a. Dose response assessments for developmental toxicity, II: Comparison of generic benchmark dose estimates with NOAELs. *Fundam Appl Toxicol* 23: 487–495.

______. 1994b. Dose response assessments for developmental toxicity, III. Statistical models. *Fundam Appl Toxicol* 23: 496–509.

Borm P, Klaessig FC, Landry TD, Moudgil B, Pauluhn J, Thomas K, Trottier R, Wood S. 2006. Forum Series. Research strategies for safety evaluation of nanomaterials, V: Role of dissolution in biological fate and effects of nanoscale particles. *Toxicol Sci* 90(1): 23–32.

California DHS 2004. Groundwater recharge reuse regulations, Draft, December 1. California Department of Health Services, Sacromento, CA, Accessed at: http://www.dhs.ca.gov/ps/ddwem/publications/waterrecycling/rechargeregulationsdraft-12-01-04.pdf

California OEHHA (Office of Environmental Health Hazard Assessment). 2006. Symposium: The aging population as a sensitive subgroup, Asilomar, CA, April 9–11, 2006.

Cambell PR. 2006. State population projections, U.S. Census Bureau. Accessed at: http://www.census.gov/population/www/pop-profile/stproj.html.

Cohen SM, Klaunig J, Meek ME, Hill RN, Pastoor T, Lehman-McKeeman L, Bucher J, Longfellow DG, Seed J, Dellarco V, Fenner-Crisp P, Patton D. 2004. Evaluating the human relevance of chemically induced animal tumors. *Toxicol Sci* 78: 181–186.

Colborn T, vom Saal FS, Soto AM. 1993. Developmental effects of endocrine-disrupting chemicals in wildlife and humans. *Environ Health Perspect* 101: 378–384.

Crump K. 1995. Calculation of benchmark doses from continuous data. *Risk Anal* 15: 79–89.

Crump KS, Allen B, Faustman E. 1995. The use of the benchmark dose approach in health risk assessment. Prepared for the U.S. EPA Risk Assessment Forum.

Cunningham ML. 2006. Putting the fun into functional toxicogenomics. *Toxicol Sci* 92: 347–348.

Cunningham ML, Bogdanffy MS, Zacharewski TR, Hines RN. 2003. Forum. Workshop overview: use of genomic data in risk assessment. *Toxicol Sci* 73: 209–215.

Daughton CG, Ternes TA. 1999. Pharmaceuticals and personal care products in the environment: Agents of subtle change? *Environ Health Perspect* 107:6: 907–938.

De Rosa CT, El-Asri HA, Pohl H, Cibulas W, Mumtaz MM. 2004. Implications of chemical mixtures in public health practice. *J Toxicol Environ Health B* 7: 339–350.

Dobrev ID, Andersen ME, Yang RSH. 2002. In silico toxicology: simulating interaction thresholds for human exposure to mixtures of trichloroethylene, tetrachloroethylene, and 1,1,1-trichloroethane. *Environ Health Perspect* 110: 1031–1039.

Donaldson K, Aitken R, Tran L, Stone V, Duffin R, Forrest G, Alexander A. 2006. Carbon nanotubes: a review of their properties in relation to pulmonary toxicology and workplace safety. *Toxicol Sci* 92(1): 5–22.

Dreher KL. 2004. Toxicological highlight. Health and environmental impacts of nanotechnology: toxicological assessment of manufactured nanoparticles. *Toxicol Sci* 77: 3–5.

Fan AM, Howd RA. 2006. California's public health goals to protect infants and children from drinking water contaminants. Office of Environmental Health Hazard Assessment, California Environmental Protection Agency, Oakland/Sacramento, CA. Roundtable discussion, Society of Toxicology 45th Annual Meeting, San Diego, CA, March 5–9.

Fan AM, Howd R, Alexeeff G. 2005. Consideration of sensitive populations for risk assessment of chemicals in drinking water. Presented at the Society of Toxicology 44th Annual Meeting, New Orleans, LA, March.

Faustman EM, Allen BC, Kavlock RJ, Kimmel CA. 1994. Dose response assessment for developmental toxicity, I: Characterization of database and determination of NOAELs. *Fundam Appl Toxicol* 23: 274–486.

Folmar LC, Denslow ND, Rao V, Chow M, Crain DA, Enblom J, et al. 1996. Vitellogenin induction and reduced serum testosterone concentrations in feral male carp (*Cyprinus carpio*) captured near a major metropolitan sewage treatment plant. *Environ Health Perspect* 104: 1096–1101.

Fox GA. 1991. Practical causal inference for ecoepidemiologists. *Toxicol Environ Health* 33: 359–373.

Geller AM, Zelnick H. 2005. Aging and the environment: A research framework. *Environ Health Perspect* 113: 1257–1262.

Hemminger P. 2005. Damming the flow of drugs into drinking water. *Environ Health Perspect* 113: A679–A681.

Hill AB. 1965. The environment and disease: association or causation. *Proc Royal Soc Med* 58:295.

IARC (International Agency for Research on Cancer). 2005. *Monographs on the Evaluation of Carcinogenic Risks to Humans*, IARC Monographs, Vol 92, *Air Pollution*, Part 1; Some non-heterocyclic polycyclic aromatic hydrocarbons and some related exposures. World Health Organization, Lyon, France, October 11–18.

——. 2006. *IARC Monographs on the Evaluation of Carcinogenic Risks to Humans*. Preamble. World Health Organization, Lyon, France.

ICPS (International Programme on Chemical Safety). 2003. Toxicogenomics and the risk assessment of chemicals for the protection of human health. Workshop Report, Summary. World Health Organization. Held at Federal Institute for Risk Assessment, Berlin, November 17–19. (ICPS/Toxicogenomics/03/1).

Inoue T, Pennie W, Eds. 2003. *Toxicogenomics*. Springer-Verlag, Tokyo.

Kaiser J. 1996. Scientists angle for answers. *Science* 274: 837–1838.

Kavlock RJ, Daston GP, DeRosa C, Fenner-Crisp P, Gray LE, Kaattari S, Lucier G, Luster M, Mac MJ, Maczka C, Miller R, Moore J, Rolland R, Scott G, Sheehan DM, Sinks T, Tilson HA. 1996. Research needs for the risk assessment of health and environmental effects of endocrine disruptors: a report of the U.S. EPA–sponsored workshop. *Environ Health Perspect* 104(suppl 4): 715–740.

Kolpin DW, Furlong ET, Meyer MT, Thurman EM, Zaugg SD, Barber LB, Buxton HT. 2002. Pharmaceuticals, hormones, and other organic wastewater contaminants in U.S. streams, 1999–2000: a national reconnaissance. *Environ Sci Technol* 36(6): 1202–1211. Accessed at: http://toxics.usgs.gov/pubs/FS-027-02/index.html.

Lam CW, James JT, McCluskey R, Hunter RL. 2004. Pulmonary toxicity of single-wall carbon nanotubes in Mice 7 and 90 days after intratracheal instillation. *Toxicol Sciences* 77: 126–134.

Lamb JC, Herman BWM, Koeter H, Becker R, Gies A, Davies L, Inoue T, Jacobs A, Lyons G, Matsumoto M, Timm G. 2003. IUPAC Workshop 6.1. Risk management options for endocrine disruptors in national and international programs. Pure Appl Chem 75 (11–12):2549–2553.

Liebig M, Moltmann JF, Knacker T. 2006. Evaluation of measured and predicted environmental concentrations of selected human pharmaceuticals and personal care products. *Environ Sci Pollut Res Int* 13(2): 110–119.

Meek ME, Bucher JR, Cohen SM, Dellarco V, et al. 2003. A framework for analysis of information on carcinogenic modes of action. *Crit Rev Toxical* 33: 591–653.

Mileson BE, Chambers JE, Chen JE, Dettbarn W, Ehrich M. 1998. Common mechanism of toxicity: a case study of organophosphorous pesticides. *Toxicol Sci* 41: 8–20.

NAS (National Academy of Sciences). 1983. *Risk Assessment in the Federal Government: Managing the Process*. NAS, Washington, DC.

_____. 2005. *Health Implications of Perchlorate Ingestion*. Committee to Assess the Health Implications of Perchlorate Ingestion. National Academies Press, Washington, DC.

NRC (National Research Council). 1999. *Hormonally Active Agents in the Environment. Commission on Life Sciences*. National Research Council, Washington, DC.

_____. 2002. *Small Wonders, Endless Frontiers: A Review of the National Nanotechnology Initiative*. National Academies Press, Washington DC, p 12. Accessed at: www.nano.gov.

NSF (National Science Foundation) and U.S. EPA (U.S. Environmental Protection Agency). 2003. Nanotechnology Grand Challenge in the Environment Research Planning Workshop, Session E, Nanotechnology Implications in Health and the Environment, Arlington, VA, May. Accessed at: http://es.epa.gov/ncer/publications/nano/nanotechnology4-20-04.pdf.

Oberdorster G, Maynard A, Donaldson K, et al. 2005. Principles for characterizing the potential human health effects from exposure to nanomaterials: elements of a screening strategy. *Particle Fibre Toxicol* 2:8. Accessed at: http://www.particleandfibretoxicology.com/content/2/1/8.

Seed J, Carney EW, Corle RA, Crofton KM, De Sesso JM, Foster PM, Kavlock R, Kimmel G, Klaunig J, Meek M, Preston RJ, Slikker W Jr, Tabacova S, Williams GM, Wiltse J, Zoeller RT, Fenner-Crisp P, Patton DE. 2005. Overview: Using mode of action and lifestage information to evaluate the human relevance of animal toxicity data. *Crit Rev Toxicol* 35(8–9): 664–672.

Sonich-Mullin C, Fielder R, Wiltse J, Baetche K, et al. 2001. IPCS conceptual framework for evaluating a MOA for chemical carcinogenesis. *Regul Toxicol Pharmacol* 34: 146–152.

Straif K, Baan R, Gross Y, Secretan B, El Ghissassi F, Cogliano V. 2006. Carcinogenicity of polycyclic hydrocarbons. *Lancet Oncol* 6: 931–932.

Sumpter JP. 1995. Feminized responses in fish to environmental estrogens. *Toxicol Lett* 82–83: 737–742.

Sumpter JP, Jobling S. 1995. Vitellogenesis as a biomarker for estrogenic contamination of the aquatic environment. *Environ Health Perspect* 103(7): 173–177.

Thomas K, Aguar P, Kawasaki H, Morris J, Nakanishi J, Savage N. 2006a. Research strategies for safety evaluation of nanomaterials, VIII: International efforts to develop risk-based safety evaluations for nanomaterials. *Toxicol Sci* 92(1): 23–32.

Thomas T, Thomas K, Sadrieh N, Savage N, Adair P, Bronaugh R. 2006b. Research strategies for safety evaluation of nanomaterials, VII: Evaluating consumer exposure to nanoscale materials. *Toxicol Sci* 92(1): 14–19.

Ting D, Howd RA, Fan AM, Alexeeff GV. 2006. Development of a health-protective drinking water level for perchlorate. Office of Environmental Health Hazard Assessment, California Environmental Protection Agency. *Environ Health Perspect* 114: 881–886.

U.S. EPA (Environmental Protection Agency). 1991. *Guidelines for Developmental Toxicity Assessment*. Risk Assessment Forum, EPA/600/FR-91/001. U.S. EPA, Washington, DC.

———. 1997a. *Special Report on Environmental Endocrine Disruption: An Effects Assessment and Analysis*. EPA/630/R-96/012. Risk Assessment Forum, U.S. EPA, Washington, DC.

———. 1997b. *Guidance on Cumulative Risk Assessment*, Part 1: *Planning and Scoping*. Office of the Administrator, U.S. EPA, Washington, DC.

———. 1998. Guidelines for Neurotoxicity Risk Assessment. *Fed Regi* 63(93): 26926–26954.

———. 1999. *Guidance for Conducting Health Risk Assessment of Chemical Mixtures*. Risk Assessment Forum, U.S. EPA, Washington, DC.

———. 2000a. *Supplementary Guidance for Conducting Health Risk Assessment of Chemical Mixtures*. EPA/630/R-00/002. Risk Assessment Forum, U.S. EPA, Washington, DC.

———. 2000b. *Benchmark Dose Technical Guidance Document*. External Review Draft. EPA/630/R-00/001. U.S. EPA, Washington, DC.

———. 2000c. *Methodology for Deriving Ambient Water Quality Criteria for the Protection of Human Health*. Technical Support Document, Volume 1, *Risk Assessment*. EPA/822/B-00/005. Office of Water, Office of Science and Technology, U.S. EPA, Washington, DC.

———. 2002a. PPCPs as environmental pollutants. frequently asked questions from the public, media, and scientific communities. Accessed at: http://www.epa.gov/esd/chemistry/pharma/faq.htm.

———. 2002b. *Potential Implications of Genomics for Regulatory and Risk Assessment Applications at EPA*. EPA/100/B-04/002. Science Policy Council, U.S. EPA, Washington, DC.

———. 2003a. *Framework for Cumulative Risk Assessment*. EPA/630/P-02/001F. Risk Assessment Forum, U.S. EPA, Washington, DC.

_____. 2003b. *A Framework for a Computational Toxicology Research Program*. EPA/600/R-03/065. Office of Research and Development, U.S. EPA, Washington, DC.

_____. 2004a. *Pharmaceuticals and Personal Care Products (PPCPs) as Environmental Pollutants: Pollution from Personal Actions, Activities, and Behaviors*. National Exposure Research Laboratory, U.S. EPA, Research Triangle Park, NC. Accessed at: http://www.epa.gov/nerlesd1/chemistry/pharma/index.htm.

_____. 2004b. *Interim Policy on Genomics*. Science Policy Council, U.S. EPA, Washington, DC.

_____. 2005a. *Guidelines for Carcinogen Risk Assessment*. EPA/630/P-03/001F. Risk Assessment Forum, U.S. EPA, Washington, DC, March.

_____. 2005b. *A Framework for Assessing Health Risks of Environmental Exposures to Children*. NCEA-W-17960 National Center for Environmental Assessment, Office of Research and Development, U.S. EPA, Washington, DC.

_____. 2005c. *Supplemental Guidance for Assessing Susceptibility from Early-Life Exposure to Carcinogens*. EPA/630/R-03/003F. Risk Assessment Forum, U.S. EPA, Washington, DC.

_____. 2005d. *Aging and Toxic Response: Issues Relevant to Risk Assessment*. EPA/600/P-03/004A. Office of Research and Development, National Center for Environmental Assessment, U.S. EPA, Washington, DC.

_____. 2006a. *Endocrine Disruptor Screening Program (EDSP)*. U.S. EPA, Washington, DC. Last updated March 29, 2006. Accessed at: http://www.epa.gov/oscpmont/oscpendo/.

_____. 2006b. *Definition of endocrine disruptor*. Memorandum from Fenner-Crisp P, Maciorowski A, and Timm G, to EDSTAC members. U.S. EPA, Washington, DC. Last updated February 23, 2006. Accessed at: http://www.epa.gov/scipoly/oscpendo/edsp-archive/2-3attac.html.

_____. 2006c. *Aging Initiative*. U.S. EPA, Washington, DC. Last updated April 24, 2006. Accessed at: http://www.epa.gov/aging/agenda/index.htm.

_____. 2006d. Organophosphate cumulative risk assessment; notice of availability. *Fed Regi* 71(148): 43740–43743, August 2. 2006.

_____. 2006e. National Center for Computational Toxicology. U.S. EPA, Washington, DC. Accessed at: http://www.cpa.gov/comptox/.

Warheit DB, Laurence BR, Reed KL, DH Roach, Reynolds GAM, Webb TR. 2004. Comparative pulmonary toxicity assessment of single-wall carbon nanotubes in rats. *Toxicol Sci* 77: 3–5.

Wilkinson CF, Christoph GR, Julien E, Kelley JM, Kronenberg S, McCarthy J, Reiss R. 2000. Assessing the risk of exposure to multiple chemicals with a common mechanism of toxicity: How to cumulate? *Regul Toxicol Pharmacol* 31: 30–43.

Yang RSH, El-Masri HA, Thomas RS, Dobrev ID, Dennison JE Jr, Bae DS, Campain JA, Liao KH, Reisfeld B, Andersen ME, Mumtaz M. 2004. Chemical mixture toxicology: from descriptive to mechanistic, and going on to in silico toxicology. *Environ Toxicol Pharmacol* 18: 65–81.

INDEX

Risk Assessment for Chemicals in Drinking Water, Edited by Robert A. Howd and Anna M. Fan